W9-AUL-645

CONCEPTS AND METHODS IN DISCRETE EVENT DIGITAL SIMULATION

GEORGE S. FISHMAN

Associate Professor
Department of Administrative Sciences and
Institution for Social and Policy Studies
Yale University

A Wiley-Interscience Publication

JOHN WILEY & SONS,
New York · London · Sydney · Toronto

Library of Congress Cataloging in Publication Data

Fishman, George S.
Concepts and methods in discrete event digital simulation.

Bibliography: p.
1. Digital computer simulation. 2. System analysis.
I. Title.

T57.62.F57 001.4′24 73-5713
ISBN 0-471-26155-6

Printed in the United States of America

10 9 8 7 6 5 4 3 2 1

PREFACE

Although discrete event digital computer simulation has been with us as a research tool for almost two decades, the absence of a balanced and comprehensive account of its essential features has made the teaching of this form of simulation almost entirely a matter of the teacher's choice. No doubt the diversity of topics which discrete event simulation comprises has contributed to this situation. For example, modeling, programming languages, and statistical considerations all represent essential features of discrete event simulation; however, who can claim both equal expertise and equal interest in each of these areas?

This variety of choices has led to a situation in which one cannot tell what a student who claims to have taken a course in simulation knows about the subject. Does he know the alternative ways of modeling a queueing environment? Does he know the differences between GPSS and SIMSCRIPT? Is he familiar with the problem of statistical reliability in evaluating a simulation experiment?

One way to reduce this variability in course content is to provide a textbook that deals with modeling, programming languages, and statistical considerations in some depth. This I have attempted to do in this volume. However, I must quickly add that, although the book includes topics in each area that are necessary to a useful knowledge of simulation methodology, my own interest in statistical considerations has led to a considerably deeper discussion in this area than in the other two. Moreover, I believe that this imbalance in favor of statistics is justified on the grounds that the ability of the average user of simulation to make statistically valid statements about his simulation result is considerably more in doubt than are his abilities to model and program his simulation.

In writing this book I have, wherever possible, described alternative techniques for accomplishing a particular objective. The importance of having alternative methods available cannot be underestimated in simulation work, where the nature of the situation often dictates the methods that can be employed. In addition, I have included flowcharts for modeling and for computational algorithms to speed the transition from theoretical ideas to practical use. The reader should realize that these algorithms serve a descriptive purpose and that he is free to alter them to accomplish the same objectives in more efficient ways.

The material in this book has been used in both undergraduate and graduate semester courses at Yale. The undergraduate course concentrated on modeling and programming considerations along with selected topics in the statistical area. For an undergraduate course with a prerequisite knowledge of a programming language and an introductory course in statistics, Chapters 1 through 5, Sections 7.1 through 7.6, 8.1 through 8.5, 10.1 through 10.8, and Chapter 12 are recommended. For a graduate course with the same programming requirement but a comprehensive probability and statistics prerequisite, the entire book should be used. Because of the volume of material, an instructor may wish to make Chapters 1 and 2 a reading assignment and begin his lectures with Chapter 3.

I began this book in September 1970 and completed the present draft in August 1972. I am grateful to Professor Robert Fetter, Mr. Philip Kiviat, Mr. Arnold Ockene, and Professor Alan Pritsker, who kindly read a preliminary draft and provided me with their critical comments. The book has benefited substantively from their contributions. My thanks go also to Mr. Joseph Faulkner of UNIVAC, who provided me with the correct coding of the SIMULA example in Chapter 5 and to Mr. William Eddy, who wrote the FORTRAN ANALYS subroutine in Appendix B.

Appreciation is due to Mrs. Linda Oestreich and Mrs. Lynne Black, who typed the major part of the manuscript, and to Mrs. Elizabeth Walker and Mrs. Irene Loukides for their typing contributions. Mrs. Ann Docherty also deserves my gratitude for her programming assistance.

This work was supported by Grant Number T01-HS-00090 from the National Center for Health Services Research and Development as part of the Health Services Research Training Program, Institution for Social and Policy Studies, Yale University.

GEORGE S. FISHMAN

New Haven, Connecticut
December 1972

CONTENTS

Tables **xi**

Figures **xiii**

Chapter 1 Introduction **1**

Chapter 2 Systems, Models, and System Simulation **4**

2.1 System, 4
2.1.1 System Classification, 8
2.1.2 System State, 8
2.1.3 Performance, 10
2.1.4 Optimization, 10
2.2 Models, 11
2.2.1 Model Classification, 11
2.2.2 Need for and Cost of Detail, 12
2.2.3 Modeling Dangers, 13
2.3 System Simulation, 14
2.3.1 Identity Simulation, 14
2.3.2 Quasi-Identity Simulation, 14
2.3.3 Laboratory Simulation, 15
2.3.4 Computer Simulation, 16
2.4 Monte Carlo Methods, 19

Chapter 3 Discrete Event Digital Simulation **22**

3.1 Introduction, 22
3.2 Alternative Discrete Event Modeling Techniques, 24

3.3 Queueing Models, 25
3.3.1 Event Scheduling Approach, 26
3.3.2 Data Collection, 32
3.3.3 Activity Scanning Approach, 38
3.3.4 Process Interaction Approach, 40
3.4 More Complex Queueing Problems, 44
3.4.1 One Task—Many Servers, 44
3.4.2 Two Task—Many Resources, 46
3.4.3 An Inventory Problem, 49
3.5 PERT Networks, 58
3.6 Multitask-Multiresource Problem, 61
3.7 Period Modeling, 62

Chapter 4 Programming Considerations and Languages 70

4.1 Introduction, 70
4.2 Data Structures, 71
4.2.1 Identification of Objects and Characteristics, 72
4.2.2 Relationships Between Objects, 72
4.2.3 Object Generation and Manipulation, 73
4.3 Simulation Control Programs, 82
4.4 Time Flow, 85
4.5 Random Number Generation, 88
4.6 Data Collection, Analysis, and Display, 88
4.7 Initialization and Termination, 89
4.8 Error Messages and Documentation, 90
4.9 Simulation Programming Languages in Perspective, 92

Chapter 5 GPSS/360, SIMSCRIPT II, and SIMULA 98

5.1 Introduction, 98
5.2 GPSS/360, 98
5.2.1 Transaction Creation, 103
5.2.2 Assignments, 104
5.2.3 Queueing and Service, 104
5.2.4 Statistics, 105
5.2.5 Logical Testing, 106
5.2.6 Set Operations, 107
5.2.7 Extended Computing Capability, 108
5.3 GPSS Single Server Queueing Problem, 108
5.4 SIMSCRIPT II, 112
5.5 SIMSCRIPT II Single Server Queueing Problem, 121

5.6 SIMULA, 129
5.7 SIMULA Example, 131

Chapter 6 Statistical Definitions and Concepts 136

6.1 Statistical Association, 137
6.2 Stochastic Sequences, 143
6.3 Stationarity, 148
6.4 The Autocorrelation and Spectral Density Functions, 149
6.5 A Queueing Problem, 159
6.6 Autoregressive Processes, 160

Chapter 7 Random Number Generation 167

7.1 The Importance of Uniform Variates, 167
7.2 Considerations in Random Number Generation, 169
7.3 A Table of Random Numbers, 170
7.4 Pseudorandom Number Generation, 171
7.4.1 Mixed Congruential Generators, 173
7.4.2 Multiplicative Congruential Generators, 175
7.4.3 A Bad Multiplicative Congruential Generator, 176
7.4.4 An Almost Full Period Multiplicative Congruential Generator, 178
7.5 *k*-Tuples, 180
7.6 GPSS/360, SIMSCRIPT II, and SIMULA Pseudorandom Generators, 182
7.7 Tests of Independence and Uniformity, 184
7.7.1 Chi-Square Goodness-of-Fit Test, 185
7.7.2 Kolmogorov-Smirnov Test, 187
7.7.3 Serial Test, 188
7.7.4 Gap Test, 190
7.7.5 Poker Test, 190
7.7.6 Runs Test, 190
7.7.7 Tests of Correlation, 190

Chapter 8 Stochastic Variate Generation 197

8.1 Introduction, 197
8.2 Continuous Distributions, 200
8.2.1 Uniform Distribution, 200
8.2.2 Triangular Distribution, 202
8.2.3 Exponential Distribution, 203

8.2.4 *Gamma Distribution with Integral Shape Parameter*, 204
8.2.5 *Beta Distribution*, 204
8.2.6 *Gamma Distribution with Nonintegral Parameters*, 208
8.2.7 *Beta Distribution with Nonintegral Parameters*, 209
8.2.8 *Weibull Distribution*, 211
8.2.9 *Normal Distribution*, 211
8.2.10 *The Chi-Square, t, and F Distributions*, 213
8.2.11 *Lognormal Distribution*, 214
8.3 Bivariate and Multivariate Distributions, 215
8.3.1 *Bivariate Exponential Distribution*, 215
8.3.2 *Bivariate Gamma Distribution*, 215
8.3.3 *Multivariate Normal Distribution*, 215
8.4 Discrete Distributions, 216
8.4.1 *Discrete Uniform Distribution*, 219
8.4.2 *Binomial Distribution*, 220
8.4.3 *Beta-Binomial Distribution*, 221
8.4.4 *Geometric Distribution*, 222
8.4.5 *Poisson Distribution*, 224
8.4.6 *Negative Binomial Distribution*, 226
8.4.7 *Hypergeometric Distribution*, 228
8.5 Other Distributions, 228
8.5.1 *Truncated Distributions*, 228
8.5.2 *Empirical Distributions*, 232
8.5.3 *Tabled Distributions*, 233
8.6 Autocorrelated Sequences, 234

Chapter 9 Input Parameters **242**

9.1 Introduction, 242
9.2 Estimation, 244
9.3 Fitting Distributions, 249
9.4 Estimation of Autoregressive Schemes, 254

Chapter 10 Output Analysis **262**

10.1 Introduction, 262
10.2 Static Simulation Output Analysis, 263
10.3 Dynamic Simulation Output Analysis, 270
10.4 Initial Conditions, 272
10.5 Final Conditions, 276

10.6 Data Collection Errors, 278
10.7 Variance Considerations, 278
10.8 Variance Estimation, 279
10.9 Variance Estimator Based on Subsamples in a Single Time Series, 282
10.10 An Estimator from Spectrum Analysis, 283
10.11 Estimation Based on Sample Autoregressive Parameters, 286
10.12 Confidence Intervals for $\overline{X}$, 288
10.13 Nonparametric Confidence Intervals, 293
10.14 The Autocorrelation Function and Spectrum, 295
10.15 Determination of Sample Size, 297
10.16 The Sample State Time Approach, 299

Chapter 11 The Design of Experiments 310

11.1 Introduction, 310
11.2 Prior Information, 312
11.3 Importance Sampling, 317
11.4 Antithetic Sampling, 319
11.5 Stratified Sampling, 322
11.6 Control Variates, 324
11.7 Comparison of Experiments, 325
11.8 Validation, 328
11.9 More than Two Treatments, 330
11.10 2^k Factorial Experiments, 331
11.11 Fractional Factorial Designs, 334
11.12 Response Surfaces, 335
11.13 The Value of Foresight, 341

Chapter 12 Questions and Procedures 347

Appendix A 351

Appendix B 362

References 367

Index 379

TABLES

1. Examples of Systems 5
2. Examples of Models 12
3. Tasks and Resources 44
4. Simulation Programming Languages: Identification Methods 73
5. Simulation Programming Languages: Relationship Concepts 74
6. Simulation Programming Languages: Object Generation Methods 75
7. Simulation Programming Languages: Modeling Approaches 85
8. Debugging Features 92
9. GPSS Block Types 100
10. GPSS Standard Numerical Attributes 101
11. GPSS/360 Sample Output 114
12. Single Server Queueing Problem: SIMSCRIPT II Report Routine Statistics 128
13. Single Server Queueing Problem: SIMSCRIPT II Analysis Routine Statistics 130
14. Independence, Dependence, and Correlation 143
15. Example of a Stochastic Sequence 146
16. Definitions in Stochastic Sequences 147
17. Definitions and Properties of Stationarity 150
18. Properties of the Time and Frequency Domains 159
19. Number Generation Example 172
20. Test Results for a Multiplicative Congruential Generator 186
21. Continuous Distributions 201
22. Discrete Distributions 218
23. Maximum Likelihood Estimators for Selected Distributions 246
24. Selected Critical Values of $\chi^2_{j,\,1-\alpha}$ 250
25. Intervals for Chi-Square Test of the Exponential Distribution 252
26. Intervals for Chi-Square Test of the Normal Distribution 252
27. Sample Tableau for Chi-Square Test of the Normal Distribution 253

28. Sampling Properties of Autoregressive Estimates 256
29. Determination of Autoregressive Order 256
30. Formulae for Determining Sample Autoregressive Scheme 258
31. Population Descriptors of Probabilistic Behavior 264
32. Estimation of m 289
33. Confidence and Probability Intervals for Queueing Problem 292
34. Sample Sizes for Normal, Chebyshev, and Unimodal Assumptions 294
35. Confidence Intervals for Queueing Problem 305
36. Variance Comparisons for Two Estimators 317
37. Statistics for the 2^3 Factorial Experiment 332
38. Graphical Techniques 337
39. Treatments for Second-Order Polynomial with Two Factors 341
A1. Maximum Likelihood Estimates of $\hat{\alpha}$, $\Delta(\hat{\alpha})$, and $\psi(\hat{\alpha})/\Delta(\hat{\alpha})$ for the Gamma Distribution 352
A2. Percentage Points l_γ for the Maximum Likelihood Estimator $\hat{\alpha}/\alpha$ for the Weibull Distribution 354
A3. Percentage Points t_γ for the Maximum Likelihood Estimator $\hat{\alpha} \log (\hat{\beta}/\beta)$ for the Weibull Distribution 356
A4. Unbiasing Factors for the Weibull Maximum Likelihood Estimator of α 358
A5. Complete Sample Maximum Likelihood Estimates of Parameters of the Beta Distribution 359
A6. Modified Critical Values of the Kolmogorov-Smirnov Test for Normality with the Maximum Likelihood Estimates Used for Parameters 360
A7. Modified Critical Values of the Kolmogorov-Smirnov Test of the Exponential Distribution with the Maximum Likelihood Estimate Used for the Parameter 361

FIGURES

1. Input, System, and Output 6
2. Examples of System Response 9
3. Event, Process, and Activity 24
4. Flowchart Symbols Used in Simulation Models 27
5. Primitive Arrival Event: Event Scheduling Approach 28
6. Single Server Queueing Model, Form 1: Event Scheduling Approach 29
7. Single Server Queueing Model, Form 2: Event Scheduling Approach 30
8. Single Server Queueing Model, Form 3: Event Scheduling Approach 33
9. Data Collection Event 35
10. Single Server Queueing Model: Activity Scanning Approach 39
11. Single Server Queueing Model: Process Interaction Approach, Concept 1 41
12. Single Server Queueing Model: Process Interaction Approach, Concept 2 43
13. Multiserver Queueing Model: Event Scheduling Approach 45
14. Two Task—Multiresource Queueing Model: Event Scheduling Approach 47
15. Two Task—Multiresource Queueing Model: Activity Scanning Approach 50
16. Two Task—Multiresource Queueing Model: Process Interaction Approach, Concept 1 52
17. Inventory Model: Event Scheduling Approach 56
18. PERT Network 59
19. PERT Network Model: Event Scheduling Approach 60
20. Multitask-Multiresource Problem: Event Scheduling Approach 62
21. Blood Bank Simulation Model 65

22. Successor Relationships 78
23. Timing Routine: Event Scheduling Approach 84
24. Simulation Control Program: Process Interaction Approach 86
25. Initialization and Termination Routines: Event Scheduling Approach 91
26. GPSS Block Diagram Symbols 99
27. GPSS/360 Coding Form 102
28. GPSS Block Diagram of Single Server Queueing Problem 109
29. GPSS Coding of Single Server Queueing Problem 110
30. SIMSCRIPT II Coding of Single Server Queueing Problem 122
31. SIMULA Coding of Supermarket with 2 Clerks and 1000 Customers 132
32. Independence and Correlations 140
33. Waiting Times 144
34. Two Replications of Queue Length 148
35. Autocorrelation Function, $\rho_\tau = \alpha^{|\tau|}$ 151
36. $\rho_\tau = 0.25(-0.95)^{|\tau|} \cos(3\pi\tau/2) + 0.75(0.75)^{|\tau|} \cos(6\pi\tau/7)$ 153
37. Spectral Density Function Corresponding to Fig. 36 156
38. Estimated Queue Length Correlogram and Spectrum, Constant Service Time 157
39. Theoretical and Sample Autocorrelation Functions of Queue Length 163
40. Theoretical and Sample Spectra of Queue Length 164
41. Pseudorandom Number Plot of U_{i+1} and U_{i+2} 178
42. Item Demand through Time 199
43. Beta Variate Generation from $\mathscr{Be}(a, b)$ 208
44. Gamma Variate Generation from $\mathscr{G}(\alpha, \beta)$ 210
45. More Efficient Beta Variate Generation from $\mathscr{Be}(a, b)$ 212
46. Computation of Lower Triangular Matrix C for Generation of Multivariate Normal Variates 217
47. Binomial Variate Generation from $\mathscr{B}(n, p)$ 221
48. Geometric Variate Generation from $\mathscr{G}e(p)$ 223
49. Poisson Variate Generation from $\mathscr{P}(\lambda \Delta t)$ 225
50. Negative Binomial Variate Generation from $\mathscr{N}\mathscr{B}(r,p), r = 1,2,\ldots$ 227
51. Hypergeometric Variate Generation from $\mathscr{H}(n_1, n_2, x)$ 229
52. Examples of Histogram H_i and Sample Cumulative Distribution Function $F_x(x)$ 266
53. k and m_3 for Queueing Problem in Section 6.5 285
54. k and m_3 for Waiting Time Sequence 286
55. Flowchart for Specified Statistical Accuracy Constraint 298
56. Iterative Scheme for Estimating μ with Specified Accuracy 300

CHAPTER 1

INTRODUCTION

Discrete event digital computer simulation is now moving into its third decade. As a tool of analysis, it has been used to study many problems. Although queueing-oriented problems account for a considerable share of these, simulation has been used to study subjects as diverse as the turtle population on the Australian Great Barrier Reef [5] and the resolution of conflicts between nations [4]. A review of problems to which simulation has been applied indicates that success in applying it as a research technique depends, to a great extent, on the user's familiarity with simulation modeling concepts, programming language options open to him, and statistical techniques needed to produce desired input behavior and to analyze output behavior. Although the relative importance of each consideration varies with the problem to be studied, experience has shown that all three play important roles.

The purpose of this book is to introduce readers to the modeling, programming language, and statistical aspects of discrete event digital computer simulation as it applies to the study of systems. The book emphasizes methods of accomplishing individual steps in designing, running, and evaluating a simulation. The need for a discussion of simulation methodology was first noted by Conway, Johnson, and Maxwell in their 1959 paper [1]. However, efforts to provide such a discussion have usually been descriptive in character with a relatively low content on the evaluation of alternative methods. The papers by Kiviat on modeling [2] and simulation programming languages [3] are exceptions, being unusually helpful to both the experienced simulation user and the novice.

Before one can understand the meaning of system simulation in perspective, he must be familiar with the formal concepts of a system and a model. Chapter 2 begins with a description of these concepts, followed by a discussion of the alternative forms of system simulation, with special emphasis on computer system simulation.

Chapter 3 discusses system modeling, especially queueing-oriented systems. It describes three alternative simulation modeling methods [3]; the event scheduling approach, the activity approach, and the process interaction approach. Although the first and third of these are more commonly employed in practice, a comparison of the features of all three provides useful insight into discrete event modeling.

While the modeling of a system provides a major step toward organizing one's understanding of it, the actual simulation of the system, using the model on the computer, is the major focus of attention in a simulation study. Many programming considerations must be borne in mind, to facilitate both the programming of the model and its execution on a computer. Chapter 4 describes these programming considerations, and Chapter 5 presents limited discussions of GPSS/360, SIMSCRIPT II, and SIMULA, three commonly employed simulation programming languages. In addition to describing statements in the three languages, Chapter 5 illustrates their use.

Statistical considerations pervade almost all aspects of a system simulation. Since the passage of time also plays an integral role in system simulation, one cannot hope to view most of the statistical aspects of a simulation unless these aspects are cast in appropriate perspective with regard to time. To provide this perspective, Chapter 6 gives an introductory discussion of stochastic sequences. In particular, it describes the concepts of the autocorrelation function and spectrum and illustrates their value for understanding statistical phenomena evolving through time.

Chapter 7 describes the generation of pseudorandom numbers, which form the basis for producing random variates from different distributions during a simulation. The history of pseudorandom number generation is reviewed, presently employed methods are discussed, and testing procedures to reduce the possibilities of nonrandomness and dependence are described. The chapter also suggests procedures to follow, when using pseudorandom numbers, that reduce the chance of their misuse.

Algorithms for generating random variates from a variety of continuous and discrete probability distributions are presented in Chapter 8. Flowcharts are included for most of these. The chapter also gives an account of how to generate autocorrelated time series from moving average and autoregressive representations of stochastic sequences. These time series generation techniques are especially useful in econometric simulations.

Every simulation has input parameters. Sometimes these parameters assume hypothetical values because few, if any, data are available about their true values in the real system. Whenever data are available, however, it behooves the investigator to use them to estimate the parameter values. Naturally some methods of estimation provide more accurate estimates than others do, and consequently we would like to use the "better" methods. Chapter 9 describes methods of deriving maximum likelihood estimators of the parameters of most probability distributions for which generation algorithms are given in the previous chapter. These estimators have many desirable statistical properties. However, they are sometimes difficult to compute. Accordingly, the chapter describes computational algorithms and gives references to many useful tables, some of which are contained in Appendix A. Chapter 9 also describes estimation methods for autoregressive parameters for use in generating time series, as described in Chapter 8, and in evaluating simulation output, as described in Chapter 10.

Among the statistical topics relevant to computer simulation, the analysis of simulation results has traditionally been one of the least organized. No doubt, this situation has occurred partly because of the diversity of statistical measures that are encountered in simulation. Chapter 10 reviews many of these measures and describes inferential methods applicable to them. In particular, the chapter devotes special attention to estimating how accurate a measure the sample mean of an autocorrelated stochastic sequence is of the true mean. This constantly recurring problem in simulation can be solved in one of the several ways that Chapter 10 describes.

Whether we realize it or not, every study that we undertake has an implicit experimental design associated with it. This design relates to how the study is conducted, and it is not difficult in most instances to identify one experimental design that meets the requirements of the study better than others do. Chapter 11 discusses design considerations in the peculiar environment of computer simulation. In particular, it dwells on methods of making individual experiments efficient with regard to providing useful results and also stresses the need to select the experiments to be performed with an eye toward the same efficiency considerations. Chapter 12 provides a guide of important steps for the conscientious simulation user to follow.

As the reader may perceive by this time, discrete event digital computer simulation of systems covers many topics. For each of these an established literature exists. By including these topics here we hope to give the investigator wishing to use simulation a background that has both breadth and depth.

CHAPTER 2

SYSTEMS, MODELS, AND SYSTEM SIMULATION

System simulation on a digital computer offers one method for studying the behavior of a system. Although systems vary in their characteristics and complexities, the synthesis of model building, computer science, and statistical techniques that this type of simulation represents provides a useful set of tools for learning about these characteristics and complexities and imposing structure on them. Before we can understand the technical features of this approach and apply them to a real problem, we need to familiarize ourselves with the concepts that describe a *system* and a *model*. In addition, a review of the general concept of *system simulation* will add perspective to the reader's understanding of this topic as a method of scientific investigation. This chapter describes the basic concepts of a system and a model and discusses the development of system simulation in general and system simulation using a digital computer in particular.

2.1 SYSTEM

By a *system* we mean a collection of related *entities*, each characterized by *attributes* that may themselves be related. For example, an automobile and a driver form a system. The automobile entity has the attributes of type and cost, whereas the driver entity has the attributes of age and income. The cost of the automobile is related to its type, which in turn may be related to the driver's income. When speaking of relationships we say that the cost of the automobile is a *function* of its type and its type may be a function of its driver's income. The generality of the definition readily brings many systems to mind, those in Table 1 being but a few.

Table 1 Example of Systems[a]

System	Entities	Attributes
Fire department	Stations	Location, size
	Battalions	Location, size
	Companies	Location, size
	Firemen	Location, speciality
	Fire-fighting equipment	Location, specific use
	Fires	Location, type, detection time, expected time duration
Hospital	Inpatient unit	Number of beds
	Outpatient unit	Number of clinics
	X-ray unit	Number of X-ray machines
	Laboratory	Specific capabilities
	Clinics	Size, speciality
	Physicians	Specialty, quantity
	Nurses	Skill, quantity
	Outpatients	Appointment status (scheduled or unscheduled), consultation time, quantity
Maintenance shop	Servers	Skill
	Machines	Type
	Jobs	Arrival rate, service time

[a] These conceptualizations are hypothetical and do not describe a particular fire department, hospital, or maintenance shop.

A number of relationships exist among the entities within each of the three systems in Table 1. For example, in a fire department:

1. a station has battalions assigned to it,
2. a battalion has companies,
3. a company has men and equipment,
4. a station is responsible for fighting fires in areas in its geographical proximity, and
5. responses to calls and alarms occur in the order in which a station receives them.

We denote relationships 1 through 3 as *class* relationships that exist among certain entities. They and 4 are *static* in character. In contrast, relationship 5 describes *dynamic* behavior or the way in which the system responds to work demands as time evolves.

Concentrating on the outpatient unit in a hospital, we note that:

1. an outpatient unit has clinics,
2. a clinic has physicians and nurses assigned to it,
3. a clinic may use the services of the laboratory and X-ray unit,
4. a clinic may process patients according to an appointment schedule,
5. a physician may belong to several clinics,
6. a nurse may belong to only one clinic, and
7. a patient may have appointments at more than one clinic but not at the same time.

Here 1, 2, 5, and 6 denote class relationships and, together with 7, are static in character. Relationships 3 and 4 describe dynamic behavior.

Collectively the attributes of an entity define its *state*, and the states of critical entities define the state of the system. For example, station A has one battalion with three companies. Companies 1 and 2 each have six men and a pumper; company 3 has three men and a ladder truck. These numbers describe a static state. The number of active fires that station A is now servicing and the numbers of committed men and equipment provide a dynamic picture of the system state. In particular, the numbers of busy men and engines play a significant role in deciding how many men and engines to commit to the next fire call.

The objectives in studying one or several phenomena in terms of a system are to learn how change in state occurs, to predict change, and to control change. Most studies combine these objectives with varying emphasis. One particular combination of these objectives, called the *evaluation of alternatives*, concerns the relationship between the *input* to and the *output* from a system, as depicted in Fig. 1. Input refers to stimuli external to a system that induce changes in the system state. Output refers to measures of these state changes.

Three variants of the evaluation of alternatives commonly occur. First, we can undertake a straightforward analysis wherein we specify the input and the system (or a representation of it) and then measure the output. For

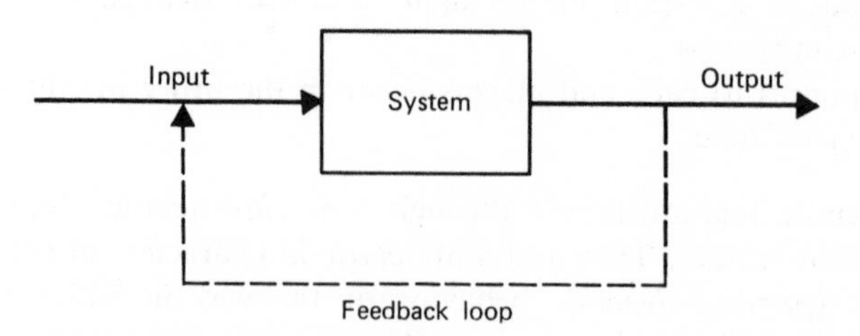

Fig. 1 Input, system, and output.

example, we may specify a stream of fire alarms (input) and a collection of fire stations with their respective manning and engine levels (system) and then measure the mean response time (output) between an alarm and the dispatch of the resources required to fight the fire.

A second, broader purpose is to evaluate the desirability of alternative system *designs* when the input is given and certain desirable characteristics for the output are specified. For example, in the fire department system the objective may be to try alternative rules (system design) for dispatching fire equipment in response to fire calls (input) in order to determine which rule gives the minimum response time (output).

A second example is also instructive here. A health center has a walk-in clinic with three physicians to serve the neighborhood inhabitants. Patients (input) are currently processed in the order of their arrival (system). The center plans to institute a new processing system that classifies a patient in group 1 if his expected consultation time is short and in group 2 if his expected consultation time is long. One physician will be assigned to process patients in group 1 on a first-come-first-served basis. The remaining two physicians will do likewise for patients in group 2. The principal concern is the extent to which the new system will lead to a shorter mean waiting time for group 1 patients (output).

The third variant in the evaluation of alternatives relates to *control.* Here we specify the system and we wish to determine the input that produces a desired output. It is in the control problem that *feedback* between response and stimulus plays a critical role. When a system has the capability of reacting to changes in its own state, we say that the system contains feedback. A *nonfeedback* or *open-loop* system lacks this characteristic. Fire department and outpatient clinic operations offer good examples of feedback. A dispatcher may send more men and equipment to a fire, once the men he has already sent reach the scene and report to him the need for additional resources. In the health center a group 2 physician may treat a group 1 patient if no group 2 patient is waiting. In these two examples the added firemen help to bring the fire under control more quickly and the flexible physician behavior contributes to the reduction of the waiting time of group 1 patients. This control of input resources is the essence of the benefit of feedback.

Every system exhibits three features. It has *boundaries*, exists in an *environment*, and has *subsystems*. An environment is the set of surroundings in which a model is embedded, whereas boundaries distinguish the entities in a system from the entities that make up its environment. For example, the continental United States defines the boundaries of Amtrak, the U.S. railway network. It operates in several environments of which economic climate and weather are two. It also exists as a subsystem, along with airlines and bus service, in a national transportation network and itself contains subsystems such as the Penn Central and Southern Pacific Railroads.

2.1.1 System Classification

We can classify systems in a variety of ways. There are *natural* systems and *man-made* systems. Examples of both come quickly to mind. When a system can exist only in a particular environment, we call it an *open* system in contrast to a *closed* one, which can exist in alternative environments. An *adaptive* system reacts to changes in its environment, whereas a *nonadaptive* system does not. The distinctions between open and closed and between adaptive and nonadaptive take on special importance when a system is under analysis. Inferences drawn about an open system should be carefully qualified in terms of the environment in which they are valid. Moreover the analysis of an adaptive system requires a description of how the environment induces a state change.

Suppose that over a period of months the number of local inhabitants who come to the health center increases. As a result patient waiting time lengthens. If the center adds more staff to handle the increased workload, we say that the center is an adaptive system. In terms of the dichotomous classifications mentioned, we would describe the center as being a man-made, adaptive feedback system.

2.1.2 System State

When we speak of the state of a system, we mean the values of the attributes of its entities. The analysis of a system involves a study of its state changes as time elapses. If a system can assume a finite (infinite) set of states, then over any finite time period the system state exhibits one of a finite (infinite) set of possible sequences of states. The longer the period is, the greater the set of possible sequences is. In any arbitrarily small interval of time there is a likelihood of finding a system in a particular state and also a likelihood of the state changing to one of the remaining states. These probabilistic characteristics are discussed in greater depth in Chapter 6.

Two elements describe state change in time. One is *magnitude*; the other is *delay*. The magnitude of a change refers to the absolute difference in an attribute's value over a specified time period compared to its value before the change. If X_t and X_{t+1} denote the states at times t and $t + 1$, respectively, then $|X_{t+1} - X_t|$ denotes the magnitude of the state change from t to $t + 1$.

In most systems an external stimulus induces a change in system state. This change may occur when the stimulus is first received, after some time has elapsed, or over a period of time following the stimulus. The time lapse between stimulus and state change is termed the delay. We call the characteristics of magnitude change and delay *system response*. Figure 2 describes five possible system responses. Figures 2*a*, 2*b*, and 2*c* illustrate *stable* responses since in each case the state of the system moves either at once or over some time to a permanent new *equilibrium* or *steady state* with finite value in response to a single stimulus. We say that a system in process of moving from one equilibrium state to another is in a *transient state*. Figure 2*d* exhibits an unstable response that continues to grow without ostensible convergence to a new finite value. Figure 2*e* describes a system in which an equilibrium level *r* exists but system response oscillates around this level without evidence of convergence. We denote a system with this response as unstable but non-explosive, in contrast to a system with response, as in Fig. 2*d*.

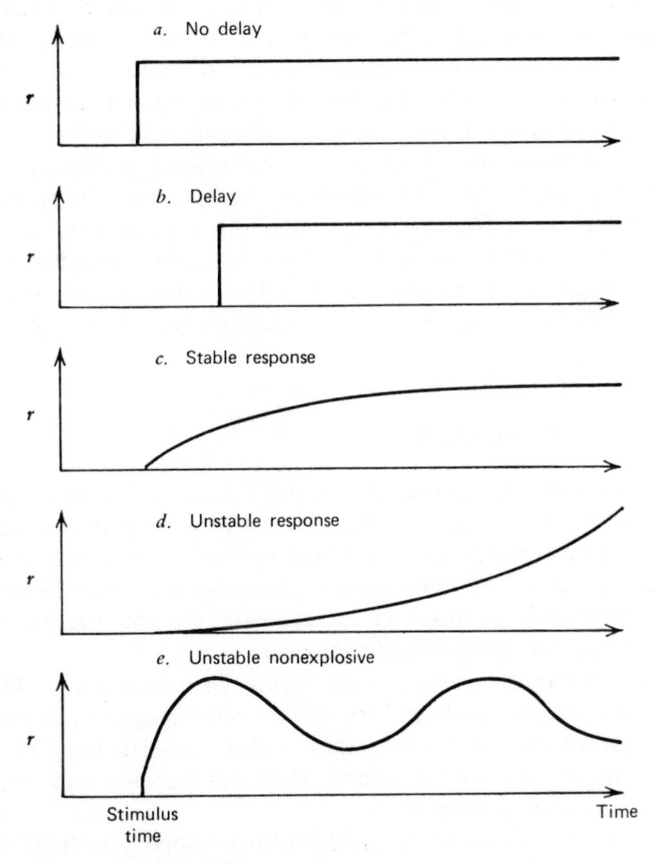

Fig. 2 Examples of system response.

2.1.3 Performance

Earlier we noted that three reasons for analyzing a system were to understand how change occurs, to predict change, and to control change. These reasons follow from a desire to improve system *performance* in some sense. System performance refers to the sequence of states that a system assumes over a specified time interval. Usually these states are averaged in some way to provide a single performance measure.

The concept of what constitutes a performance measure varies among systems. In a business enterprise annual profits may be the measure. In the fire department example mentioned earlier the average response time is the critical consideration. In some systems more than one performance measure is important. For example, in an outpatient clinic the average number of patients who wait for service (queue length) and the average length of a wait are critical. It is entirely possible for the average queue length to be short but the average waiting time to be long because a patient without an appointment is given service only after all scheduled patients receive attention. In the fire department example, we may be interested in studying average response time under alternative rules for assigning men and equipment to fires.

In structuring the analysis of an ill-defined system we often can say only that we want to get a "feel" for the system. It is precisely this type of effort that can benefit most from a priori thoughtful consideration about what performance measures enable an investigator to obtain the understanding he needs. Failure to devote adequate time to this activity before analysis can easily lead to confusion later in trying to determine the direction that analysis should take.

2.1.4 Optimization

The ideal objective is to *optimize* system performance. This involves controlling some aspect of a system so that the best possible performance can be realized. Generally certain aspects of the system are beyond the analyst's control, and these aspects often impose *constraints* on system behavior that preclude unrestricted optimization. In such cases the objective becomes one of optimizing performance subject to the constraints.

These constraints can take many forms. In a business establishment financial considerations and resource ceilings often restrict optimization. In a large organization sheer size makes a clear understanding of how all subsystems relate virtually impossible. Then optimization may occur with regard to individual subsystems.

In practice we must often be content with developing methods of simply *doing better* than before analysis begins [12, p. 1]. Since there are many levels of doing better, it behooves an analyst to bear in mind constantly the desirability of approaching the unrestricted optimum as closely as possible. Then simply doing better does not become a source of criticism when others are evaluating his results.

2.2 MODELS

The first step in studying a system is to build a *model*. A model can be a formal representation of theory or a formal account of empirical observation. Usually, however, it is a combination of both. The purposes for using a model are many. In particular, it does the following:

1. enables an investigator to organize his theoretical beliefs and empirical observations about a system and to deduce the logical implications of this organization.
2. leads to improved system understanding,
3. brings into perspective the need for detail and relevance,
4. expedites the speed with which an analysis can be accomplished,
5. provides a framework for testing the desirability of system modifications,
6. is easier to manipulate than the system is,
7. permits control over more sources of variation than direct study of a system would allow, and
8. is generally less costly.

2.2.1 Model Classification

Models can be classified in many ways. There are *physical* models such as a model airplane or, more generally, a scaled-down replica of a system. There are *schematic* models that include diagrams, maps, and charts. There are *symbolic* models, of which those based on mathematics or computer code are the most common. As we see shortly, schematic and symbolic models play major roles in the design of a computer system simulation study.

Since many types of mathematical models are used in practice it is instructive to distinguish them according to certain characteristics. One dichotomy separates *analytical* and *numerical* models. In an analytical model it is possible to deduce a solution to the problem under study directly from its mathematical representation. Ohm's law is an example. A numerical characterization implies that a solution is not feasible but that numerical methods can provide a solution for specified numerical values of the parameters of the model. Numerical integration offers an example.

Some models are *static*; others are *dynamic*. A static model either omits a recognition of time or describes a snapshot of the state of a system at a moment in time. By contrast a dynamic model explicitly acknowledges the passage of time. In addition to providing a sequence of snapshots of system state through time, some dynamic models specify relationships between the states of a system at different points in time. A *Markov probability* model is an example of a dynamic model that exhibits time passage and state dependence.

One other distinction concerns *deterministic* versus *stochastic* models. In a deterministic model all entities bear fixed mathematical or logical relationships to each other. As a consequence these relationships completely determine the solutions. In a stochastic model at least part of the variation is random in nature. Hence an investigator can, at best, obtain *average* solutions by using stochastic models to solve problems. This book concentrates exclusively on stochastic models.

The three sets of classification enable us to define eight possible descriptions of a model, examples of which are given in Table 2. Notice that a discrete event simulation is numerical in character.

Table 2 Examples of Models

	Analytical	Numerical	Static	Dynamic	Deterministic	Stochastic
Ohm's law	X		X		X	
Normal probability law	X		X			X
Newton's laws of motion	X			X	X	
Markov process	X			X		X
*p*th-Order polynomial[a] $p > 4$		X	X		X	
Non-linear differential equations		X		X	X	
Maximum likelihood[a] estimation procedure		X	X			X
Job-shop simulation		X		X		X

[a]This model occasionally permits analytical solution.

2.2.2 Need for and Cost of Detail

When building a model an investigator constantly faces the problem of balancing the need for structural detail with the need to make the model amenable to problem-solving techniques. Being a formalism, a model is necessarily an abstraction. However, the more detail that a model includes explicitly the better we think the model resembles reality. An additional reason for including detail is that if offers increased opportunities for studying system response when a structural relationship within the model is altered for investigative purpose. First, more combinations of structural changes can be considered and, second, more aspects of the response can be studied.

On the other hand, detail generally makes solution of problems more difficult. Often added detail shifts the method for solving a problem from an analytical to a numerical one, so that the generality of an analytical solution is lost. Detail may also increase the cost of solution. However, the most limiting factor in the use of detail is that we often do not know enough about the system under study to specify more than its most ostensible characteristics.

Every model must limit detail in some respect. To fill in a system description in place of detail we make *assumptions* about system behavior. Since we do not want these assumptions to conflict with the observable behavior of the system, we test them against observation whenever we can. Chapter 9 describes methods for conducting this testing with regard to the characterization of simulation input parameters.

2.2.3 Modeling Dangers

Although modeling clarifies many relationships in a system, there are at least three warnings than at investigator should always bear in mind. First, no guarantee exists that the time and effort devoted to modeling will return a useful result and satisfactory benefits. Occasionally failure occurs because the level of resources is too low. More often, however, the investigator has relied more on method and not enough on ingenuity when the proper balance between both will lead to the greatest probability of success.

The second warning concerns the tendency of an investigator to defend his particular depiction of a problem as the best representation of reality. This situation often occurs after he has spent much time and effort and expects some useful results.

The third warning concerns the use of the model to predict beyond the range of its applicability without proper qualification. For example, a model may be developed to forecast system behavior one time period ahead. If the same model is used to predict two periods ahead, an explicit qualification should be given to users of these predictions that the two period prediction is not as accurate as the one period prediction. Neglect of appropriate qualification with regard to extrapolating model results is perhaps the single greatest cause of model misuse in practice. In Chapter 10 we describe methods for estimating how statistically accurate the results of a simulation run are and how they can be presented to decision makers with appropriate qualification using confidence intervals.

2.3 SYSTEM SIMULATION

In the study of a system the problems to be solved determine to a great extent the method of solution. Whenever possible, analytical solutions are desirable since they can be evaluated for a range of input parameter values, usually with minimal effort. More often than not, however, the need for detail to reproduce system behavior realistically makes analytical solution impossible. Then one approach for studying a system is *system simulation*.

The concept of *system simulation* became a reality in the early 1950s when a shift in emphasis occurred from looking at parts of a problem to examining the simultaneous interactions of all parts. This shift was partially due to the fact that system simulation experiments had become feasible on electronic computers which themselves were undergoing order-of-magnitude advances in speed. Simulation made it possible to carry out fully integrated system analyses which were generally far too complex to be made analytically. This was especially true for studies of the interactions among parts of a system.

A description of alternative forms of system simulation is helpful at this point. Hereafter we refer to system simulation simply as simulation, unless otherwise indicated.

2.3.1 Identity Simulation*

For present purposes *system simulation* means the act of representing a system by a symbolic model that can be manipulated easily and that produces numerical results. The range of system simulation models is broad. At one extreme we may use a system itself as a model to gain insight into its behavior. Although this *identity* simulation sounds like a direct and simple approach, it ignores certain fundamental rules of modeling. It is usually expensive, is seldom feasible, and permits little or no control over phenomena that affect response. It is especially unappealing when answers are needed in a time period short compared to the length of time required for an actual study using the true system.

2.3.2 Quasi-identity Simulation

One step removed from an identity simulation is a modeling effort that attempts to preserve as many aspects of the real system as possible but excludes elements whose presence would make an identity simulation impossible. For example, the U.S. Air Force may wish to test this nation's air

* These classifications of system simulation partially follow the account in reference 11.

defenses. One approach would be to use SAC bombers to play the role of enemy bombers in a trial run. Rather than having the Air Defense Command proceed with an actual attack on these aircraft, the study would seek to determine the extent of penetration of the SAC bombers into the U.S. air defense network before detection and response could be accomplished. We refer to this approach as *quasi-identity* simulation and note that it is also an expensive method for learning about the adequacy of an air defense posture.

2.3.3 Laboratory Simulation

A *laboratory simulation* offers a method of analysis that is more feasible and less expensive than identity and quasi-identity simulations, while retaining the essential features of the system in question. The laboratory components consist of such diverse elements as people, computers, hardware, operating procedures, mathematical functions, and probability distributions. This approach includes the ability to reproduce some aspects of a system literally and to substitute symbolic representations for other aspects. Two types of laboratory simulation deserve mention. One is *operational gaming*; the other, *man-machine simulation.*

In operational gaming a computer is used to collect, process, and produce information that human players, usually adversaries, need to make decisions about system operation. Each player's objective is to perform as well as possible, given specified information. Moreover, each player's decisions affect the information that the computer provides as the game progresses through simulated time. The computer can also play an active role by initiating predetermined or random actions to which the players respond.*

War games and business management games have attracted the most attention in the operational gaming literature. For example, the Naval Electronic Warfare Simulator developed in the 1950s consisted of a large analog computer designed primarily to assess ship damage and to provide information to two opposing forces regarding their respective effectiveness in a naval engagement [11, pp. 15–16]. The exercise, which is one form of *simulation gaming*, was a training device for naval fleet officers in the final stages of their training.

Man-machine simulation brings to the fore another concept in investigative techniques. Here gaming need not play a role. In addition to the computer some of the subjects in the laboratory perform the data reduction and analysis. For example, the Rand Systems Research Laboratory employed

* See reference 18 for a bibliography on gaming and reference 19 for a discussion of the scope of gaming.

simulation to generate stimuli for the study of information processing centers [11, p. 16]. The principal features of a radar site were reproduced in the laboratory, and by carefully controlling the synthetic input to the system and recording the behavior of the human detectors it was possible to examine the relative effectiveness of various man-machine combinations and procedures.

In 1956 Rand established the Logistics System Laboratory under U.S. Air Force sponsorship [9]. The first study in this laboratory involved the simulation of two large logistics systems in order to compare their effectiveness under different management and resource utilization policies. Each system consisted of men and machines, together with policy rules for the use of such resources in simulated stress situations such as war. The simulated environment required a specified number of aircraft in flying and alert states, while the system's capability to meet these objectives was limited by malfunctioning parts, procurement and transportation delays, and the like. The human participants represented management personnel, while higher echelon policies in the utilization of resources were simulated in the computer. The ultimate criteria of the effectiveness of each system were the number of operationally ready aircraft and the dollar cost of maintaining this number.

Although the purpose of the first study in this laboratory was to test the feasibility of introducing new procedures into an existing air force logistics system and to compare the modified system with the original one, the second laboratory problem had quite a different objective. Its purpose was to improve the design of the operational control system through the use of simulation.

2.3.4 Computer Simulation

If we remove the people and hardware from the laboratory simulation concept and retain the computer, operational rules, mathematical functions, and probability distributions, we have the essential features of an *all computer simulation*, given that a method of analytical solution is not readily available.* All behavior is reduced to programmable, logical-decision rules and operations. Computer simulation, which is the topic of this book, has been applied to many diverse systems. As examples, it has been used to study the operations of the New York City Fire Department [6], hospital resource allocation problems [8], urban transportation systems, and inventory systems.†

* The term *computer simulation* should not be confused with simulation study of a computer system. Such studies have grown in number recently and are becoming major users of computer simulation, as defined here.

† Reference 13 provides a bibliography of computer simulation through 1968, and references 7 and 14–17 contain accounts of applications to diverse areas.

Computer simulation offers many conveniences that make it an attractive tool for analysis. It can *compress* time so that several years of activity can be simulated in minutes or, in some cases, seconds. This ability enables an investigator to run through a variety of operational designs that he may wish to study in a very small fraction of the time required to try each on the real system.

Computer simulation can also *expand* time, in effect. By arranging for statistics of interest to be produced over small intervals of simulated time, an investigator can study by computer simulation the detailed structure of change in a real system which cannot be observed in real time. This figurative time dilation is especially helpful when few data are available on change in the real system.

One important consideration in any experiment is the ability to identify and control *sources of variation*. This ability is especially important when a statistical analysis of the relationship between independent (input) and dependent (output) factors in an experiment is to be performed. In the real world the ability is to a major extent a function of the system under study. In a computer simulation an investigator, of necessity, must explicitly specify the sources of variation and the degree of variation due to each source in order to make his simulation run. This requirement enables him to eliminate unwanted sources of variation simply by omitting them from his simulation. However, this ability also requires an investigator to devote sufficient attention to a system in order to derive an adequate understanding of how to describe quantitatively the sources of input variation that he wishes to include. Chapter 9 describes inferential methods for deriving such a description from quantitative data.

In a field experiment, *errors of measurement* inevitably occur when results are recorded. This occurrence is due to the fact that no perfect measuring device exists for field use. By contrast, no measurement errors occur in a computer simulation since the programmed simulation produces numbers free of any superimposed variation due to external and uncontrollable sources. There are, of course, roundoff errors due to finite word length in a computer, but a modicum of care on the investigator's part can make such errors relatively negligible.

In the course of an experiment it is occasionally desirable to *stop* the experiment and *review* the result to date. This means that all phenomena associated with the experiment will have to retain their current states until resumption of the experiment begins. In field experiments a complete halt of all processes is seldom possible. Computer simulation, however, offers this convenience, provided that the termination part of the program contains instructions for *recording* all relevant states. When the experiment resumes, the terminal states become the initial states so that no loss of continuity occurs.

The ability to restore the state of the simulation provides another benefit. At the end of a computer simulation run *analysis* of the results may reveal a situation explainable by additional data that were not collected. Here the investigator can reprogram the data collection package to record the desired data and can run the simulation again with the same initial conditions. The simulation runs identically as before with the additional data being made available. Since a rerun at least doubles the running time, however, this reproducibility feature should in no way reduce the importance of taking sufficient time before experimentation to decide which data can be most instructive.

Computer simulation also enables us to *replicate* an experiment. Replications means rerunning an experiment with selected changes in parameters or operating conditions being made by the investigator. For example, an independent replication is one in which the simulation model is unchanged but, say, the sequence of random numbers used to drive the model in the replication is independent of the sequence employed in the original run.* In addition, computer simulation often allows us to induce correlation between these random number sequences to improve the statistical analysis of the output of a simulation. In particular, a negative correlation is desirable when the results of two replications are to be summed, whereas a positive correlation is preferred when the results are to be differenced, as in the comparison of experiments. Sections 11.4 and 11.5 describe correlation-inducing techniques in detail.

While the technical benefits of computer simulation are substantial, the method also offers great flexibility for studying a variety of problems. It provides the ability to experiment, to test and evaluate new systems or proposed changes to existing systems in advance of having to invest effort, money, and time in their physical development. It also makes it possible to consider hypothetical systems that may be dangerous or impossible to try in any other way. For example, the Cornell Aeronautical Laboratory first used a computer simulation to demonstrate that the Mark I perceptron for the automatic recognition of simple patterns could be built [11, p. 10].

As mentioned earlier, computer simulation becomes a legitimate research tool when known analytical methods cannot supply a solution to a problem. Once simulation is adopted, however, certain admonitions made earlier in regard to modeling must be re-examined. Principal among these considerations is the issue of detail.

Our earlier remarks stressed the fact that the amount of detail in a model is generally inversely related to our ability to obtain an analytical solution. However, if the *minimal model*, which contains the least detail needed for

* See Section 6.2.

useful study, is not amenable to analytical solution and we adopt a computer simulation approach, we have the prerogative of building as much detail into our model as we like without concerning ourselves about the absence of analytical solution. It is precisely this descriptive ability that holds the most attraction for modelers, for the greater the amount of detail, the more realistic the model is and consequently the closer they expect the results of a simulation run to conform to reality. However, we find in practice that judicious restraint in regard to detail is often the better policy, for at least three reasons.

First, to include detail we must devote effort and time to preliminary observation of the individual characteristics of the system under scrutiny. This attention induces a cost, the justification for which a modeler has to weigh with respect to his objective. This issue arises in most simulations, and no universal course of action fits all contingencies.

A second cost of detail arises when programming the model for eventual running. Added detail requires added programming, especially for the contingencies that detail specifies. Moreover, the inclusion of great detail at the outset of a modeling effort makes the job of locating the sources of error in the resulting computer program especially difficult because of the many potential trouble spots.

Third, because of the increased number of functions that a computer program based on a detailed model must perform, we expect a concomitant increase in running time and, consequently, cost. Testing for special situations, along with the need to update and manipulate system attributes, contributes notably to this cost. This third reason for restraint is not always apparent until actual simulation begins and we see the computing cost quickly mounting.

These sobering remarks provide a useful perspective for evaluating the degree of detail worth having. Often investigators using computer simulation get a gross model running first and then introduce detail where this model provides inadequate answers. This bootstrapping approach makes for a more judicious allocation of a modeler's time and effort and reduces debugging and running times. When changes in a simulation model are anticipated, a modeler is advised to organize his simulation program so that minimal effort is required. The choice of computer language also plays a major role here. We discuss program organization and language selection in Chapter 4.

2.4 MONTE CARLO METHODS

So far we have used "computer simulation" to denote a method for studying a system evolving in time. However, the term is also occasionally used to denote a computer-based sampling experiment in which time plays no substantive role and which has historically been called *Monte Carlo* computer simulation. The systematic development of Monte Carlo methods, which concern experiments with random numbers, began during World War II when they were applied to problems related to the atomic bomb. The work involved direct simulation of behavior concerned with random neutron diffusion in fissionable material. Shortly thereafter Monte Carlo methods were used to solve certain integral equations occurring in physics that were not amenable to analytical solution but for which corresponding probabilistic problems could be solved. Reference 10 gives a comprehensive account of these methods.

Although Monte Carlo methods play a major role in most computer system simulation, their importance is matched by other factors. These include a capability to manipulate a large data base, an ability to structure logical as well as mathematical operations in a model, and a need to accommodate model development as well as production runs as an integral part of an investigation. Because of this distinction between Monte Carlo and computer system simulation, we reserve the term computer simulation to denote a comprehensive system study in which "Monte Carlo simulation" designates the sampling experiment contained therein.

EXERCISES

1. Describe a local supermarket as a system in terms of entities, attributes, and class and dynamic relationships.
2. a. Describe a bank as a system of entities, attributes, and class and dynamic relationships.
   b. Describe (i) the environment in which the bank operates, (ii) its boundaries, (iii) its subsystems, (iv) the system in which it is a subsystem, (v) its feedback mechanism, (vi) its adaptive mechanism.
   c. Which variables constitute the state of the bank?
   d. Identify the bank's performance measure from the point of view of the stockholders.
   e. Identify the bank's performance measure from the point of view of the federal government agency that underwrites loans to low income families for home purchasing.
   f. The bank wants to shorten its customers' waiting time. Its manager thinks that establishing an express line in each branch for people making deposits only can shorten this time. Would you study this problem as an identity simulation, a quasi-identity simulation, or a computer simulation? Explain your answer.

3. a. Describe a library in terms of entities, attributes, and class and dynamic relationships.
   b. Describe the library in terms of the elements listed in Exercise 2b.
   c. Which variables constitute the state of the library?
   d. Identify the library's performance measure from the point of view of its users.

4. A new skyscraper is to be built. The first five floors above the street level are to serve as parking space for automobiles. In addition to entering and leaving the building on the street floor, people can enter and leave on a basement level that connects to an underground arcade. Describe how you would go about formulating a study to determine how many elevators to install in the new building and how to program the elevators to operate. (A visit to a local skyscraper, if possible, would be helpful.)

CHAPTER 3

DISCRETE EVENT DIGITAL SIMULATION

3.1 INTRODUCTION

This chapter presents concepts germane to the construction of a discrete event digital simulation model. The approach is to describe alternative, but equivalent, ways of representing behavior. The presentation also includes a collection of model descriptions of increasing complexity. By observing the evolution of a simple model to a more complex one in terms of flowcharts, the reader can begin to recognize the intrinsic value of using the flowchart approach as a first step in simulation modeling.

Changes in the states of a system's entities occur continuously as time evolves or at discrete moments in time. Both *continuous event* and *discrete event* systems abound in practice, and well-established methodologies exist for studying each type of change by computer simulation. Here an *event* denotes a change in the state of a system entity. Since modeling techniques differ for formalizing the characteristics of continuous event and discrete event systems, we emphasize that the techniques described in this chapter apply to discrete event systems or to continuous event systems that can be reasonably approximated by discrete event models.

The fire department, outpatient clinic, and maintenance shop described in Table 1 exemplify three discrete event systems. An alarm received at the fire department's dispatching center represents a discrete event that activates the fire-fighting system, which responds by committing (discrete event) apparatus that eventually arrives (discrete event) at the scene of the fire. A patient arrives at the outpatient clinic; a physician eventually sees the patient, who then leaves after treatment is completed. Here the arrival denotes a discrete event that adds one more patient to the number in the clinic at that moment, the beginning of treatment constitutes a discrete event that reduces the number of patients waiting for treatment by one, and treatment completion denotes a discrete event that reduces the number of patients in the clinic by one. The maintenance shop example follows the same format as the outpatient clinic.

Discrete event digital simulation concerns the modeling on a digital computer of a system in which state changes can be represented by a collection of discrete events. Here the appropriate technique for modeling a particular system depends on the nature of the *interevent intervals*. These time intervals may be random or deterministic. When random interevent intervals occur the modeling technique must allow for intervals of varying lengths. When the intervals are deterministic they may vary according to a plan or they may all be of equal length. When they vary the modeling techniques must again acknowledge the nonuniform nature of the interevent intervals. When they are constant we can employ a considerably simpler modeling technique to structure change in a model. The techniques described in Sections 3.2 through 3.6 describe modeling for nonuniform interevent intervals. Section 3.7 describes modeling for constant intervals.

In a discrete event system change occurs when an event occurs. Since the states of entities remain constant between events, there is no need to account for this inactivity time in our modeling. Accordingly all modern computer simulation programming languages use the *next event* approach to time advance. After all state changes have been made at the time corresponding to a particular event, simulated time is advanced to the time of the next event, where required state changes are again made. Then simulated time is again advanced to the time of the next event, and the process is repeated. In this way a simulation is able to skip over the inactive time whose passage in the real world we are forced to endure.

Although the next event approach is the most commonly used, the reader should note from the outset that it is not the only method available for advancing time and processing events. In Section 4.4 we describe several alternative methods and their advantages for particular problems. Unfortunately use of one of these alternatives necessitates programming it—hence the appeal of the next event approach, which is a standard feature of most simulation programming languages.

In modeling a system for eventual computer simulation, two different structures play significant roles. One includes the *mathematical* relationships that exist between variables (attributes) associated with the entities. For example, if Q is the number of patients waiting in the outpatient clinic, then we set $Q = Q + 1$ when a patient arrives and $Q = Q - 1$ when a patient begins treatment. Sometimes the specification of the mathematical relationships for a system serves to describe completely the way in which state changes occur, as, for example, in a model of a national economy system.

Logical relationships comprise the other set of structures that are used to describe a system. In a logical relationship we check to see whether a condition holds. If it does, we take a certain action. If it does not hold, we take an alternative action. For example, when a server becomes available in a maintenance shop, he checks to determine whether a job is waiting for service. If one is, he begins the service. If no job is waiting, then the server remains idle. If several jobs are waiting, the server selects one according to an established *logical operating* or *job selection* rule. A logical operating rule describes a set of mutually exclusive conditions with corresponding actions. For example, if several jobs are waiting, the rule may be to select the job that has been waiting for the longest time. If there are n jobs waiting, this rule is formally: For $i = 1, \ldots, n$, if job i has been waiting for the longest time, select job i for service.

Our purpose in discussing time passage and relational questions in this introduction is to establish the importance of these topics in modeling a discrete event system. The rest of this chapter describes modeling methods that account for these concepts in varying ways.

3.2 ALTERNATIVE DISCRETE EVENT MODELING TECHNIQUES

The concepts of *event*, *process*, and *activity* are especially important when building a model of a system. As already defined, an *event* signifies a change in state of an entity. A *process* is a sequence of events ordered on time. An *activity* is a collection of operations that transform the state of an entity. To illustrate the relationships among these concepts we consider a job arriving at a maintenance shop with two tasks to be performed. Figure 3 shows the arrival and the eventual service that each task receives.

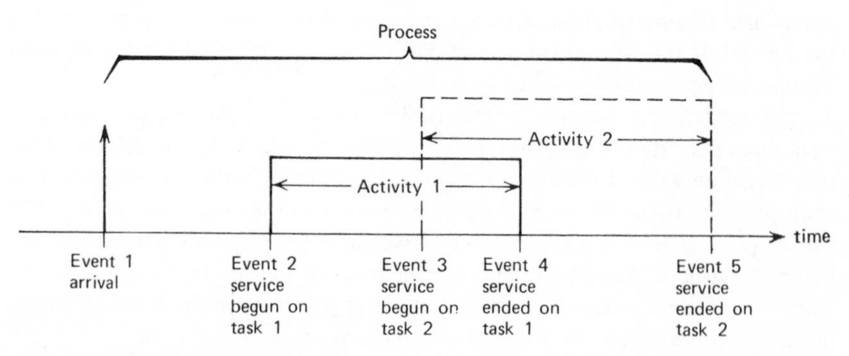

Fig. 3 Event, process, and activity.

These three concepts give rise to three alternative ways of building discrete event models [23]. The *event scheduling approach* emphasizes a detailed description of the steps that occur when an individual event takes place. Each type of event naturally has a distinct set of steps associated with it. The *activity scanning approach* emphasizes a review of all activities in a simulation to determine which can be begun or terminated each time an event occurs. The *process interaction approach* emphasizes the progress of an entity through a system from its arrival event to its departure event. The development of the three concepts is related to the development of discrete event computer simulation programming languages. In particular, SIMSCRIPT and GASP use the event scheduling approach; GPSS and SIMULA, the process interaction approach; and CSL, the activity scanning approach.* Each approach has advantageous features. In Section 3.3 we offer several examples for comparison.

3.3 QUEUEING MODELS

In the early days of discrete event digital simulation, modelers quickly realized that many of the systems they encountered in practice resembled *queueing* or *waiting line* problems. In a queueing problem an arrival occurs and demands that a service be performed. The system responds by performing the service if it can, or by keeping the demand waiting until it can perform it. For example, a demand in the fire department example of Table 1 is a request to send equipment to a fire. In the outpatient clinic patients wish to see a physician. In an inventory system an order waits for filling until the requested items are in stock.

Recognition of this commonality among problems led to a formalization of their basic queueing features into a *job-shop simulation* model.† At least two simulation programming languages, GPSS and SIMSCRIPT, are oriented around this concept. Because of the central position that queueing models hold, it is instructive for anyone contemplating the use of a discrete event digital simulation model to familiarize himself with the formalisms of a job-shop environment.

Three considerations play roles in the study of problems using queueing-oriented models. The first consideration concerns the nature of the jobs to be performed. This includes their *frequency* of occurrence, the *number* of tasks per job, the *resource requirements* per task, and the *service* time per task. The second consideration relates to resources, for example, questions about the

* See Chapter 5.

† See reference 29 for a recent list of queueing problems studied by simulation methods.

number and skill *types* of available resources, the *assignment* of resources by station, and the assignment of resources to tasks. The third consideration concerns the way in which tasks are *selected* for service or, as already mentioned, *logical operating rules*. Several *selection* rules are defined shortly, and their data requirements are discussed.

Usually the objective in queueing-oriented problems is to learn how system performance varies in response to changes in workload, resource characteristics, or task selection rules. For a given computer simulation experiment each of these issues regarding jobs, resources, and selection must be resolved and specified explicitly in the programming logic. To provide a more general setting, however, we describe the modeling considerations that apply regardless of the forms these three specifications assume.

3.3.1 Event Scheduling Approach

In the simplest form of a queueing problem a job requiring service arrives at a shop that has one server. If the server is idle he services the job; if he is busy the job is placed in a *queue* or *waiting line* to await later service. We define the state of the system as the number of jobs in the shop at a given moment of time. This queue length is identical with the sum of the number of jobs waiting for service plus the number in service.* A state change occurs every time a job arrives and every time a job departs. Their arrivals and departures are events. In modeling terms a job is an object or entity. The server is also an entity. The queue is a *set* to which jobs may belong.

Two additional events occur in this simple queueing system: one, when the server becomes busy; the other, when he becomes idle. These events, however, are *conditional* on the occurrence of an arrival or departure event. For example, an arrival when the server is idle causes him to become busy. A departure, when no other jobs are waiting, causes the server to become idle.

A symbolic flowchart provides a convenient way of describing the problem. Using the conventions given in Fig. 4, we describe the arrival event in Fig. 5. Subsumed therein is the conditional event of becoming busy. Here the server has the characteristic or attribute of status, which assumes either a busy or an idle mode.

Although Fig. 5 tells us how an activity starts, it provides no information about the jobs that wait nor does it describe how completions leave the shop. When a completed job departs it is no longer responsible for any activity in the shop; hence a departure is also an event.

* Some investigators prefer to define queue length as the number of jobs waiting for service. Our definition coincides with that used in most books that treat queueing problems analytically. See, for example, reference 34.

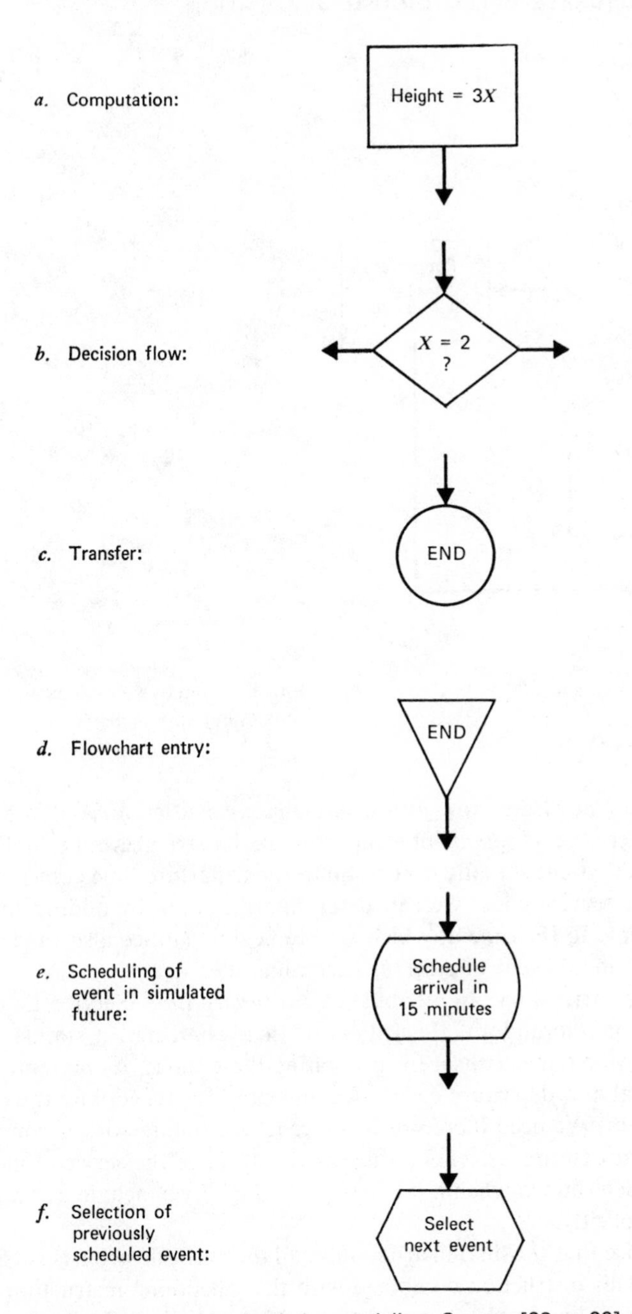

Fig. 4 Flowchart symbols used in simulation modeling. *Source:* [23, p. 28].

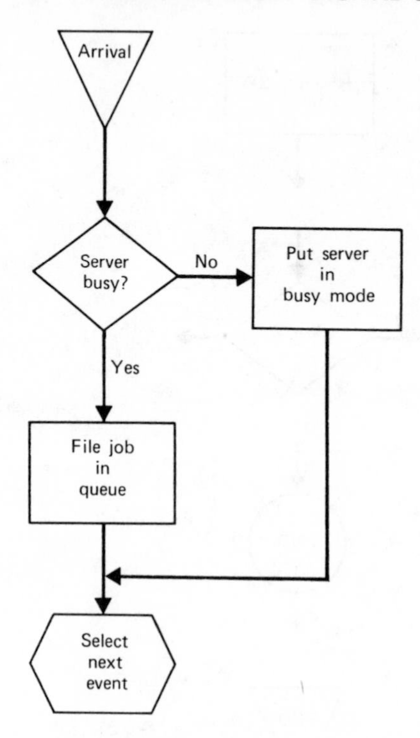

Fig. 5 Primitive arrival event: event scheduling approach.

To acknowledge how jobs eventually receive service and depart we use Fig. 6. Here the *scheduling* of a departure in the arrival event establishes when a departure event actually occurs. Since the departure time coincides with the time that service ends, we can determine this time by adding the arrival's service time to the time at which service begins. Notice also the inclusion of the conditional event of a server becoming idle.

Every arrival has an attribute called *service time*. Service times may be random or nonrandom. Regardless of their character, a simulation model must provide a mechanism for generating these times. As presently depicted, the arrival and departure events account only for scheduling the occurrence of an event. We need therefore to precede scheduling with a computational block that determines service time or else to have the service time attribute already determined when a job arrives. In Fig. 7 we include a computational block explicitly.

Notice that the final instruction in all the event flowcharts is "Select next event." This instruction combined with the scheduling instruction forms the basis for making a discrete event simulation work. Whenever an event is scheduled, a *record* identifying the event and the time at which it is to occur

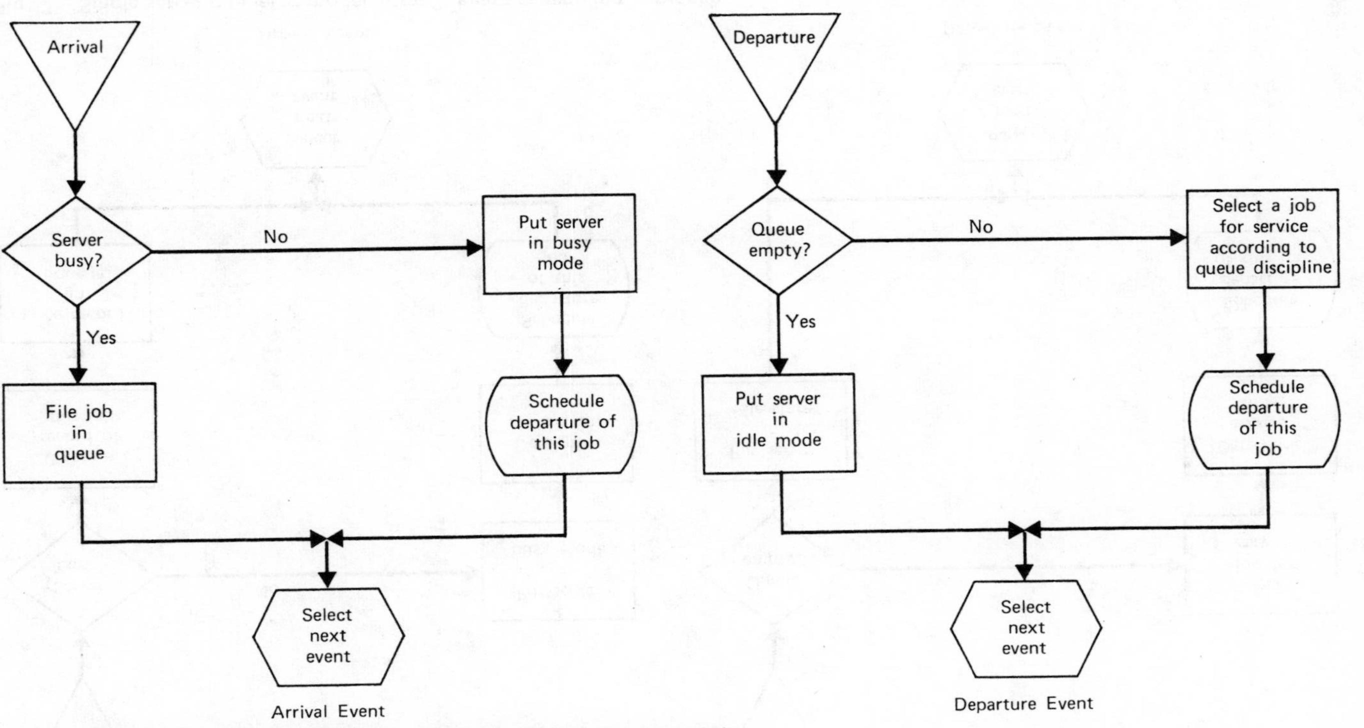

Fig. 6 Single server queueing model, form 1: event scheduling approach.

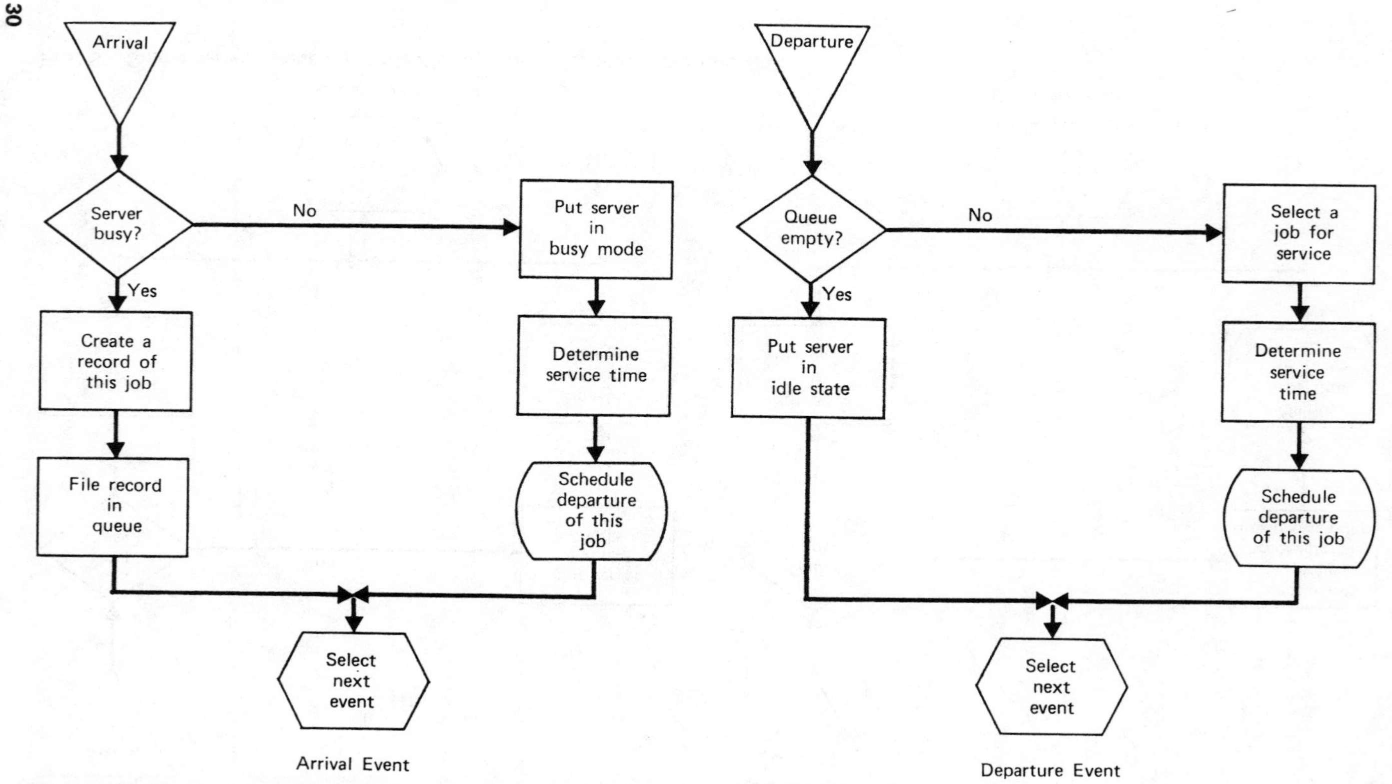

Fig. 7 Single server queueing model, form 2: event scheduling approach.

is filed in a special list. When the instruction to select the next event is encountered, the computer simulation *searches* this list to find and perform the event with the earliest scheduled time. Then simulated time is advanced to this scheduled time, thus skipping the "dead" time. This procedure is the essence of the *next event approach* to simulation.

The scheduling and event selection functions provide insight into the desirability of incorporating the conditional events that affect the server into the arrival and departure events. If, for example, the server becoming busy were formalized as a separate event, we would have to schedule such an event every time we find the server idle in the arrival event. Scheduling the event causes record creation and filing; moreover an additional search occurs in the event selection process. It is easy to see that the alternative method described in Fig. 7 eliminates the need for this additional record creation, filing, searching, and selection. The same considerations hold true when the server becomes idle in regard to the departure event. The ability to incorporate conditional events in this way can lead to a relatively large cost saving in applications.

Since ties may occur in scheduling events, a precedence rule is needed to establish order among them. For example, we may specify that if an arrival and a departure are scheduled for the same time the arrival event is to be effected before the departure event.

Whenever the event scheduling approach to simulation modeling is used, a computer program is needed to conduct a search of the list of scheduled events to determine which is the next to be executed. This *simulation control* program has many titles, among which the name *timing routine* is one of the most common. One convenience of using a simulation programming language is that it provides a timing routine based on the next event approach. Section 4.3 discusses this topic in detail.

As mentioned earlier, the queue is a set of jobs that are waiting for service. We may think of the queue as a *list* from which arrivals are selected for service according to a rule called the *queue discipline*. For example, jobs may be selected as follows:

1. in the order of their arrivals (first-come-first-served),
2. in the reverse order of their arrivals (last-in-first-served),
3. in the order of shortest service time,
4. in the order of longest service time, or
5. according to a priority number that each job has for service.

A queue discipline is a rule by which the job shop operates. In a computer simulation this rule is translated into a logical operating rule whose form depends on the queue discipline adopted. For example, adoption of the first-come-first-served (FCFS) discipline means that records are filed in the queue in order of their arrival times. The next job to receive service is then the first job in the queue. Figure 7 provides for this eventual search by including the creation of a record for each job that waits.

Adoption of the second queue discipline means that the jobs receive service in inverse order of their arrival times. If job records are filed in order of their arrival times in the queue, the next job to receive service is the last job in the queue.

If discipline 3 or 4 is adopted service time would have to be determined as the first computation in the arrival event and would have to enter into the record of jobs that queue. Incidentally, it is entirely possible for jobs to arrive with already determined service times, but, as Section 7.4 discusses, certain procedures should be avoided when generating arrival and service times at the same time in a simulation.

If selection is a function of a priority number (discipline 5), then this attribute must be part of the job record. If a tie occurs during selection, a second selection rule is needed to break the tie. For example, we may select the earliest arrival from among jobs with the same priority number.

There remain several procedural features which should be added to the arrival and departure events to enrich them. One fetaure concerns arrivals themselves. Suppose that the first arrival is scheduled externally to the arrival flowchart. Then one way to make the model self-generating is to determine the time of the next arrival and schedule it after the processing of the first arrival begins. This is done in Fig. 8.

Another feature pertains to recordkeeping. As Fig. 7 presently appears, all records of arrivals that wait remain in the model. Clearly the list of records grows as a function of simulated time and occupies increasing space in the computer. One way to limit this growth is to destroy a record when the corresponding job receives service, as shown in the departure event of Fig. 8. This step is discretionary, however, since the investigator may wish to save all records for a subsequent purpose. The idea of creating and destroying records and searching lists is one of the principal concepts on which discrete event digital simulation is based.

3.3.2 Data Collection

So far our model development has concentrated on the depiction of all changes that occur as time evolves. Since the number of such changes naturally increases with elapsed time, an investigator faces the problem of how to *summarize* the effects of all these changes into an account that is sufficiently compact to be manageable, yet sufficiently comprehensive to be useful. Since such a compact, yet comprehensive, analysis of behavior is an imperative for understanding any system, it is certainly a double imperative

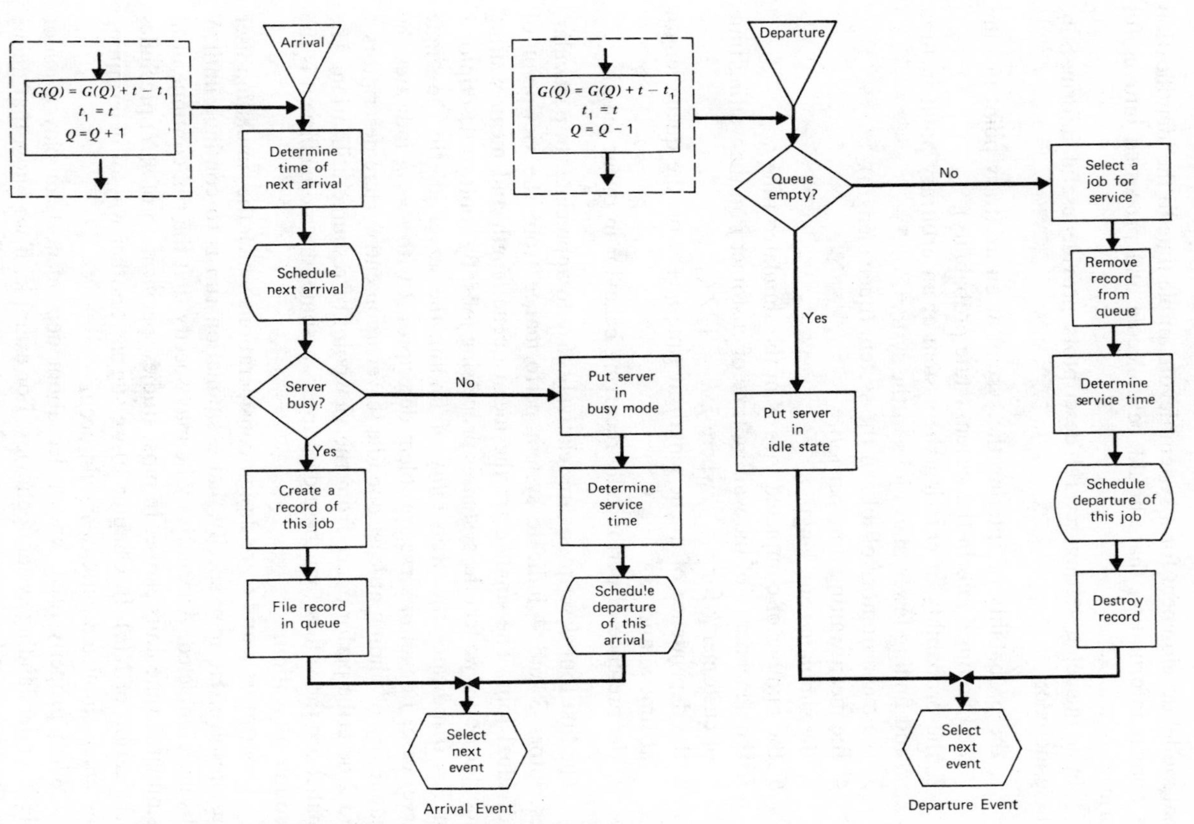

Fig. 8 Single server queueing model, form 3: event scheduling approach.

that model development for a system devote ample time to the identification of critical information that should be collected and processed into useful form.

In a queueing setting several descriptors provide useful information. These include:

1. the probability of entering the system at an arbitrary point in time and finding i jobs in the system (state probability),
2. the probability of entering the system at an arbitrary point in time and finding fewer than i jobs in the system,
3. the mean number of jobs in the system (queue length),
4. the mean waiting time per job,
5. the simulated run length n,
6. the number of completed jobs N in the simulated time t,
7. the dependence of the waiting time of customer j on the waiting time of customer k for $j > k$ where $j, k = 1, 2, \ldots,$
8. the dependence of queue length at time $u + s$ on the queue length at time u, and
9. the presence of periodically recurring behavior in queue length.

The first four descriptors are relatively easy to appreciate. In particular, descriptors 3 and 4 indicate system performance from the viewpoint of individual jobs. The smaller are the mean queue length and mean waiting time, the better we feel the system is processing jobs. By contrast descriptor 1 for $i = 0$ indicates the probability of finding the server idle in the single server case. For an m server problem descriptor 2 with $i = m$ indicates the probability of finding at least one idle server upon entry. Here descriptors 1 and 2 describe performance from the viewpoint of resource utilization. The smaller are the values that these descriptors assume, the more efficient is the resource allocation.

Descriptors 5 and 6 are design considerations which the simulation user must resolve. He may specify that a simulation run is to continue until N jobs are completed. Alternatively he may specify that the run continue until n simulated time units elapse. In most studies specification of $n(N)$ precludes specification of $N(n)$. In Chapter 10 we suggest another way of determining how much simulated activity should occur.

When properly used, knowledge about dependence can play a major role in understanding system behavior. For example, if we know that queue length changes little over time intervals of length s, and s is large, we would regard the system as being relatively unresponsive to its waiting workload. The presence of periodically recurring components in queue length alerts us to the presence of a source of variation that we may wish to treat in a special way, rather than regarding it as a component in an otherwise probabilistic system. We defer our account of the role of dependence in simulation work to Chapter 6. However, we wish to emphasize at this point that an understanding of the nature of dependence in a simulation is a prerequisite to any later claim of analyzing simulation results appropriately. The reader is advised to have this statement in mind in all simulation work.

At least two methods exist for collecting and processing data for descriptors 1 through 4. In method 1 we create a *collection event* to record queue length periodically at intervals Δt. Figure 9 shows this event with $\Delta t = 1$. Notice that the collection event generates itself recursively.

To understand how queue length data can be used to provide estimates of descriptors 1 through 4, we consider a simulation of specified time length n in which the queue lengths $X_1, \ldots, X_n$ are collected. Let p_i denote the state probability as defined by descriptor 1. Then

$$H_i = \frac{1}{n} \sum_{t=1}^{n} Y_{it},$$

where

$$Y_{it} = \begin{cases} 1 & X_t = i \\ 0 & X_t \neq i, \end{cases}$$

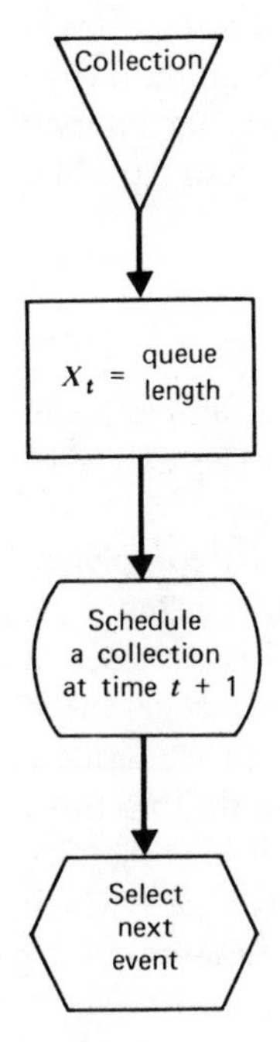

Fig. 9 Data collection event.

is an estimate of p_i and $\sum_{j=0}^{i-1} H_j$ is an estimate of descriptor 2. Since the mean number of jobs in the system is $\sum_{i=0}^{\infty} ip_i$, we estimate this quantity (descriptor 3) by

$$\bar{X} = \sum_{i=0}^{\infty} iH_i = \frac{1}{n}\sum_{i=0}^{\infty} i \sum_{t=1}^{\infty} Y_{it} = \frac{1}{n}\sum_{t=1}^{n} X_t.$$

A moment's reflection will convince the reader that, for m servers in parallel,

$$\bar{W} = \frac{n}{N}\sum_{i=m+1}^{\infty} (i - m)H_i = \frac{1}{N}\sum_{t=1}^{n} Z_t, \tag{3.1}$$

where

$$Z_t = \max(X_t - m, 0),$$

provides an estimate of mean waiting time (descriptor 4). Here nH_i is the sample mean time that the system spends in state i. Then $n(i - m)H_i$ is the sample total waiting time that occurs for state $i > m$. Summing $n(i - m)H_i$ over all $i > m$ and dividing by N as in (3.1) thus leads to a sample mean waiting time. Since n is the determinant of run length, (3.1) overstates the sample mean service time since it includes waiting time for incomplete jobs which have not been added to N. However, this influence should diminish as n increases.

An additional problem arises from the use of N in the denominator of (3.1) when the simulation has a stochastic input. Since a run length of n time units has been specified, the number of completed jobs N in that time interval is a random variable. Consequently (3.1) is the ratio of two random variables $n\bar{Z}$ and N where

$$\bar{Z} = \frac{1}{n}\sum_{t=1}^{n} Z_t.$$

In the present context, the form of (3.1) induces a bias in the sample mean waiting time. Section 10.5 describes one method of reducing the extent of bias in (3.1).

If an investigator elects to run his simulation until N jobs are completed, he may use the following method to compute the sample mean waiting time. Before the "queue empty?" block in the departure event of Fig. 8, insert a block to accumulate the completed job count N. After the job selection block, insert a block to record waiting time. At the end of the run accumulate these waiting times and divide by N to derive the sample mean waiting time. Notice, however, that we have not included the waiting time of jobs that are queueing at termination. This leads to an underestimate, albeit one which diminishes as N increases. This estimation of mean waiting time based on specified N avoids the bias produced by (3.1), when N is random.

The second method of computing relevant descriptors is to record the amount of time that the system spends in each possible state. For state i we denote that time as t_i. Clearly

$$n = \sum_{i=0}^{\infty} t_i.$$

Then the sample state probabilities are

$$H_i = \frac{t_i}{n} \qquad i = 0, \ldots, \infty.$$

Moreover the sample mean queue length is

$$\bar{Q} = \frac{1}{n} \sum_{i=0}^{\infty} i t_i$$

by substitution of the t_i/n for the true state probabilities in the expression for a mean expectation. Furthermore the sample mean waiting time for m servers in parallel is

$$\overline{W} = \frac{1}{N} \sum_{i=m+1}^{\infty} (i - m) t_i. \tag{3.2}$$

If the simulation contains a stochastic input, the H_i and $\bar{Q}$ are biased if N is the specified parameter for running the simulation; conversely $\overline{W}$ in (3.2) is biased if n is the specified parameter.

By means of this approach we may record the relevant data, using the blocks in dashed lines in Fig. 8. Here Q denotes queue length, and $G(Q)$ the accumulated time spent in state Q. At the end of the run $G(Q) = t_Q$, and we may use the preceding formulae to compute the statistics of interest. In subsequent flowcharts we denote the block with $Q = Q + 1$ (see Fig. 8) as $+$ and the block with $Q = Q - 1$ as $-$. Their inclusion will remind the reader of the need to provide for data collection.

Although the estimation of the population parameters mentioned above is straightforward, we wish to emphasize at this point that these conventionally employed data collection methods leave much to be desired in most queueing models. Because of the inherent randomness of the arrival times, service times, or both, the sample summary quantities that we compute are themselves subject to random variation. To make use of these quantities for decision making of any kind we must know how well they represent the true population parameters. Bias is one aspect of such an evaluation. Variance consideration is the other. In Chapters 10 and 11 we discuss how statistical analysis of the adequacy of results can be performed.

3.3.3 Activity Scanning Approach*

The single server queueing problem contains one activity, the servicing of a job. Accordingly the activity scanning approach calls for the modeling of this service activity. Figure 10 presents the model. Here there are two channels by which service can begin. Either a new arrival occurs when the server is idle, or else at the completion of a job at least one other job is awaiting service. Notice that no scheduling of events occurs here. After an event passes through the model, simulated time is advanced to the time of the next event. To perform this step, it is necessary that each entity that changes state have a *clock*. This holds for the next arriving entity and for the server entity. By checking these clocks the computer logic used in the activity scanning approach determines when the next event occurs, advances time accordingly, and recycles through the service activity. Then it is necessary to determine whether the event is a job arrival or completion. This contrasts with the event scheduling approach, in which a detailed list of scheduled events with their occurrence times are kept and the simulation control program selects the next event. In effect, the activity scanning approach substitutes logical tests in the model for the event scheduling and next event selection features of the event scheduling approach [24, p. 333].

Since the single server queueing model contains only one activity, we cannot adequately contrast the event scheduling and activity scanning approaches by means of this model. Nevertheless, it is instructive to conceptualize the difference now for the reader. The essential feature of activity scanning appears when there are several activities. Then, whenever time is advanced to the next event, *all* activities are scanned to determine which can be begun or ended. The event scheduling approach requires that the beginning and end of each activity be events. Consequently, they must be scheduled for execution. As the number of activities grows, so do the number of scheduled events and, correspondingly, the amount of computer time spent in creating records of these events, filing them in the list of scheduled events, selecting them for execution, and destroying them once they have been executed. The activity scanning approach, however, substitutes less time-consuming, logical checking in the model at each time advance for the required event scheduling steps.

When there are few activities but many arrivals, the appeal of the event scheduling approach in relation to the activity scanning approach becomes more evident. Fewer activities call for fewer events; however, the limited information available in the activity scanning approach requires repeated scans to guarantee that all possible state changes occur. Many arrivals mean many scans. The reader can begin to appreciate this point when studying Fig. 10. By contrast, the recursive generation of arrivals in Fig. 8 makes the

* Reference 28 contains interesting examples of the activity scanning approach.

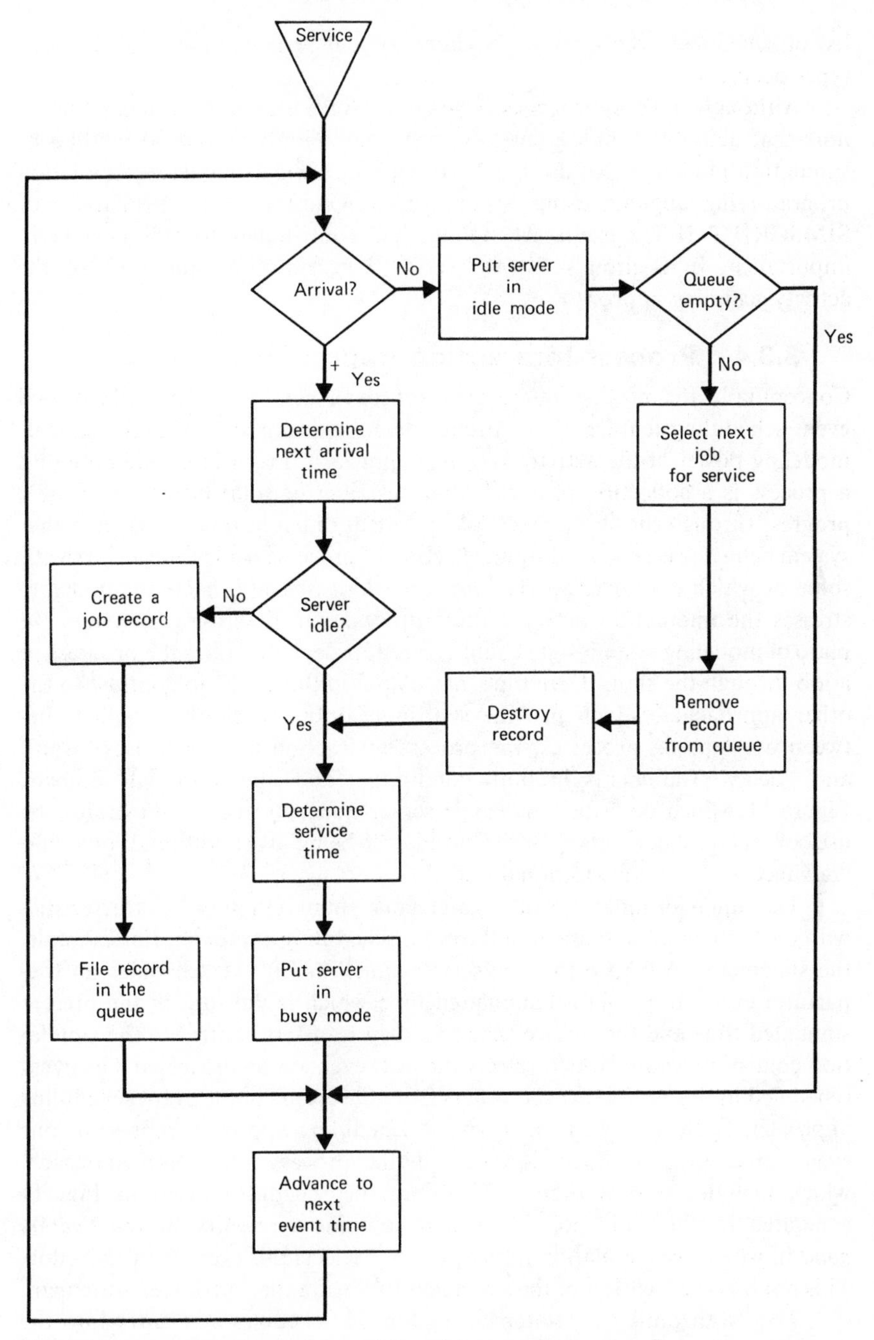

Fig. 10 Single server queueing model: activity scanning approach.

list of scheduled events relatively short, even if arrivals of several different types occur.

Although both approaches have attractive features, it is interesting to note that activity scanning has not been a commonly employed method of simulation modeling. No doubt, the absence of a widely available simulation programming language using this approach, in contrast to the availability of SIMSCRIPT II for event scheduling, has contributed to this decline in importance. In Section 4.4 we describe a recent re-examination of the activity scanning approach.

3.3.4 Process Interaction Approach

Conceptually the process interaction approach combines the sophisticated event scheduling feature of the event scheduling approach with the concise modeling power of the activity scanning approach. From Fig. 3 we note that a process is a collection of events that describe the total history of a job's progress through the shop. Since jobs arrive at different times, it is clear that system behavior is described by a collection of processes (one for each arrival), some of which may overlap. The *process interaction* approach to modeling stresses the interaction between these processes in describing a system. In place of modeling system state change, this approach describes the progress of a job through the shop. Here time may elapse in the model in contrast to the other approaches, which provide snapshots only. To resolve conflicts between overlapping processes, the process interaction approach uses "wait" and "delay" statements in both conditional and unconditional contexts. Figure 11, which describes the single server queueing model, illustrates the use of such statements, "wait until . . . " being a conditional one and "advance . . . " an unconditional one.

The unconditional "wait" statements share common characteristics with the "schedule" statement of the event scheduling approach. For example, the statement "Advance time by service time" in Fig. 11 schedules an (departure) event at a point in simulated time which is the sum of the present simulated time and the service time and then transfers control to the simulation control program, which selects the next event to be processed The event scheduled in this way, however, differs from the events of the event scheduling approach. Each flowchart in the event scheduling approach represents one event, in contrast to each flowchart in the process interaction approach, which includes several events. Therefore the scheduled event in Fig. 11 generated by the "advance" statement contains a re-entry or *reactivation point* to which the simulation returns when it selects this event for processing. This point is the location of the statement following the "advance" statement.

The "wait until . . . " statement in Fig. 11 schedules a re-entry into the flowchart when the job is selected for service. This conditional scheduling procedure cannot assign an occurrence time when the "wait until . . . "

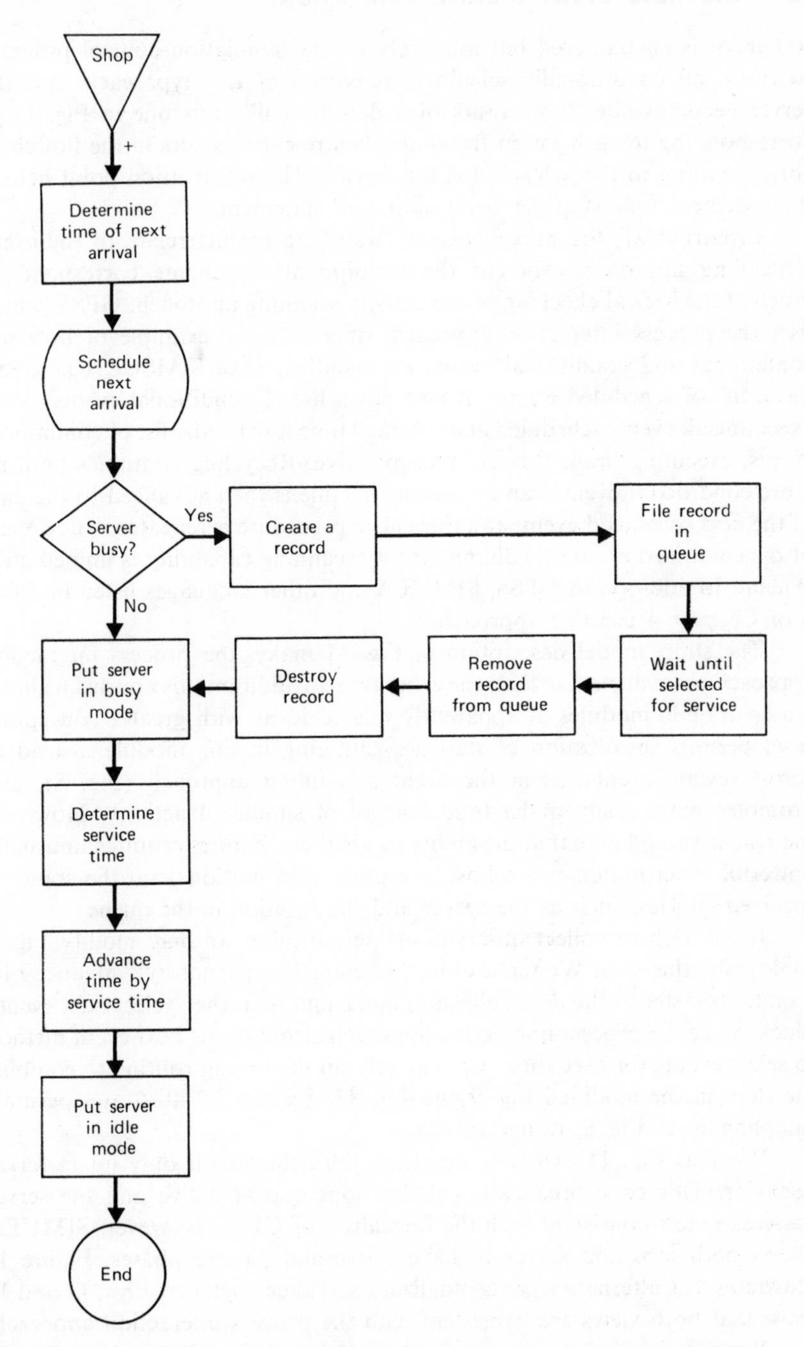

Fig. 11 Single server queueing model: process interaction approach, concept 1.

statement is encountered but must rely on the simulation control program to check all conditionally scheduled re-entries of this type each time the server becomes idle. If we think of a flowchart like this one in Fig. 11 as corresponding to each job in the shop, then re-entry occurs in the flowchart corresponding to the job selected for service. The reactivation point here is the statement following the "wait until . . . " statement.

In particular, the unconditional "wait" statements relate to the event scheduling approach, whereas the conditional statements correspond to much of the logical checking of the activity scanning approach. GPSS, which uses the process interaction approach, offers a good example of how unconditional and conditional events are handled. Like SIMSCRIPT, GPSS has a list of scheduled events. It also has a list of conditional events. After executing all events scheduled at a specified time it scans the list of conditional events, executing those that are then possible. Recycling continues until no more conditional events can be executed. Time is then advanced to the date of the next scheduled event, and this entire procedure is repeated. The appeal of the combined event scheduling-activity scanning capability is immediately evident. In addition to GPSS, SIMULA and other languages listed in Table 7 of Chapter 4 use this approach.

The short model description in Fig. 11 makes the process interaction approach more attractive than the other two. In addition this approach allows a user to build modules of apparently related ideas with greater conceptual ease, permits localization of statistics gathering in one module instead of across several events, as in the event scheduling approach (Fig. 8), and promotes more easily understood control of simulated activity. However, the reader should note that the ability to use these features requires unusually powerful macrostatements whose execution is conditional on the states of specified entities, such as the server and the location in the queue.

If we wish to collect observations periodically, we may modify Fig. 9 to describe this step. We replace the "schedule" statement by "advance t by 1 unit," transfer to the data collection block and omit the "select next event" block. Since the process interaction approach also uses the next event method to select events for execution, we may rely on the timing routine to combine the steps in the modified Fig. 9 and Fig. 11. Section 5.3 illustrates periodic sampling, as in Fig. 9, using GPSS/360.

Whereas Fig. 11 explicitly describes job behavior, it only infers server behavior. This conceptualization makes jobs appear active and the server passive, a view consistent with the formalism of GPSS. However, SIMULA allows both jobs and server to have active and passive phases. Figure 12 illustrates this alternative conceptualization. Taken together, Figs. 11 and 12 show that both views are consistent with the process interaction approach.

Reactivation points are an inherent feature of Fig. 12 as well as Fig. 11. In Fig. 12 the reactivation point for a job is the block in which the record is

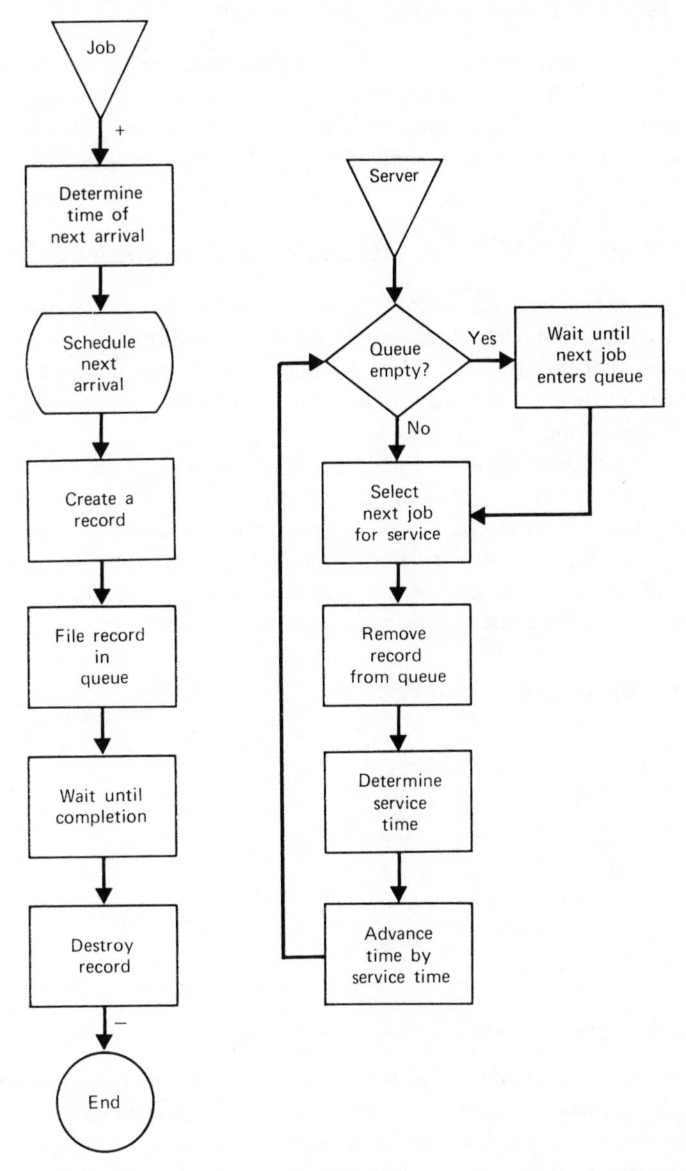

Fig. 12 Single server queueing model: process interaction approach, concept 2.

destroyed. The server has two reactivation points, one being the check for an empty queue and the other being the selection of a job. Notice that in order for the server to function he must be activated somewhere outside the model, just as the first job must be scheduled externally to its flowchart.

3.4 MORE COMPLEX QUEUEING PROBLEMS

Section 3.3 introduced the concepts of scheduling; record creation, selection, and destruction; and list and list search in a relatively simple queueing model. Since most queueing problems met in practice are more complex than our illustration, we want to extend the use of the above concepts to handle these more difficult problems.

One way of looking at queueing problems is in terms of the *tasks* that require service in any one job and the *resources* to be used in performing any one task. Table 3 provides a convenient array for depicting six possible arrangements. Here we assume that a distinct task requires a unique combination of resources. Added to this task-resource relationship is the choice of a rule for selecting tasks to be performed by a specific resource or server.

Table 3 Tasks and Resources

	Distinct Tasks		
Required Resources	One	One of Several Alternatives	Many
One	X		
One of several alternatives	X	X	
Many	X	X	X

3.4.1 One Task—Many Servers

A job consists of a single task, as before, but there are many servers who can provide the necessary service. Here we form a roster (list) containing a record for each server. This list is of constant length in contrast to the queue. Each server's record contains at least two attributes; one is status, and the other is name or identification. When a single task arrives, the roster is searched for an available server. Here a preference ordering is necessary to choose among servers. If arrivals have no preference, a selection rule based on an attribute unique to each server is necessary. For example, if servers are numbered 1, 2, 3, . . . in the name attribute, then one selection rule is to pick the available server with the lowest number.

Figure 13 shows one way of describing this many server queueing

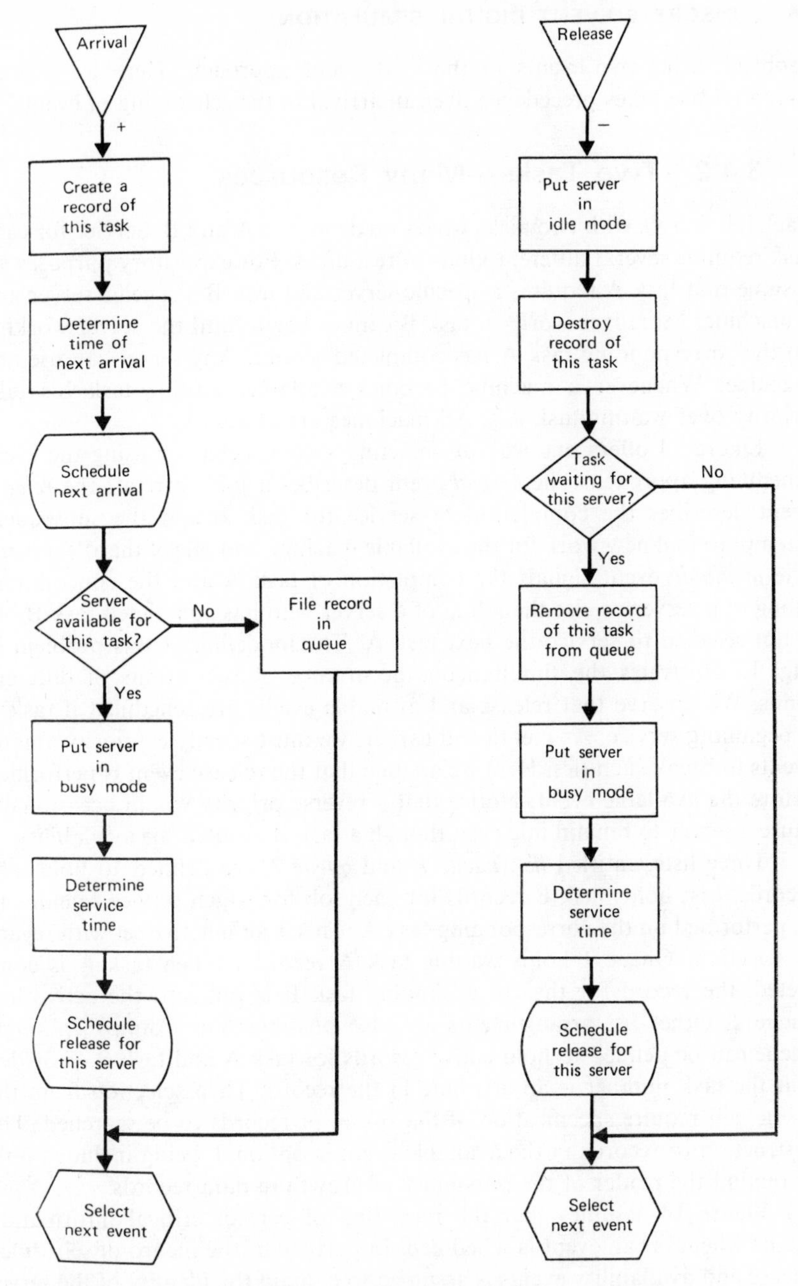

Fig. 13 Multiserver queueing model: event scheduling approach.

problem, using two events in the next event approach. Here we assume that a release takes precedence over an arrival in the scheduling of events.

3.4.2 Two Tasks—Many Resources

Each job arrives with two tasks which we denote as A and B. Service for each task requires several different kinds of resources. For expository purposes we assume that task A requires a specific server and task B a specific server and a machine. Moreover work on task B cannot begin until the server working on the corresponding task A has completed service. Any server can use any machine. Whenever a machine becomes available, waiting task B's take priority over waiting task A's. All machines are alike.

Figure 14 offers one way of depicting system behavior using the event scheduling approach. The *arrival* event describes a job's arrival; the *release* event describes the completion of service for task B and the subsequent attempt to find new work for the available machine and one of the idle servers. The *available* event signals the completion of task A and the concomitant idling of a server or else the idling of a server who has completed task B but is not selected to service the next task A. The modeling of this problem in Fig. 13 illustrates the simultaneous occurrence of two events of different kinds. We observe that release and available events are scheduled if task B is beginning service. As mentioned earlier, we must specify a priority among events to break such ties. Here we assume that the release event is performed before the available event. Notice that a reverse priority would occasionally cause a server to remain idle even though a task B awaited his availability.

Three lists, entitled *list*, *queue 1*, and *queue* 2, are defined to hold task records. List holds task B records for each job for which service remains to be performed on the corresponding task A. This is an inactive set with regard to selection. Queue 1 holds waiting task A records. When task A is completed, the record for the corresponding task B is put into the active list, queue 2. Other list arrangements are also possible. For example, a single queue can be defined to hold active records for task A and task B, provided that the task number is an attribute in the record. Then selection from the queue will require specification of the subset of records to be searched. The destruction of records in the available event is optional, being included only to remind the reader of the possibility of growth in data records.

Figure 14 assumes that the recording of certain critical information occurs whenever an event is scheduled. In particular, the record of scheduled release and availability events is assumed to contain the identity of the server and the machine, if one is used, for the task whose completion is being arranged. As a result, when execution of either of these events begins, the relevant server or machine is identified. Were this information lacking in the

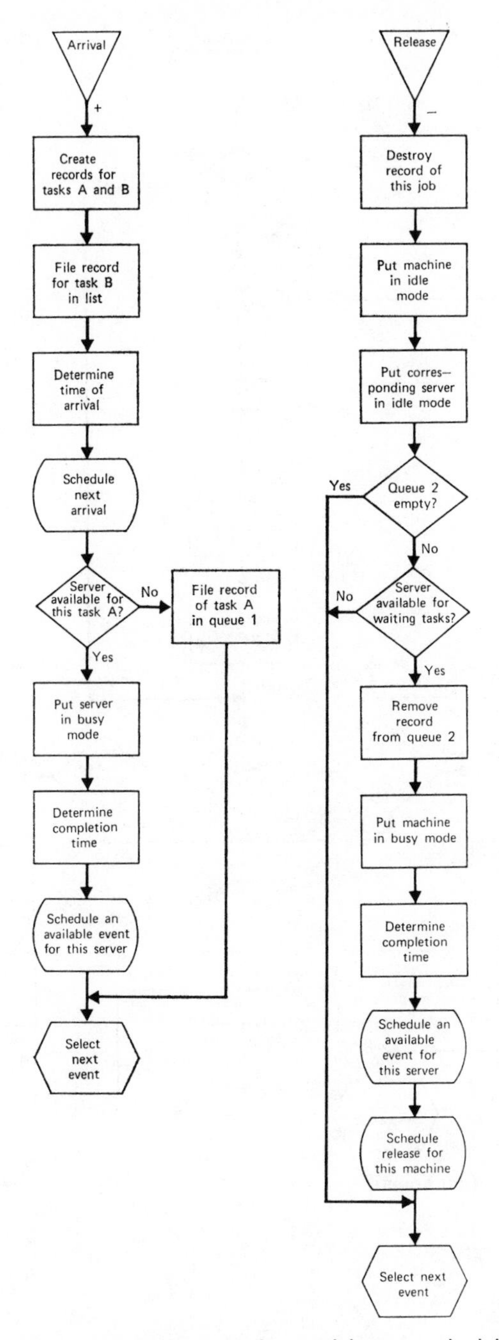

Fig. 14 Two task-multiresource queueing model: event scheduling approach.

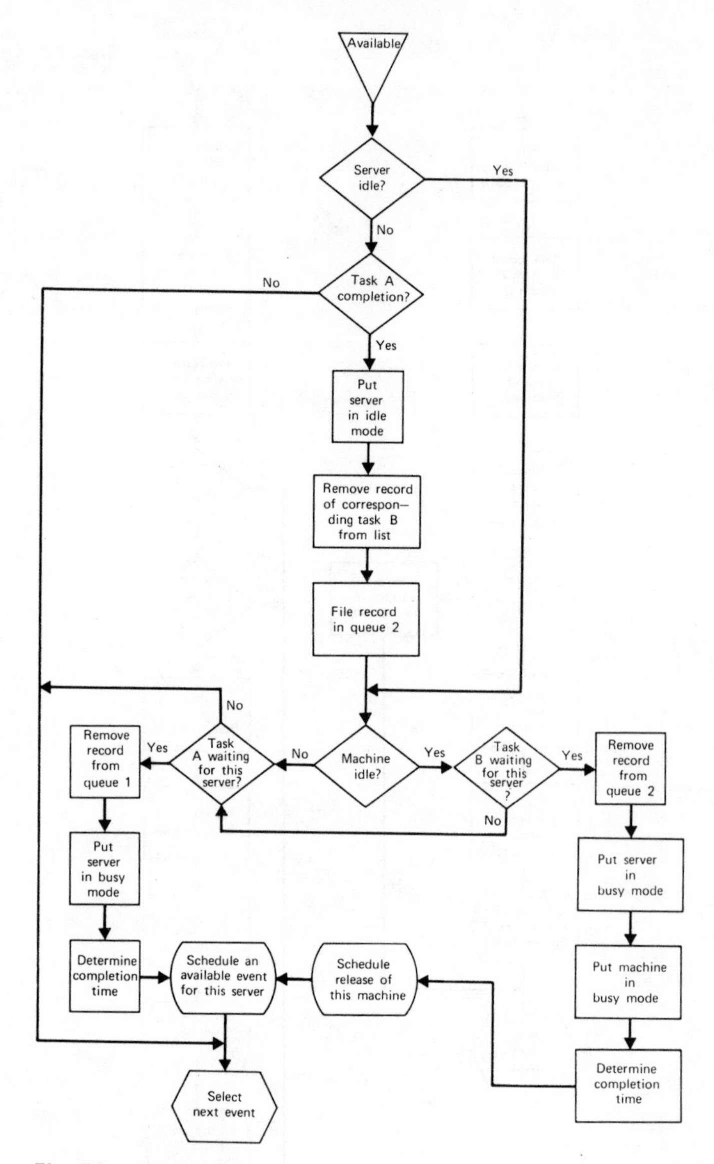

Fig. 14 (continued)

record of the scheduled event, then each server and machine would require the attribute of completion time and a search would be necessary to identify the correct server and machine.

Figure 15 illustrates the activity scanning approach to the two task-multiresource problem. Notice the considerably more concise form of the model compared to that of the event scheduling approach. However, also notice that in every pass through the model *both* activities are scanned. In practical terms this approach is less efficient than the event scheduling technique since it checks for all possible state changes at every time advance. This observation becomes more obvious as the number of activities to be scanned increases. To retain a balanced view, we also note that more activities usually mean more events and consequently the list of scheduled events becomes larger.

Figure 16*a* shows one way of representing the two task-multiresource queueing problem using the process interaction approach. Here simulated time elapses as a job progresses task by task through the shop. Conditional and unconditional time delays again appear in the "wait" and "advance" blocks, respectively.

A slightly different representation of this problem illustrates a flexibility that the process interaction approach offers but Fig. 16*a* does not show. In Fig. 16*b* we have created two flowcharts instead of one. Here progress through the second flowchart does not begin until service is completed on the corresponding task A. However, in other problems the series arrangement need not be so. This division into several flowcharts assumes special importance in problems in which jobs arrive with tasks that may be performed in parallel. In particular, the division enables the *time-oriented* blocks in each flowchart to interact with the simulation control program as becomes necessary, rather than being forced into the series arrangement that Fig. 16*a* represents. The depiction of this problem with active and passive resources as well as tasks is left to the reader as an exercise.

3.4.3 An Inventory Problem

Consider a shop that stocks one type of item. Shoppers purchase the item one unit at a time if it is in stock. Otherwise some place an order for it, whereas others cancel their demand. In developing an inventory policy the shopkeeper faces the need to balance the cost of holding inventory against the cost of lost sales. We refer to these as *inventory* and *stockout* costs, respectively. In addition he incurs a fixed *setup* cost every time he places an order.* Let time be measured in days.

* This problem has received attention in the simulation literature [21, 22] regarding desirable experimental designs. We discuss one design aspect in Section 11.13.

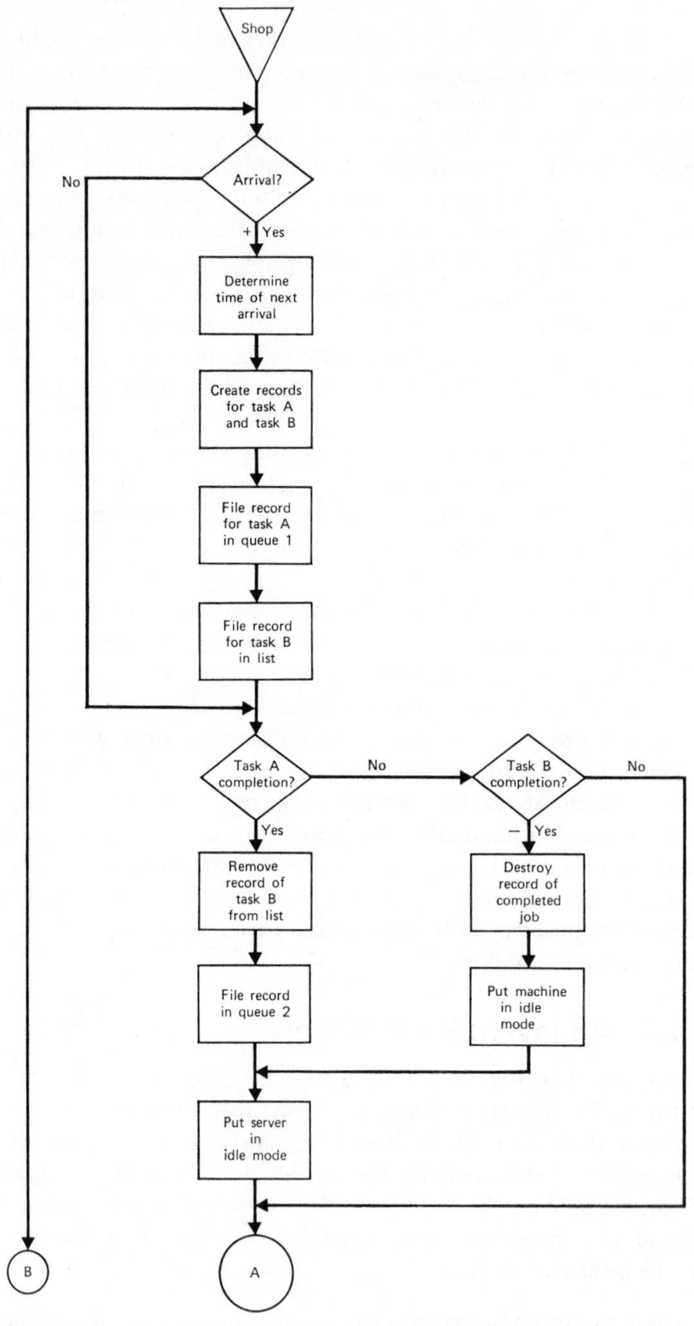

Fig. 15 Two task-multiresource queueing model: activity scanning approach.

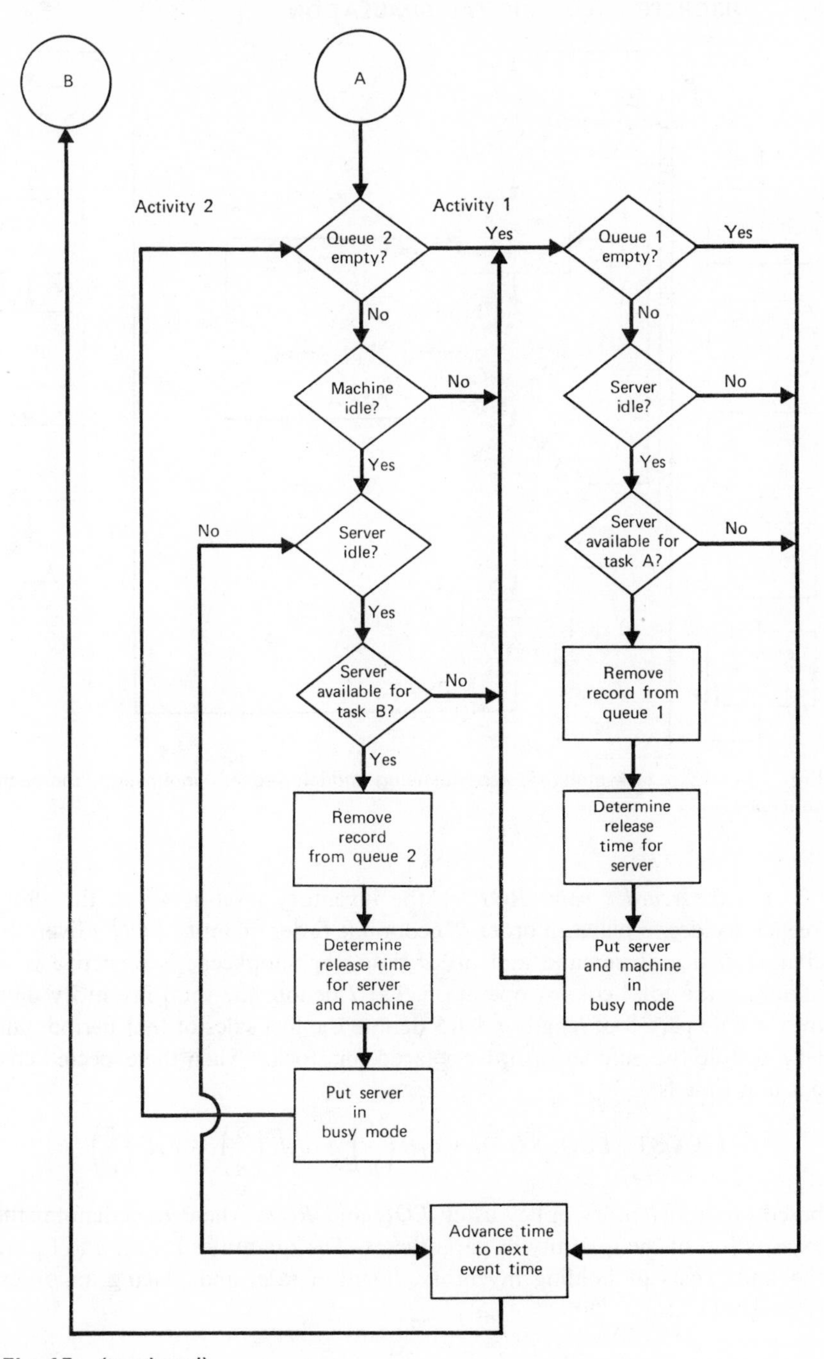

Fig. 15 (continued)

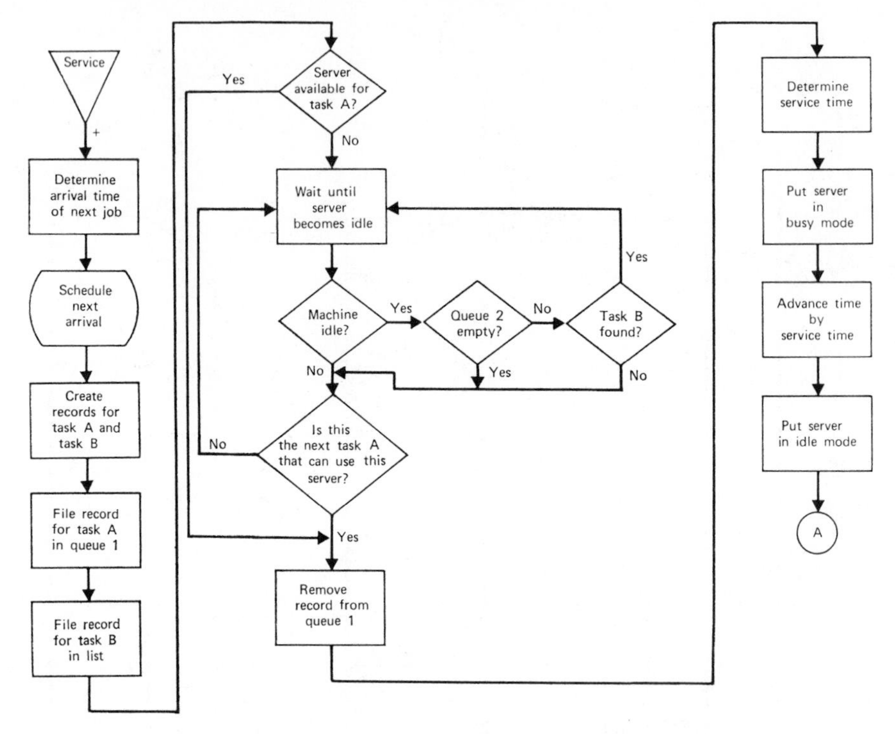

Fig. 16*a* Two task-multiresource queueing model: process interaction approach, concept 1.

Let the *reorder point ROP* be the inventory level at which the shopkeeper decides to place an order of economic order quantity *EOQ*. Given the characteristics of demand and order time, the shopkeeper's objective is to minimize the total cost of operation. Let I denote the total inventory days over a time period of length n, let S denote the lost sales of that period, and let p denote the number of orders placed for stock.* Then the expected cost per unit time is

$$E(\text{COST} \mid EOQ, ROP) = c_1 E\left(\frac{I}{n}\right) + c_2 E\left(\frac{S}{n}\right) + c_3 E\left(\frac{p}{n}\right)$$

based on an inventory policy using *EOQ* and *ROP*, where $E(\cdot)$ denotes the expectation of the quantity in parentheses. The quantities c_1, c_2, and c_3 are the units costs of holding inventory, losing a sale, and placing an order, respectively.

* Figure 17 shows how to compute inventory days.

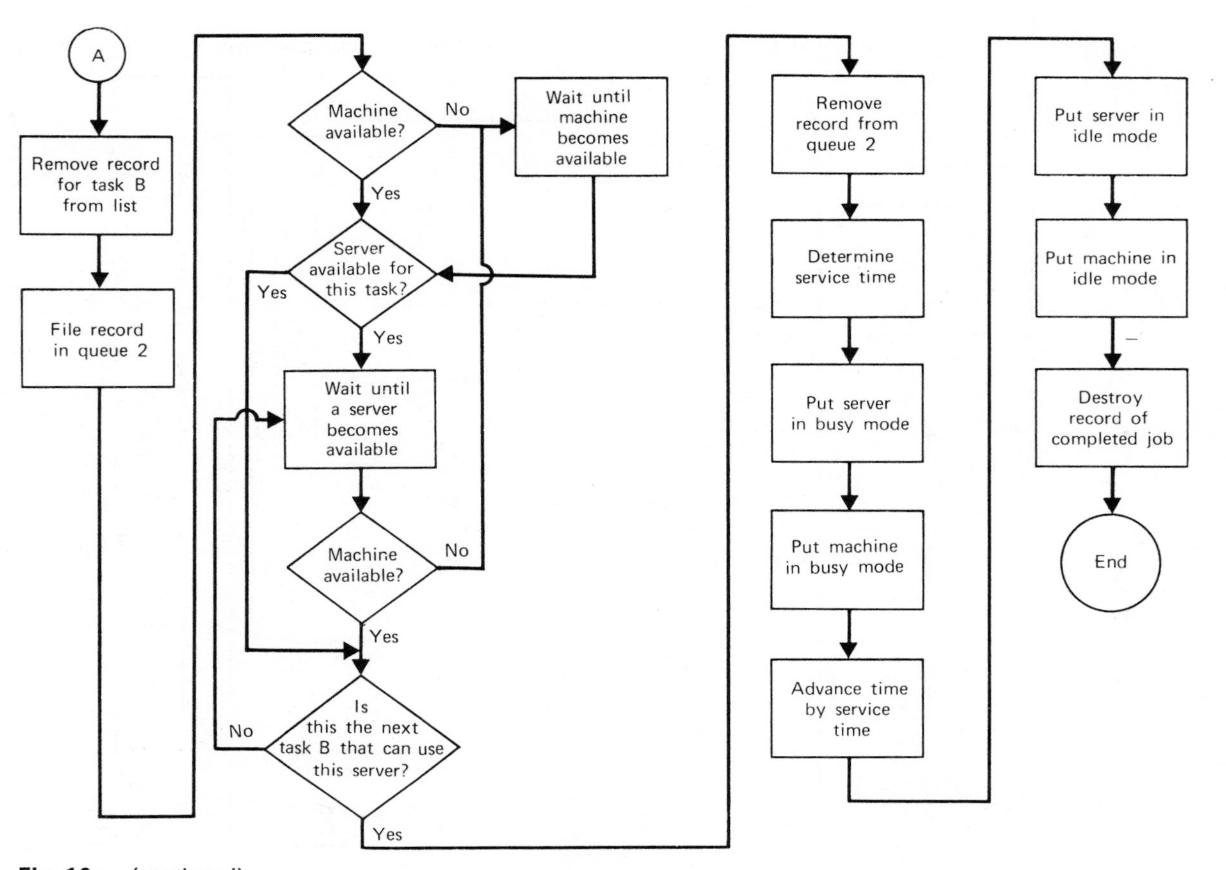

Fig. 16*a* (continued)

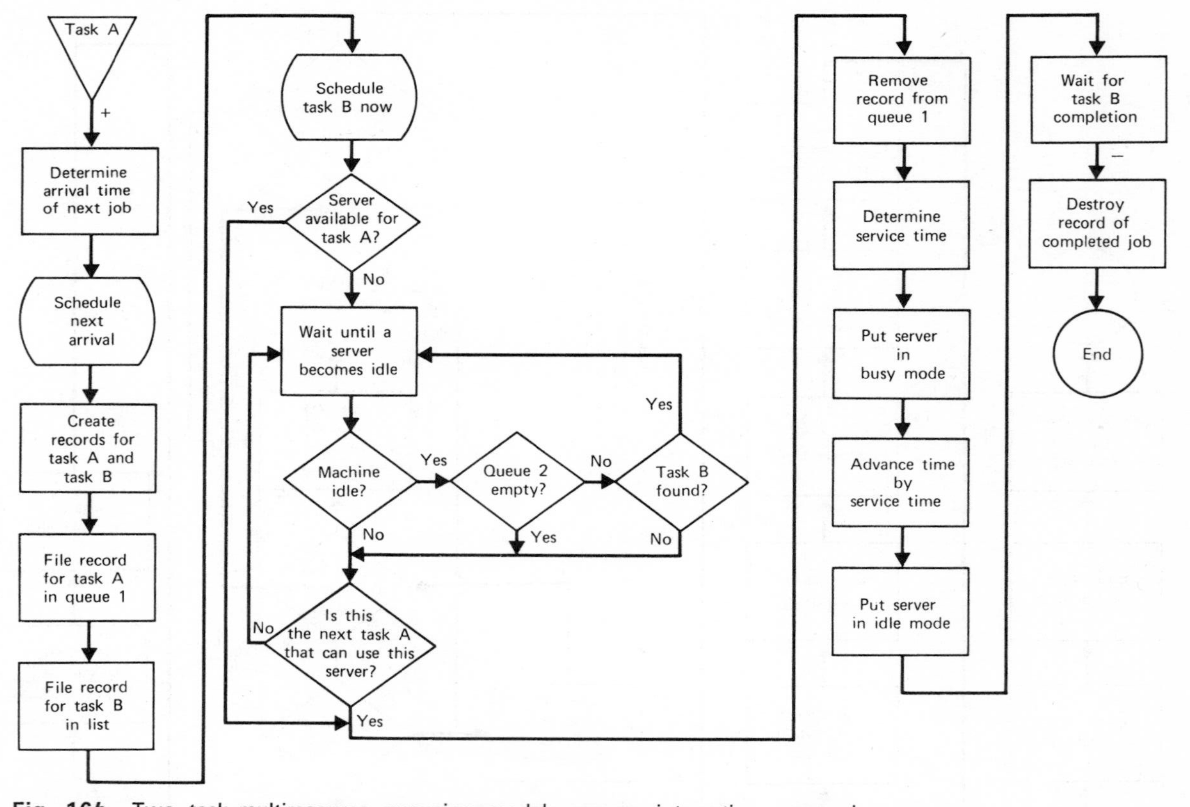

Fig. 16*b* Two task-multiresource queueing model: process interaction approach, concept 1.

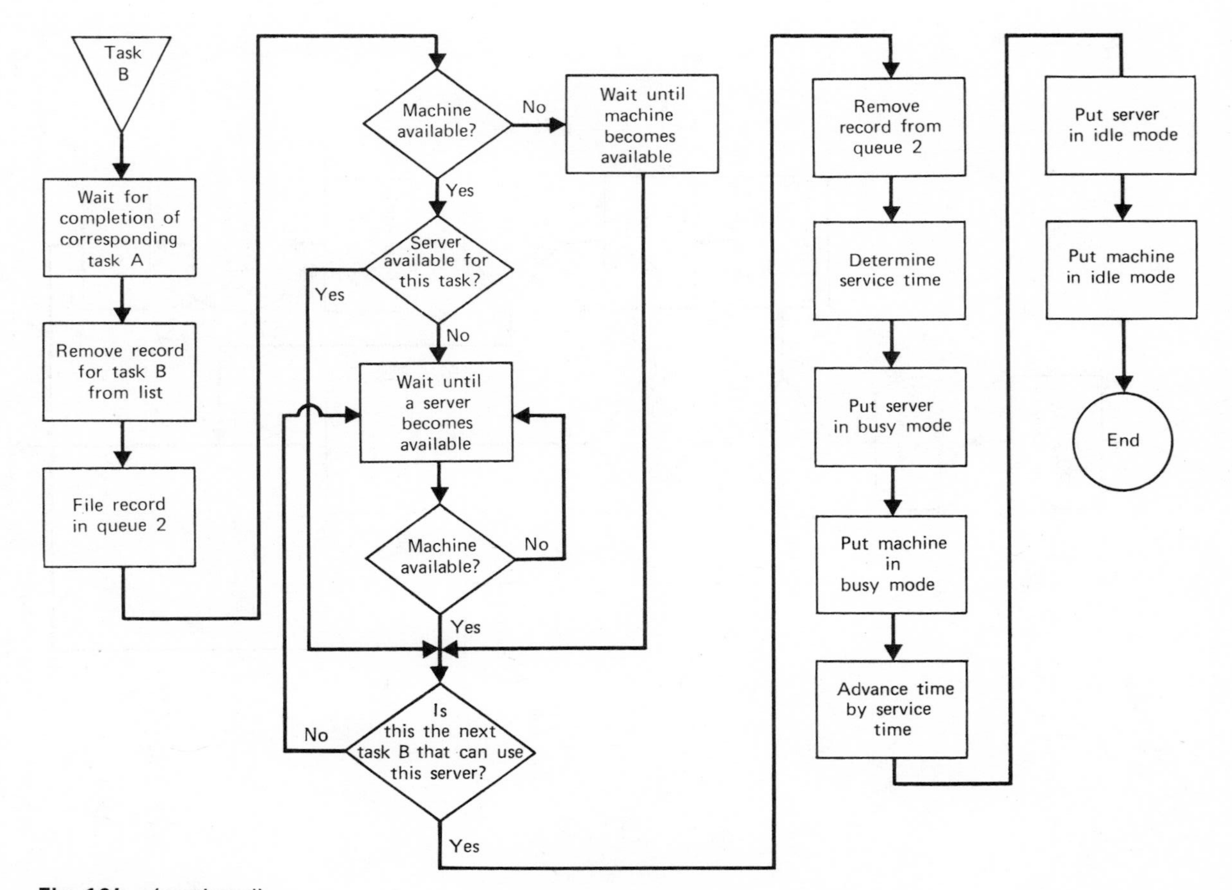

Fig. 16*b* (continued)

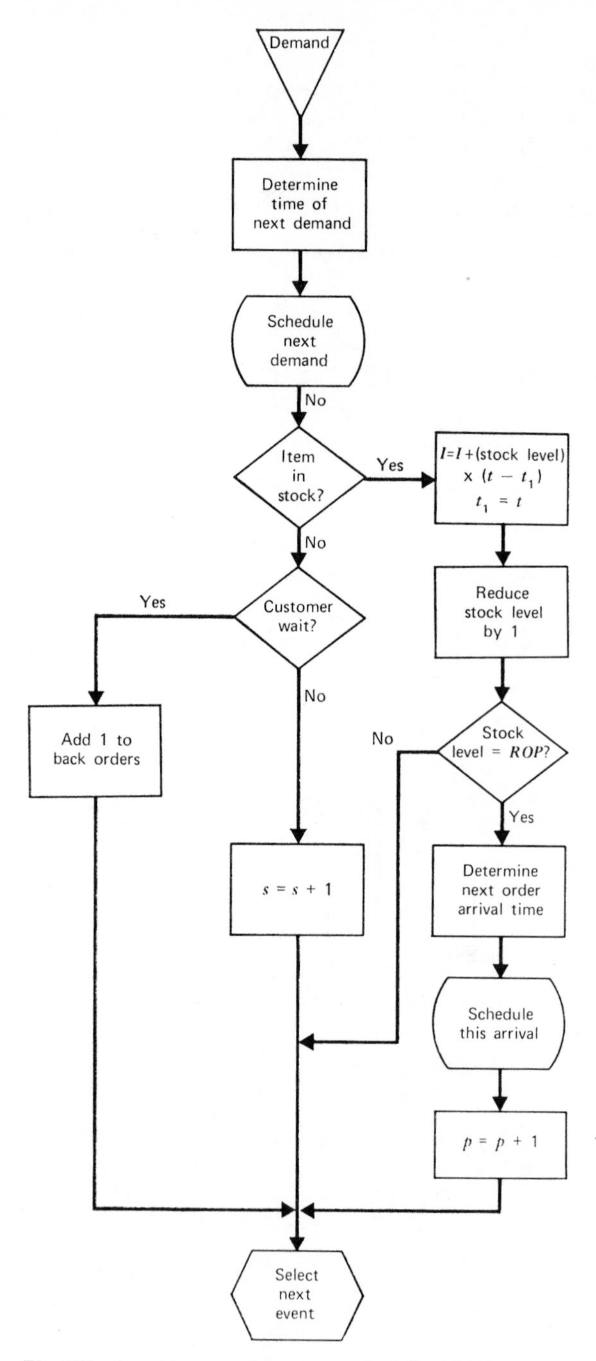

Fig. 17 Inventory model: event scheduling approach.

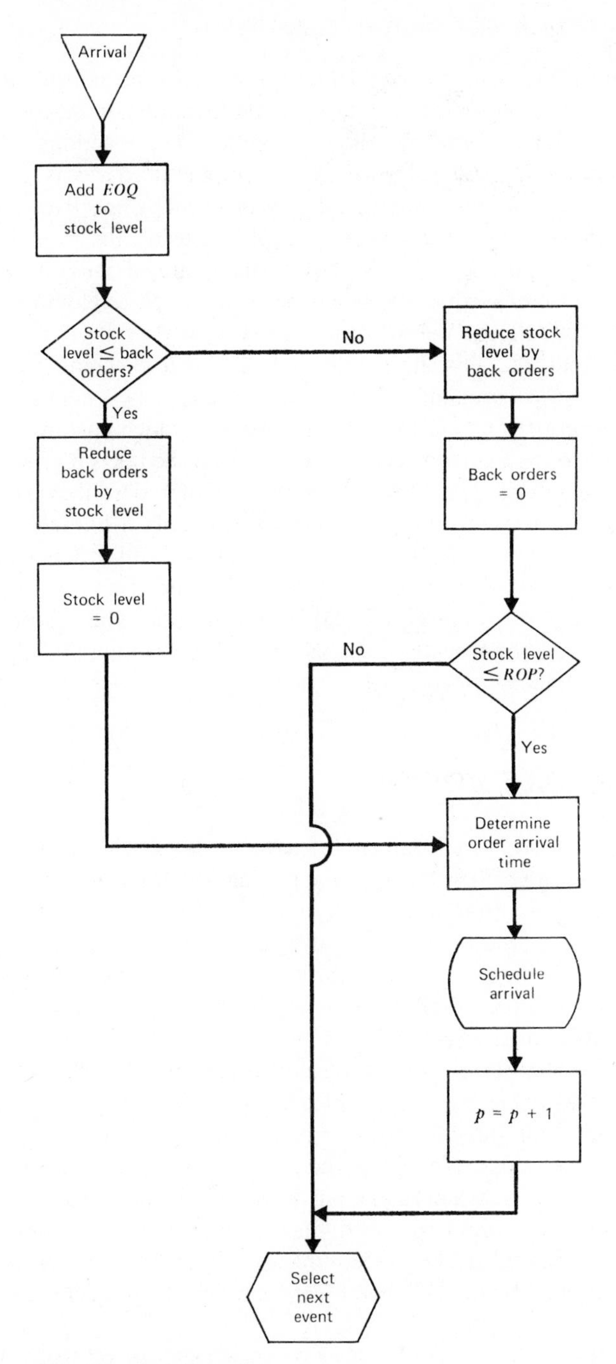

Fig. 17 (continued)

To model this inventory problem we note that three considerations contribute to cost: increases or decreases in the inventory level, loss of a sale, and placement of an order by the shopkeeper. These actions occur either when a customer demand occurs or when a shipment arrives at the store. We label these as the *demand* event and *arrival* event, respectively. Figure 17 schematically describes this inventory problem using the event scheduling approach. The quantities I, p, S, and t_1 are assumed initially set to zero. Periodic observations can easily be collected on stock level and back orders by scheduling a collection event as in Fig. 9.

Notice that the flowchart performs no recordkeeping for customer orders. This is a consequence of two facts. One is that system performance, as defined here, pays no attention to the order in which customer orders are filled. Second, once a customer has agreed to wait, he takes no further action. In contrast, Exercise 3 at the end of this chapter describes an inventory problem in which a customer is willing to wait only a limited time for his order to be filled. Then the storekeeper may elect to fill orders starting with the customer whose allowable waiting time is least. Here every order that waits will have an attribute of allowable waiting time. The creation of order records facilitates the storing of the value of this attribute and the selection of orders to be filled.

3.5 PERT NETWORKS

A powerful tool for management control of large-scale programs is the procedure known as PERT (Program Evaluation Review Technique). This system of charting into a network key milestones for the accomplishment of an objective, dependent on many diverse factors, was first developed in conjunction with the Polaris program [25]. The resulting management control contributed to the program becoming operational two years earlier than originally anticipated.

Figure 18 shows an example of a PERT network.* The *nodes* are the milestones to be achieved, and the *arcs* are the activities to be performed in achieving these milestones. The time to perform each activity may be predetermined or random. Node 1 is the *source* node, from which all activity begins. A milestone is achieved at a particular node when all activities leading into that node are completed. Activities from a node can commence only after all incoming activities are completed. Node 9 is the *terminal* node. When it is reached by the last incomplete activity, the project is complete. By

* The GASP II book, [32], by Pritsker and Kiviat describes the event scheduling program details for PERT networks.

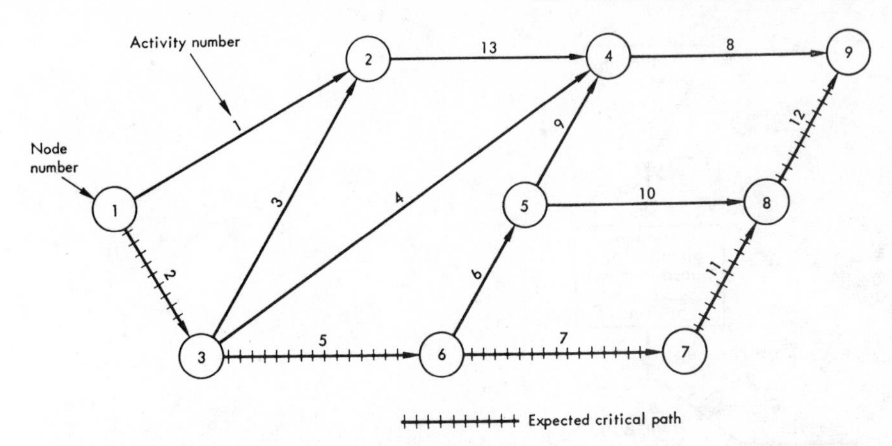

Fig. 18 PERT network. *Source:* A. Alan B. Pritsker and Philip J. Kiviat, *Simulation with GASP II: A Fortran Based Simulation Language*, © 1969, p. 199. Reprinted with permission of Prentice-Hall, Inc., Englewood Cliffs, N.J.

recording the arc number corresponding to the last activity to be completed at each node we can trace the *critical* path of the network. The critical path thus denotes the route requiring longest time to traverse the network. When the activity completion times are random, the critical path may differ on several replications of an experiment. In Fig. 18 there are six possible paths, and if we replicate the experiments many times and identify the critical path on each replication we can construct a sample frequency distribution (histogram) for the critical path. In addition we can form a histogram for the critical arc which identifies the most commonly occurring activity in the critical path. By identifying the critical path and arc we can often take action in the real project to modify activities and thus reduce the time required to traverse the network.

For planning purposes it is often desirable to know the completion time of the network. Here too we can form a histogram, but it is for a continuous random variable rather than a discrete one.*

The modeling of a PERT network using the event scheduling approach is shown in Fig. 19. Here each node has two attributes giving the numbers of activities that terminate and originate there. Each activity has the attributes of origin, destination, completion time, and specification with regard to how its completion time is determined. Notice that only the completion event is needed to carry out the entire movement, provided that each action has attributes that identify its associated actions.

* See Section 10.2.

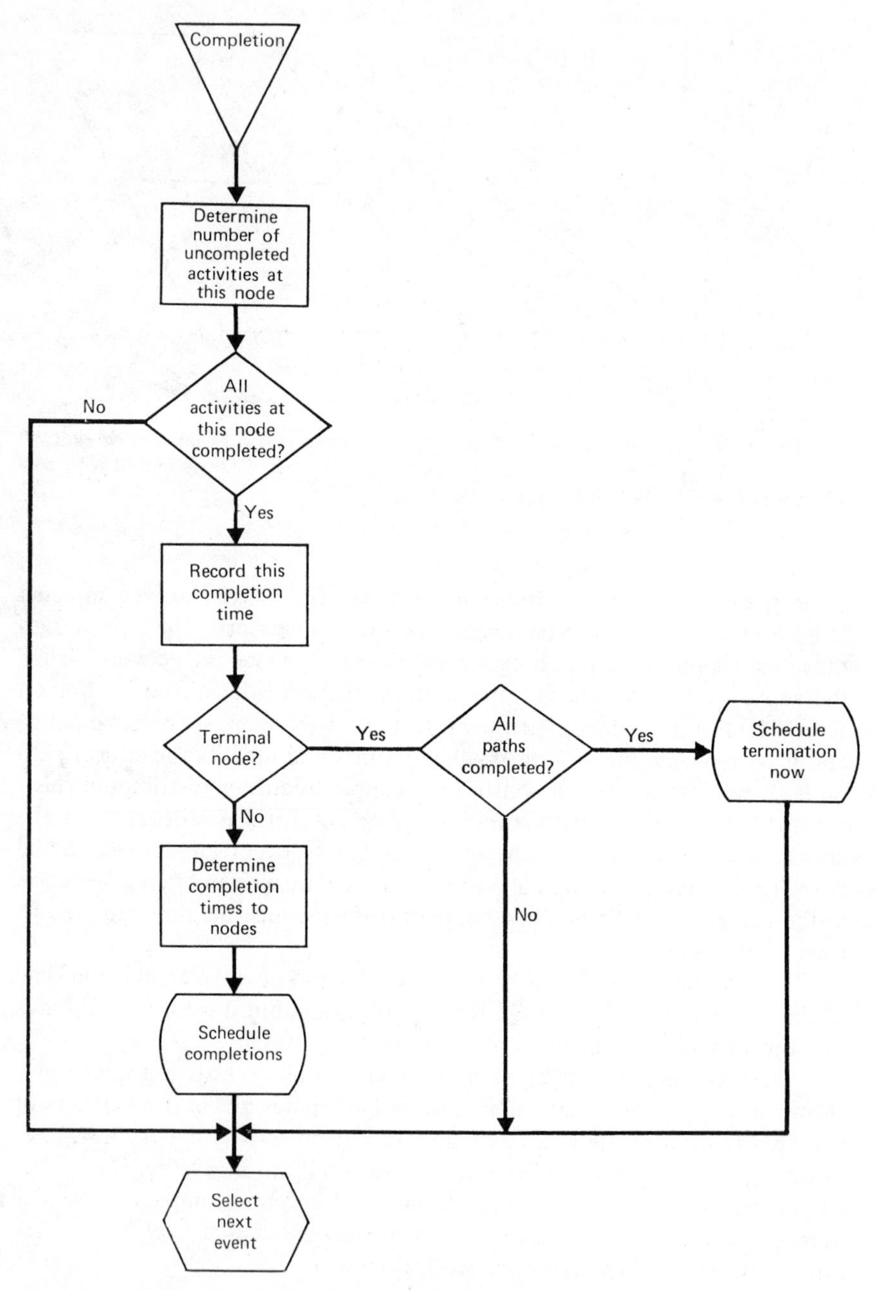

Fig. 19 PERT Network model: event scheduling approach.

A termination event is needed to make the PERT network analysis complete. This event, which is scheduled only once per run, contains the statements for providing the summary computations and report. An example of a termination event is given in Section 4.7.

3.6 MULTITASK-MULTIRESOURCE PROBLEM*

Consider a multiresource maintenance shop where each job consists of several tasks. Stations exist in the shop, each with a finite set of resources. Once the activities leading to a station are completed, work begins on the tasks to be undertaken at that station. Since generally not enough resources are available to begin all activities at a station simultaneously, it is necessary to establish a priority rule to determine the order in which activities are begun. The objective is to find priority rules that minimize a job's transit time through the shop.

One convenient way of structuring the problems is in terms of arcs and nodes as in the PERT network problem. An arc denotes an activity, and when an item reaches a node along a particular arc it has completed that activity. A node denotes a station at which activities terminate and begin. Problems of this type involve highly interactive processes for which the activity scanning approach is most appropriate. However, it would be necessary to describe each activity in a flowchart. Therefore we limit ourselves to a presentation of the event scheduling approach to this problem.

First, all tasks are put in an *initial* file. A *completion* event as in Fig. 20 is then carried out for the source node. Here records of the tasks emanating from the source nodes are removed from the initial file and placed in a queue. Those than can be undertaken immediately are removed from the queue and put in the *in-process* file. Completion events are scheduled for each activity. Whenever a completion event occurs, the record of this task is removed from the in-process file and placed in the *completion* file. When the terminal node activities are completed, a termination is scheduled.†

Figure 20 provides for no testing of components. However, we can easily add testing to the flowchart by inserting a test block between blocks 3 and 4. If an item passed the test the flow would continue as before to block 4. Otherwise the flow would go to a block that put the task record back in the queue and then transferred to the queue selection block.

* See reference 23 for an elaborate example of the analysis of a multitask-multiresource problem using the event scheduling approach. Techniques exist specifically for analyzing multitask-multiresource problems as stochastic networks. These techniques are due to A. A. Pritsker and are known as GERT. See reference 31 for an introduction to GERT.

† This model is discussed in reference 32, p. 215.

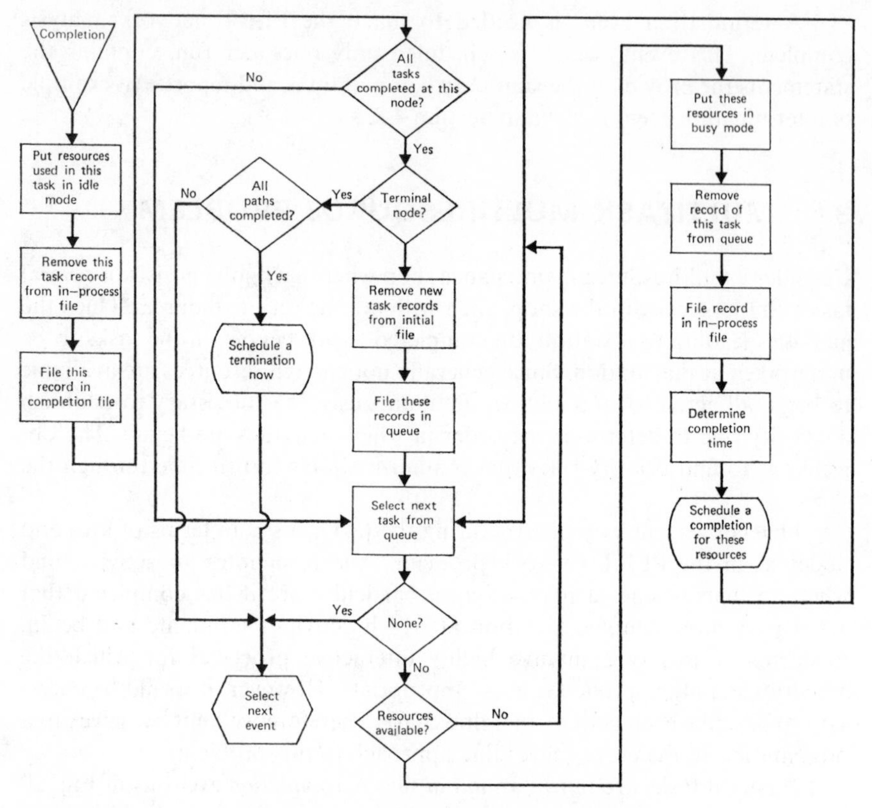

Fig. 20 Multitask-multiresource problem: event scheduling approach.

As in the PERT network example the objective would be to measure the time flow through the shop. This can again be accomplished by collecting the completion times of the last item to reach each node. The critical path and critical node can then be identified. We have many more options at our disposal for affecting flow time than are available with the PERT network. In addition to completion or service times, we also have the choice of task selection and resource dispatching rules to vary.

3.7 PERIOD MODELING

So far our discussion has emphasized the explicit modeling of all discrete state changes that occur in a system. In some systems many discrete changes occur over short time intervals, but each individual change has little effect on gross system behavior. In such a case explicit modeling of all changes can create data and model manipulation requirements that are so onerous as to jeopardize successful analysis of the system under study.

Since subaggregations of these state changes interest the analyst more than individual changes do, one convenient model adjustment is to allow time to advance in fixed increments Δt and to update the model at times $\Delta t, 2\,\Delta t, 3\,\Delta t, \ldots$ to reflect the changes that occur during the previous interval Δt between updates. For a particular system the sampling interval Δt is chosen so that successive interval changes do not fluctuate widely and so that treatment of all changes as occurring at the points $\Delta t, 2\,\Delta t, 3\,\Delta t, \ldots$ does not seriously distort the interpretation of system behavior.

Differential equations are often used to model continuous event systems. When analytical solutions are not feasible, a simulation using the differential equations is conducted on an *analog* computer. However, by selecting a sufficiently small time interval over which relatively little change occurs, the differential equations that represent a continuous event system can be replaced by corresponding difference equations, thereby *digitizing* the inter-event time and enabling a discrete event digital simulation to be undertaken.

Customarily a general purpose language such as FORTRAN is used to program these models. However, special purpose languages and packages exist, among which DYNAMO is the most notable. DYNAMO provides a concise means of programming models based on *Industrial Dynamics*. Developed at M.I.T. by Forrester and his associates,* Industrial Dynamics provides a set of concepts that facilitate study of the performance of an individual firm in response to alternative management policies. These concepts can also be applied to an entire industry.

Models are formulated in terms of *rates* of flow within the system and *decision* functions that specify the ways in which rates depend on system levels. By explicitly acknowledging feedback, Industrial Dynamics enables a user to construct a model whose output is a response to internally as well as externally generated stimuli. Although time advances continuously here, the fixed time advance feature conveniently allows a DYNAMO program to produce updated performance measures periodically.

A traffic flow network offers another example of a continuous event system in which the distribution of automobiles by location constantly changes as time passes. However, we can usually define a time interval over which the location distribution changes relatively little. Sakai and Nagao [33] created such a discrete event digital simulation and applied it to an area of Kyoto, Japan.

An economic system represents another continuous event system. Because of the sheer complexity of interactions among economic phenomena

* See reference 20 for a review and bibliography on industrial dynamics.

and the availability of data only on time intervals of constant length, economists have traditionally used discrete events or distributed lag models to describe the behavior of economic systems evolving through time. The book by Orcutt, Greenberger, Korbel, and Rivlin [27] provides an early example of the application of discrete event digital simulation to the study of socio-economic systems.

As an example of fixed time advance modeling, we consider a hospital blood bank in Fig. 21 similar to one that Malkani and Pegels studied in [26]. Every morning a central blood bank delivers a supply of blood to the hospital. This delivery, which contains units of varying ages, is added to the inventory of the hospital blood bank. During the day requests arrive for units to be transfused. In each case the age of the demanded blood is specified. If the demand can be met, the corresponding unit is removed from the inventory. If it cannot be met, an available unit with a younger age is used. If none is available, an emergency order is placed to the central blood bank. The number of emergency orders is recorded to assist in the evaluation of different inventory policies on the performance of the inventory system.

In practice it is customary to remove blood units from the inventory when they have reached a specified age. This is shown in Fig. 21 and the number of outdated units are then recorded for system evaluation purposes. For example, in Malkani and Pegels' study one objective was to study the effect of outdate age on the number of units outdated. The size of the next day's order is then determined according to the blood bank's ordering policy, time is advanced by one day, the ages of the units to be delivered are assigned, and a recycling throughout the flowchart occurs.

The simulation begins in block 1 with the specification of an initial inventory of blood units by age. Block 2 computes the frequency distribution of blood by age. Block 3 determines the number of units to be transfused on a given day by generating a random number and comparing it with the cumulative probability distribution of the number of units transfused daily.*

Block 4 identifies which of the total number of units specified in block 3 are to be transfused. The age of each unit is determined by a random draw from the blood inventory distribution. After each draw, the age distribution must be recomputed to reflect the selection of a unit with a specific age and consequently its removal from the distribution. Block 4 is repeated until the correct number of units are selected.

Suppose that the policy is not to transfuse blood that is more than m days old. Once selection is completed in block 4, any m-day-old blood that remains in the inventory will be $m + 1$ days old on the next day and hence

* Chapter 8 describes the theory behind this approach to generating stochastic events and also how this can be accomplished using a table of the cumulative probability distribution.

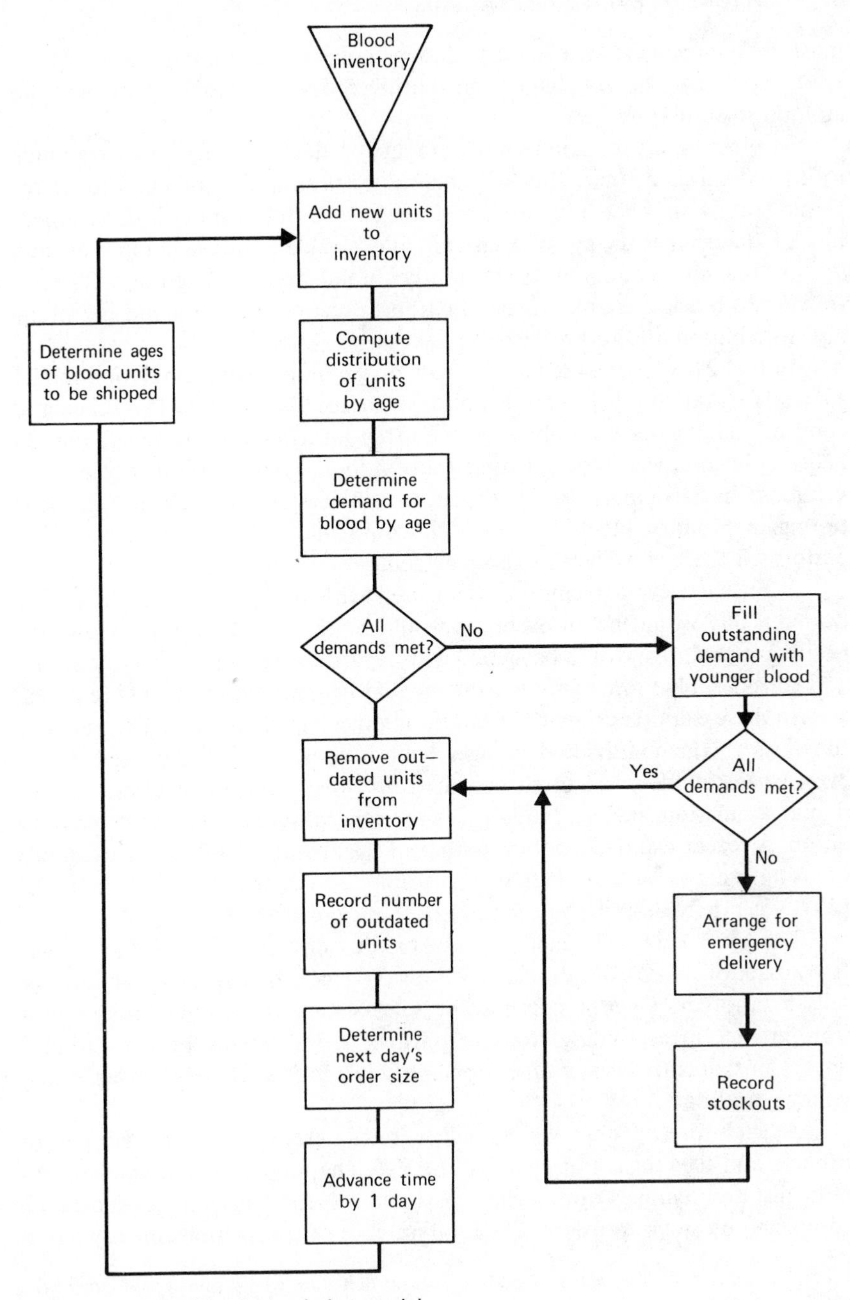

Fig. 21 Blood bank simulation model.

must be removed from inventory. Block 5 checks for outdates and, if any exist, block 10 removes them from the inventory and block 11 records the amount of outdating.

In block 6, all remaining units are aged 1 day. This action corresponds to a fixed time advance. Block 7 determines the number of units to be received from a supplier by a random draw.* Once this number is determined, block 8 determines the age of each unit, and block 9 adds the number of each age to the corresponding count in the blood bank's inventory. Then a transfer to block 2 enables the simulation to recompute the blood inventory age distribution and to produce the next day's activity.

In Fig. 21, we have used a flowchart of the blood bank simulation model similar to that in [26]. This enables us to see that our earlier schematic conventions are not the only way to convey information. However, Fig. 21 helps us to appreciate the convenience of adopting some of our earlier conventions. In particular, use of the decision flow symbol given in Fig. 4 for testing in place of block 5 in Fig. 21 would make the computation being performed there considerably more evident.

If the next event technique were used with this problem, each request during a day would be an event, as would be the blood delivery. However, nothing substantive would be gained with regard to the objective of the study, and considerably more computer time would be required to process requests.

In these examples the need for a timing mechanism no longer exists since the times at which individual changes occur hold no particular interest. However, period models can often benefit from other features of discrete event digital simulation, such as random deviate generation and data base manipulation. A recent Rand Corporation study of the operations of a stock exchange offers an example of the role that set manipulation can play in a period model [30]. We briefly describe this feature of the Rand work.†

Let $\Delta t = 1$ day. On each day brokers buy and sell stock to each other. A transaction represents the daily volume of stock that one broker buys or sells to another. Once a purchase or sale occurs, the selling broker must transfer the corresponding stock certificates to the buying broker within 5 days. Consequently every transaction entity has four attributes: buyer, seller, volume, and date.

At each time advance the model generates a transaction for every pair of brokers and files these transactions in a list. The model then reviews entries in the list and removes from it those past and present transactions that can be completed by stock transfers. The total number of transactions on any day is

* This modeled behavior seems at odds with true behavior where one would expect the blood bank to request a specific number of units.

† The presentation here is a considerable simplification of the Rand model. However, the essential features of the list processing problem are easily recognized.

a random variable, as is the date on which a particular transaction is completed. Hence the number of entries in the uncompleted transaction list on any day is also a random variable. The dynamic character of list length resembles that of the queue list in earlier sections, and we can easily recognize the benefits that accrue when a *list processing* language is used to run such a model on a computer.

The presence of random variation calls for an ability to generate random variates from specified distributions. In particular, the Rand model requires generation from normal and Bernoulli populations. In addition to providing the necessary list processing capability, the selection of SIMSCRIPT II for programming the stock transactions model enabled the investigators to generate random variables from these distributions with a minimum of effort.

EXERCISES

Explicitly specify all entities and their attributes.

Exercises Based on Examples

1. In the single server queueing problem of Section 3.3.1 a test is performed to see whether the task has been successfully accomplished. If the job passes the test, it leaves the repair facility as usual. If the job does not pass the test, it receives priority for service from the server, once he has finished the task he is currently performing. The job is then filed at the "head of the queue" to await the server.
   a. Modify Fig. 8 to reflect this testing procedure in the event scheduling approach.
   b. Modify Fig. 11 to reflect the testing procedure in the process interaction approach.
   c. Modify Fig. 12 to reflect the testing procedure in the process interaction approach where the server, as well as the jobs, can be active.
2. a. Modify Fig. 8 so that the queue discipline can depend on service time instead of on arrival time.
   b. Modify Fig. 11 so that the queue discipline can depend on service time instead of arrival time.
3. Consider the inventory problem described in Section 3.4.3. Suppose that customers are willing to wait for a specified time for an item. If the item does not arrive within this specified waiting time, the customer cancels his order. Modify Fig. 17 to show this behavior.

4. Model the inventory problem of Section 3.4.3 using the process interaction approach in two alternative ways. Identify active and passive entities.

5. Modify the flowchart in Exercise 4 to reflect the behavior described in Exercise 3.

6. Model the one task-many server problem of Section 3.4.1 using the process interaction approach.

7. Model the inventory problem of Section 3.4.3 using the period modeling approach of Section 3.7.

Exercises Based on Structured Descriptions

8. *Parallel Service:* Consider a barbershop with m barbers and n manicurists. Customers arrive one at a time. Each customer wants a haircut and may want a manicure. A haircut costs $3.25; a manicure, $2.00. If a barber is not available, some customers won't wait. When a customer's haircut and manicure, if requested, are complete, he pays his bill and departs. Haircuts take different times, but all manicures take 15 minutes. A haircut or manicure can begin any time during a customer's stay in the barbershop. For example, his manicure (haircut) can begin before his haircut (manicure) begins, after the haircut (manicure) begins, or after the haircut (manicure) is completed. However, once a customer who originally requested a manicure has his haircut completed without having his manicure begin, he may elect to leave without receiving the manicure.

   The objective here is to study the relationship between the specified inputs m and n and the revenue output. Model the problem using the event scheduling approach. *Hint:* One way is to describe three events: the arrival of a customer, the availability of a barber, and the availability of a manicurist.

9. Model the barbershop problem using the process interaction approach.

10. *Series Service:* Consider a community dental clinic staffed with m oral hygienists and n dentists. Some patients arrive by appointment; others, unscheduled. A dentist examines every new patient and determines how many visits are needed to complete dental work. If a patient has an acute problem, the dentist treats it after completing the examination. If the patient does not have an acute problem, or if he has, after he has been treated for it, he returns to the waiting room. An oral hygienist eventually cleans his teeth, and he is then given an appointment for the next visit, if required. When the patient returns for his second visit, a dentist treats him and he is given an appointment for a third visit, if required.

Patients who have had a cleaning within the last year are considered old patients. Accordingly a new patient's status changes from new to old immediately after a cleaning. Some patients do not keep their appointments. Once this occurs it is assumed that the patient will never return. Scheduled patients make their initial appointments by calling the clinic and being given an appointment time. Occasionally a patient who has had a cleaning within the last year and has no scheduled visit calls for an appointment. When he arrives at the clinic, he is treated as a new patient except that he does not receive a cleaning. When an old patient comes to the clinic for an examination without an appointment, he is also treated as a new patient except that he too receives no cleaning. With regard to patient selection, scheduled patients have priority over unscheduled ones, except when an examination reveals an acute problem, which then has highest priority.

The objective here is to determine how the appointment algorithm, the number of dentists, and the number of hygienists affect waiting time and queue length. The mean waiting time, queue length, and dentist idle times are the performance measures of interest. Model this system using the event schedule approach.

11. Model the community dental clinic using the process interaction approach and regarding patients as the only active element.

12. Model the community dental clinic using the process interaction approach and permitting patients, hygienists, and dentists to be active or passive.

Unstructured Exercises

13. Model a supermarket for the purpose of determining how many checkout counters to activate as a function of the rate of entry of customers.

14. Model a library with reference and circulation services to determine how to allocate available staff most effectively.

CHAPTER 4

PROGRAMMING CONSIDERATIONS AND LANGUAGES*

4.1 INTRODUCTION

In Chapter 3 our discussion of modeling implicitly assumed a capability to carry out a variety of instructions in a computer simulation. These included creating, filing, and destroying records; searching lists; generating random variates; collecting and analyzing data; and model initialization. All are programmable instructions. Some, however, call for more sophisticated programming techniques than others. The commonality of these and other features to most discrete event digital simulation has led to the development of special purpose *simulation programming languages* that facilitate the translation of these modeling instructions into a code recognizable by a computer.

This chapter describes the common programming features of discrete event digital simulation and the programming needs they imply. Chapter 5 discusses the way in which three simulation programming languages—GPSS, SIMSCRIPT II, and SIMULA—meet these needs. By way of introduction we note that GPSS, the most commonly used simulation programming language, has very powerful statements that relieve a user of the need to program for many of the features mentioned above. It also has a flow orientation that makes it appealing for many queueing problems. We discuss this language in Sections 5.2 and 5.3. SIMSCRIPT II, the most comprehensive simulation language available, offers all the above features in a setting that allows a user greater freedom in programming arithmetic and logical

* This chapter relies heavily on Kiviat's review [48].

operations than is afforded by GPSS. However, this language takes longer to learn and is not as generally available as GPSS. We discuss SIMSCRIPT II in Sections 5.4 and 5.5. SIMULA, an ALGOL-related simulation programming language, also has distinct features and is described in Sections 5.6 and 5.7.

GASP II, which is a package of FORTRAN subroutines especially written for discrete event digital simulation, offers many desirable features within the setting of a more widely available language [59], namely, FORTRAN. Many other simulation programming languages exist, and we also describe some of their characteristics in this chapter.

Many reasons exist for preferring simulation programming languages over languages such as ALGOL, FORTRAN, and PL/1, ease of modeling and minimal programming being the most obvious. However, the decision as to which programming language to use depends on several factors, of which knowledge and availability are the most critical. If a user has never employed a simulation programming language but knows FORTRAN, he may prefer to use it for his simulation work simply because of a reluctance to invest time in learning a new language. Sections 4.2 through 4.4, which describe the necessary features of simulation programming languages, are intended to convey the scope of the task that a user of a non-simulation-oriented language faces. For a moderate-size problem the considerations involved encourage him to learn a simulation programming language or, at least, GASP II.

Lack of availability is, unfortunately, a more difficult barrier to overcome. Because of the special character of simulation programming languages, we cannot expect them to enjoy the wide availability characteristic of more general purpose languages such as FORTRAN. As time passes, however, the availability of these special languages on different computers is increasing.

Before delving into the myriad of programming considerations that play roles in simulation, we wish to note that the examples and flowcharts in this chapter are intended to familiarize the reader with *concepts* and do not necessarily represent *features* of existing programming languages. However, by being aware of problems before they arise, the reader can progress more rapidly in learning how the language he has selected explicitly handles these problems. If it does not handle one or more of them, the conceptual framework offered here can provide a guide as to how to resolve a particular problem if it is encountered.

4.2 DATA STRUCTURES

In establishing the prerequisites for describing the static structure of a system we note our definition of system in Section 2.1: A system is a collection of related objects with characteristics or descriptors that may themselves be related. Every system contains objects that can be classified into distinct groups or *classes*. As time elapses the number of objects within a class varies. Objects are related to each other and to the environment in prescribed ways; however, the nature of these relationships can also change with time. For example, a Ford sedan belongs to a class of automobiles. It has the characteristics of age and owner, the first of which changes continuously with time and the second of which can change at discrete points in time.

In order for a simulation programming language to serve a useful purpose in simulation, it must be able to do the following [10, p. 11]:

1. define the classes of objects within a system,
2. adjust the number of these objects as conditions within the system vary,
3. define characteristics or properties that can both describe and differentiate objects of the same class, and
4. relate objects to one another and to their common environment.

These requirements are not unique to simulation programming languages; they are found also in languages and programs associated with information retrieval and management information systems.

4.2.1 Identification of Objects and Characteristics

Every simulation programming language reflects its view of the "real world" in the data structures that it provides for representing a system. These languages define a system in terms of unique classes of different kinds of objects, each class being identified by one or more distinguishing characteristics. Depending on the language, an object is referred to by such names as *entity*, *object*, *transaction resource*, *facility*, *storage*, *variable*, *machine*, *equipment*, *process*, and *element*. A characteristic of an object is referred to as an *attribute*, *parameter*, *state*, and *descriptor*. In some language all objects are passive, that is, things happen to them. In other languages objects are active, flowing through a system and initiating actions. Table 4 lists the concept names and formalisms associated with several simulation programming languages.

4.2.2 Relationships Between Objects

All simulation programming languages specify a class relationship between objects. Although a common bond may exist among objects, distinguishing characteristics enable each object to retain its uniqueness. For example, in a system containing a class of objects of the type ship, two ships may have the

Table 4 Simulation Programming Languages: Identification Methods

Language	Concepts	Example	
CSL	Entity, property	LOAD (SHIP)	LOAD OF SHIP
GASP II	Entity, attribute	AGE(I)	AGE of *I*th entity
GPSS	Transaction, parameter	P1	First parameter of current transaction
SIMSCRIPT	Entity, attribute	AGE(MAN)	AGE OF MAN
SIMULA	Process, attribute	AGE	Attribute of current process MAN

Source: [48].

names *Mary* and *Elizabeth*. The objects are related, yet different. Moreover it is often necessary to be able to relate objects, of the same class and different classes, having restricted physical or logical relationships in common. For example, it might be necessary to identify all ships of a particular tonnage or all ships berthed in a specific port.

To serve this purpose, all simulation programming languages define relational mechanisms. Names such as *set*, *queue*, *list*, *chain*, *group*, *file*, and *storage* serve to describe them. The reader may recall the use of the terms queue, roster, and list in this context in Chapter 3. Each language has operators of varying power that place objects in and remove them from these relational structures and identify objects bearing particular relationships to each other. Hence the selection of ships, say, of a certain tonnage can be accomplished by searching a list containing the names of all ships and their corresponding tonnages. The filing of tasks in the queue and their selection according to an explicitly stated queue discipline in Chapter 3 exemplify another application of such operators. Table 5 lists several relational concepts.

4.2.3 Object Generation and Manipulation

Data structure manipulation is an inherent part of any discrete event simulation. Typical operations during a simulation include:*

1. gaining access to the jth record of a list to examine or change the contents of its fields,
2. inserting a new record after the jth but before the $j + 1$st record,

* Knuth [50, p. 235] gives this list for list processing languages.

Table 5 Simulation Programming Languages: Relationship Concepts

Language	Concept	Example
CSL	Set	MAN. 3 HEAD SET(I): put the third man at the head of set I
GASP II	Set	CALL FILEM(I,NSET): put entity in *I*th set
GPSS	User chain group	LINK I, FIFO: put current transaction first in Chain I
SIMSCRIPT	Set	FILE MAN FIRST IN SET (I): insert MAN into SET(I) first
SIMULA	Set	PRCD(X,MAN): precede element X with element MAN in the set to which X belongs

Source: [48].

3. deleting the *j*th record from a list,
4. combining lists,
5. dividing one list into several,
6. determining the record count in a list,
7. ordering records in a list in ascending order based on the values stored in specified fields, and
8. searching the list for records with given values in certain fields.

We have already encountered some of these operations in Chapter 3.

Every simulation programming language contains data structures that represent objects of different classes. More specifically, the data structures consist of records of the objects in the simulation, each record containing information regarding the characteristics of a distinct object. A simulation operates on these records as simulated time elapses. To emphasize this data record manipulation we explicitly described the creation, filing, selection, and destruction of records instead of describing the corresponding objects they represented in Chapter 3.

Some languages deal only with fixed data structures that are allocated either during *compilation* or at the start of *execution.* These structures represent fixed numbers of objects of different classes. Other languages allow both fixed and varying numbers of objects.* The way in which a language handles the generation of objects is related to the way it sees a system through its world view and to the characteristic of the language as compiler oriented, as interpretive, or as a problem-oriented program package. For example, many

* See Section 4.3.

of the differences between SIMSCRIPT and SIMULA can be traced to compiler features that have little to do with simulation. The block structure-procedure orientation of SIMULA, which is rooted in ALGOL, has influenced the way processes are generated and the way they communicate with one another. The global variable-local variable subroutine orientation of SIMSCRIPT, which is rooted in FORTRAN, has similarly influenced the way entities are generated and the way they communicate with one another. In these two cases the differences are profound. A SIMULA process contains both a data structure and an activity program; a SIMSCRIPT entity holds only a data structure and is linked indirectly to an event subroutine. Table 6 describes several object generation methods.

Table 6 Simulation Programming Language: Object Generation Methods

Language	Concept	Example
CSL	Does not exist.	—
GASP II	Generate a new entity whenever one is needed.	Define ATRIB vector and file in a column of NSET
GPSS	Generate a new transaction with some specified time between successive generations.	GENERATE 10,3
SIMSCRIPT	Generate a new entity whenever one is needed.	CREATE A MAN CALLED HENRY
SIMULA	Generate a new process whenever one is needed.	HENRY: = *new* MAN

Source: [48].

Having mentioned that some languages use rigid data structures whereas others use dynamic data structure, we should discuss further the implications of these two structures for simulation. In FORTRAN a DIMENSION statement reserves storage space. For example, the statement

DIMENSION XJOB(10,20)

instructs the computer to reserve a block of $10 \times 20 = 200$ storage registers for the array called XJOB. Once the 10×20 registers are allocated to XJOB, they remain so until the program terminates.

Consider a queueing problem in which jobs arrive with varying numbers of tasks to be performed. Let XJOB(K,1) denote the number of tasks for the

Kth arrival. Suppose that the queueing discipline is based on service time, and let XJOB(K,2), . . . , XJOB(K,L) be the service times of the XJOB(K,1) tasks of the Kth arrival. If the maximum number of arrivals is I and the maximum number of tasks is $L - 1$, then the straightforward use of a dimension statement would reserve an $I \times L$ block of registers whose rows would be utilized as arrivals occur and whose columns would not be completely utilized unless $XJOB(K,1) = L - 1$ for all arrivals. Here a waste of space inevitably occurs, which is further compounded by the fact that it may not be necessary to keep arrival records of completed tasks. Therefore it is instructive to look for ways of using space more efficiently within the rigid FORTRAN data structure.

Crude Data Structure

To illustrate several ways of conserving space we use a variant of the single server queueing problem of Section 3.3. Suppose that tasks receive service in the order of minimum service time and that all jobs have nonzero service times. Then we must record the service times of all tasks that wait. Let XJOB($\cdot$) be an $L \times 1$ array for storing service times. This means we can have no more than L tasks waiting at one time. Initially we set all XJOB($\cdot$) to zero to indicate they are empty. As tasks arrive we check XJOB(1), XJOB(2), XJOB(3), . . . until we find an I ($I \leq L$) for which $XJOB(I) = 0$. We then store the service time of the new task in XJOB(I).

When a task is to be selected for service, we include a register XJOB(J) in the search only if $XJOB(J) \neq 0$. If XJOB(K) contains the minimum, we select this job for service, use the service time as required, and vacate the register by setting $XJOB(K) = 0$. This release of space allows us to use XJOB(K) again. Although the maximum number of tasks waiting at one time cannot exceed L, it is clear that the number of tasks that receive service during the simulation run can be considerably greater than L.

The benefit of space conservation in the above example is obtained, unfortunately, at the cost of added computer time. Notice that registers must be checked sequentially until an empty one is found for storage. Moreover, we must check all L registers to determine which to include in selecting a minimum. In a system with rapid state change and a long queue the computing time consumed in a search of this kind can be detrimental to the success of the project.

Crude Chain

One way to reduce the search time is to add *information* to the record of a task. Let the array XJOB($\cdot$,$\cdot$) be $L \times 2$, where XJOB(J,1) contains the service time and XJOB(J,2) contains the register *address* of the service time of the succeeding waiting arrival. For example, suppose that at a moment in simulated time four jobs await service with the following addresses:*

$$\begin{aligned} \text{XJOB}(1,2) &= 004. \\ \text{XJOB}(2,2) &= 007 \\ \text{XJOB}(3,2) &= 010 \\ \text{XJOB}(4,2) &= 000. \end{aligned}$$

These are shown conceptually in Fig. 22*a*. Since XJOB(·,2) points to an element in the list, we call it a *pointer*. The address 000 is assumed to indicate that XJOB(4,1) has no *successor*. We also assume that two memory locations are reserved to hold the *header* (first) address 001 of XJOB(1,1) and the *trailer* (last) address 010 of XJOB(4, 1). When a selection for service is to be made, the search is conducted in order of successor addresses for XJOB(·,1). The search terminates at XJOB(4,1) since XJOB(4,2) indicates no successor.

Suppose that XJOB(3,1) is the minimum arrival time. Since the corresponding task is put into service, this arrival time must be deleted from the queue. To do so we set

$$\text{XJOB}(2,2) = \text{XJOB}(3,2).$$

The new list of addresses is

$$\begin{aligned} \text{XJOB}(1,2) &= 004 \\ \text{XJOB}(2,2) &= 010 \\ \text{XJOB}(4,2) &= 000 \end{aligned}$$

so that the next search *chains* through the remaining service times only. Figure 22*b* shows the new arrangement. Moreover one can set XJOB(3,1) = 0 to indicate that XJOB(3,·) is available for use.

When an arriving task must wait, a search of XJOB(·,1) is conducted to find the first I for which XJOB(I,1) = 0. Then its service time is stored in XJOB(I,1). Suppose that I = 3. Then to enter the new service time into the chain we set

$$\begin{aligned} \text{XJOB}(4,2) &= 007 \\ \text{XJOB}(3,2) &= 000, \end{aligned}$$

and we have the arrangement in Fig. 22*c*. Here retention of the trailer address enables us to connect the new arrival record to the end of the chain.

In this second example we have shown that an increase in data storage space can reduce the amount of computer time spent in checking for data elements that should enter into the selection of a task for service. The

* This example is hypothetical and does not denote actual addresses.

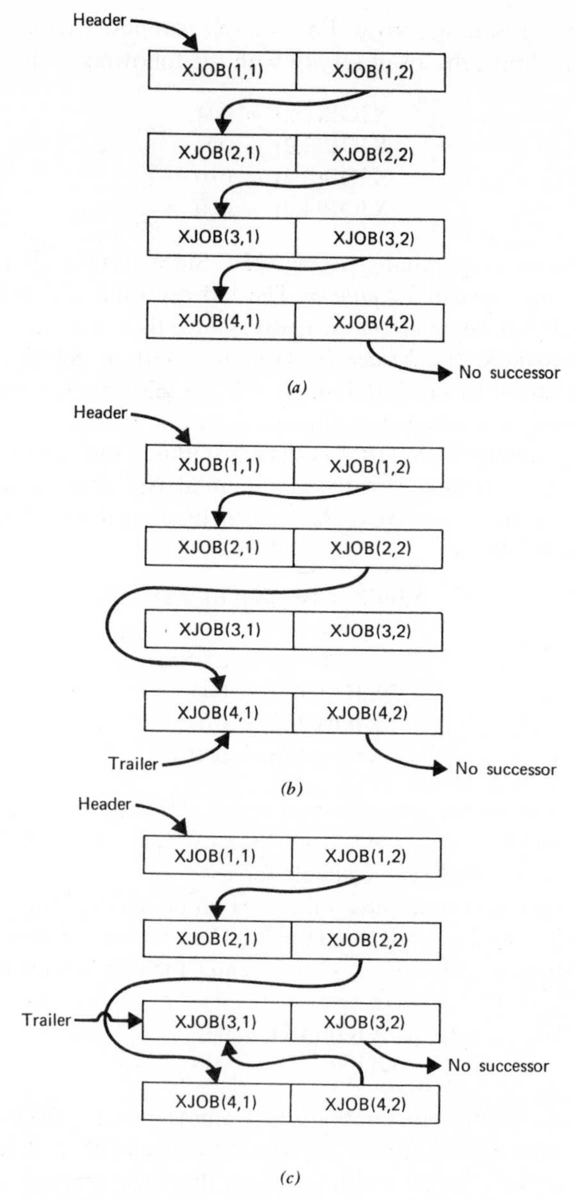

Fig. 22 Successor relationships.

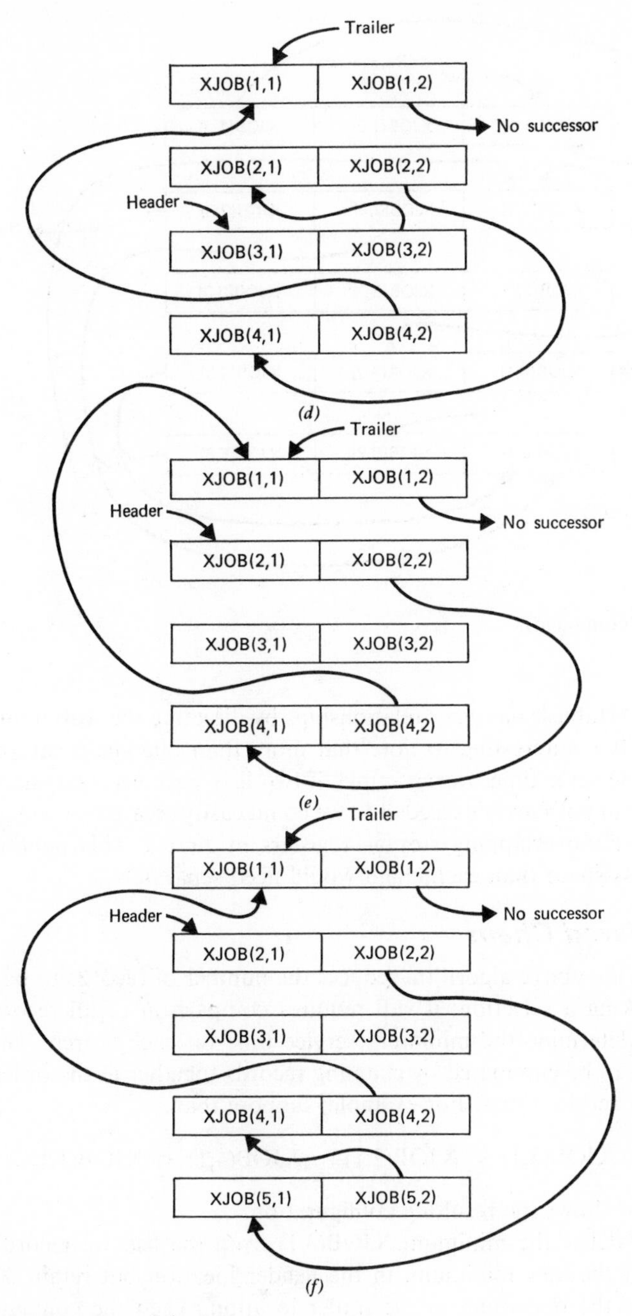

Fig. 22 (continued).

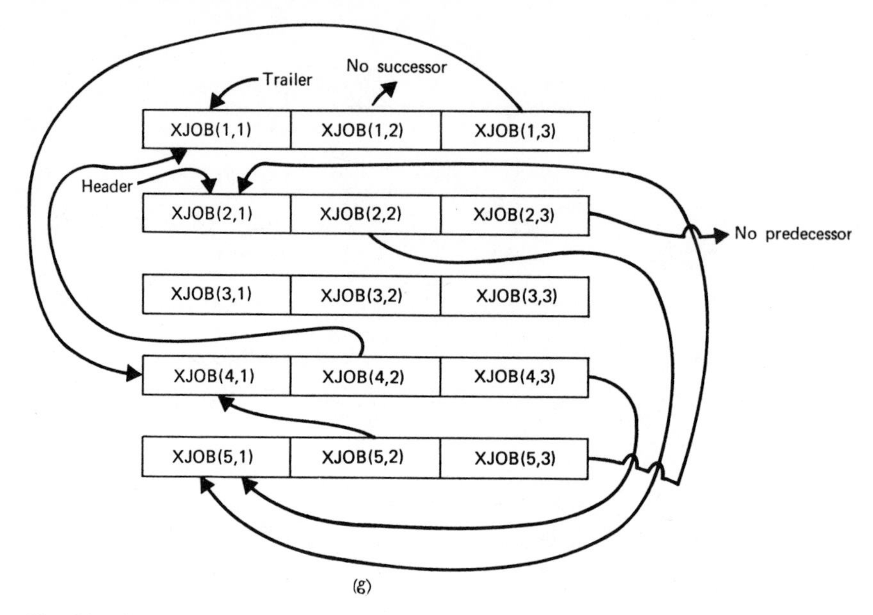

Fig. 22 (continued).

addresses establish *successor* relationships by directing the search through the data file. It is interesting to note that more than one chain can occupy an array at the same time. For example, if two lists were necessary and the sum of records in both *never* exceeded L, we could easily create two noninteracting chains in nonoverlapping storage registers in X(·,·). This pooling should require less space than each chain would need separately.

Ordered Chain

Although the above algorithm reduces the number of records to be searched when making a selection, it still requires comparison of all records in the chain to determine the minimum service time on each search. These comparisons can be eliminated by chaining records together in the order of their increasing service times. For example, suppose that

$$\text{XJOB}(3,1) < \text{XJOB}(2,1) < \text{XJOB}(4,1) < \text{XJOB}(1,1).$$

Figure 22*d* shows the resulting configuration.

If we delete the minimum XJOB(3,1) from the list, we record 004, the address of the new minimum, in the header location but retain 001 as the address of the maximum in the trailer location. Then the configuration in Fig. 22*e* obtains. Notice that no comparisons are necessary on deletion.

Let XJOB(5,1) be a new service time with address 016. To enter this service time in the chain, we must compare it with the times already in the chain. Starting with XJOB(2,1), we determine whether XJOB(5,1) < XJOB(2,1). If it is then X(5,1) is the smallest so that we change the header address to 016 and set XJOB(5,2) = 004. If, however, XJOB(2,1) > XJOB(5,2) we compare XJOB(5,1) with XJOB(4,1). If XJOB(4,1) > XJOB(5,1) then we set XJOB(2,2) = 016 and XJOB(5,2) = 010, with the results shown in Fig. 22*f*.

Suppose that our simulation uses the list of service times in two ways. In one case the minimum service time is required; in the other, the maximum service time. The successor relationship just described facilitates selection of the minimum time only. To accommodate selection of the maximum time we may expand the XJOB array to include XJOB(·,3), where we record the address of the next smallest service time for each service time. These *predecessor* addresses in addition to header, trailer, and successor addresses are also pointers. Then the user has the option of entering the list from the smallest or the largest service time.

GASP II [51] uses a considerably more sophisticated method of data management than the primitive approach described above, so that some improvement over the rigid data structure is possible in a FORTRAN program. Nevertheless the requirement for reserving space at the beginning of a program and the need not to exceed the specified array size are unattractive features. We would prefer a more dynamic data structuring capability that permits creation and destruction of records as needed without holding unutilized storage space.

The need for this greater flexibility in data manipulation, which characterizes information retrieval as well as simulation, gave rise to list processing languages in which efficiency considerations emphasize data manipulation and search, in contrast to the emphasis on arithmetic computation in FORTRAN type languages. Both GPSS and SIMSCRIPT II offer dynamic data structures. In particular, they implicitly carry out steps establishing predecessor and successor relationships, thereby relieving a user of the need to organize a search. Moreover, they carry out these steps efficiently.

To illustrate how this works in SIMSCRIPT II we consider the following statements:*

```
CREATE A TASK CALLED I
FILE THIS TASK IN THE QUEUE
REMOVE THIS TASK FROM THE QUEUE
DESTROY THIS TASK.
```

* A general reference for SIMSCRIPT II is reference 49.

With the exceptions of the *entity* TASK, the name I, and the *set* QUEUE all these words have meanings in SIMSCRIPT II. These statements must be prefaced by a declaration defining and relating TASK and QUEUE. We discuss this specification in Section 5.4. In GPSS the number of fields in a record varies dynamically as a function of the status of the corresponding transaction.*

4.3 SIMULATION CONTROL PROGRAMS

Every computer program consists of blocks of executable code. A computer executes these code blocks in the order in which it encounters them, either through sequential steps from one statement to the next or through directed transfers in the program. In a simulation program, however, a *time* is assigned to each code block for execution, and it is the ordering on these times rather than the computer's logical ordering of instructions that we want to prevail for the selection and execution of code blocks such as events. To accomplish this shift from a logical to a time ordering every simulation program, regardless of the language in which it is written, must have a *simulation control program.* All modern simulation programming languages provide a control program. Should an investigator wish to write his simulation program in a language such as PL/I, FORTRAN, or ALGOL, however, he must write a simulation control program himself. This is not an especially simple task when events can occur at random points in time.

Every simulation program has a three-level hierarchical structure in which the simulation control program or *timing routine* occupies the top level. Simulation-oriented routines such as events occupy the middle level, and housekeeping functions (e.g., computation of mathematical functions and generation of random variables) fill the lowest level. In the event scheduling approach to simulation programming a record is created every time an event is scheduled. This record identifies the event and includes the time at which it is scheduled to occur. The record is then filed in a list of scheduled events. By performing an event we mean executing a block of computer programming logic corresponding to the change in state that the event describes.

As soon as an event is performed, control returns to the simulation control program or timing routine, which selects the next event from the list of scheduled events. The event selected is the one with scheduled time closest to the current simulated time. When time advances to the scheduled occurrence time of the event, control transfers to a code block that executes the steps

* See reference 43.

comprising the event. The code block then transfers control back to the timing routine, which destroys the record of the event if so instructed. The timing routine then selects the next event. Figure 23 shows one form that a timing routine may assume.

A simulation control program in the activity scanning approach works in a different way. After all activity starts and terminations are effected at a moment in simulated time, control passes to the simulation control program, which first scans the "clocks" of the entities in the system. As mentioned in Section 3.3.3, a clock contains the incremental time from the present simulated time to the time at which the state of the corresponding entity is to change next. Once the minimal positive incremental time is determined, simulated time is advanced by this minimum, which is also subtracted from all clock times. Control then passes to the simulation program, which attempts to carry out its programmed activities as described in Section 3.3.3.

The process interaction approach offers yet a third method of shifting from a logical to a time sequencing of block execution. It combines the next event selection feature with the conditional time delay statements, as discussed in Section 3.3.4. Here we describe a control program similar to that for GPSS.

When an entity enters the system the simulation program attempts to advance it through the steps necessary for completion. If a simulated time period is to elapse during performance of the steps (e.g., when in service), control passes to a simulation control program which creates a record containing the entity's identification, the time at which the lapse ends, and the location in the program to which control is to be returned at that time. This location is called a *reactivation point*. The record is then filed in a list of scheduled "returns." The corresponding events are *unconditional*, for their execution depends only on time. The simulation control program then selects the next "return," updates simulated time to the time of this return, and passes control to the simulation program at the corresponding reactivation point. The location in the simulation program from which control passes to the control or timing routine is called the *interaction point*.

Conditional delays occur when resources are unavailable to perform a particular step. When this happens control passes to the simulation control program, which creates a record containing the entity's identification and return or reactivation point and files this record in a list containing records of other waiting entities.

After executing all unconditional events scheduled for the current time, the control program *scans* the list of conditional events to determine which ones can be performed under the conditions currently prevailing in the simulation. Each such event is executed. The control program continues to

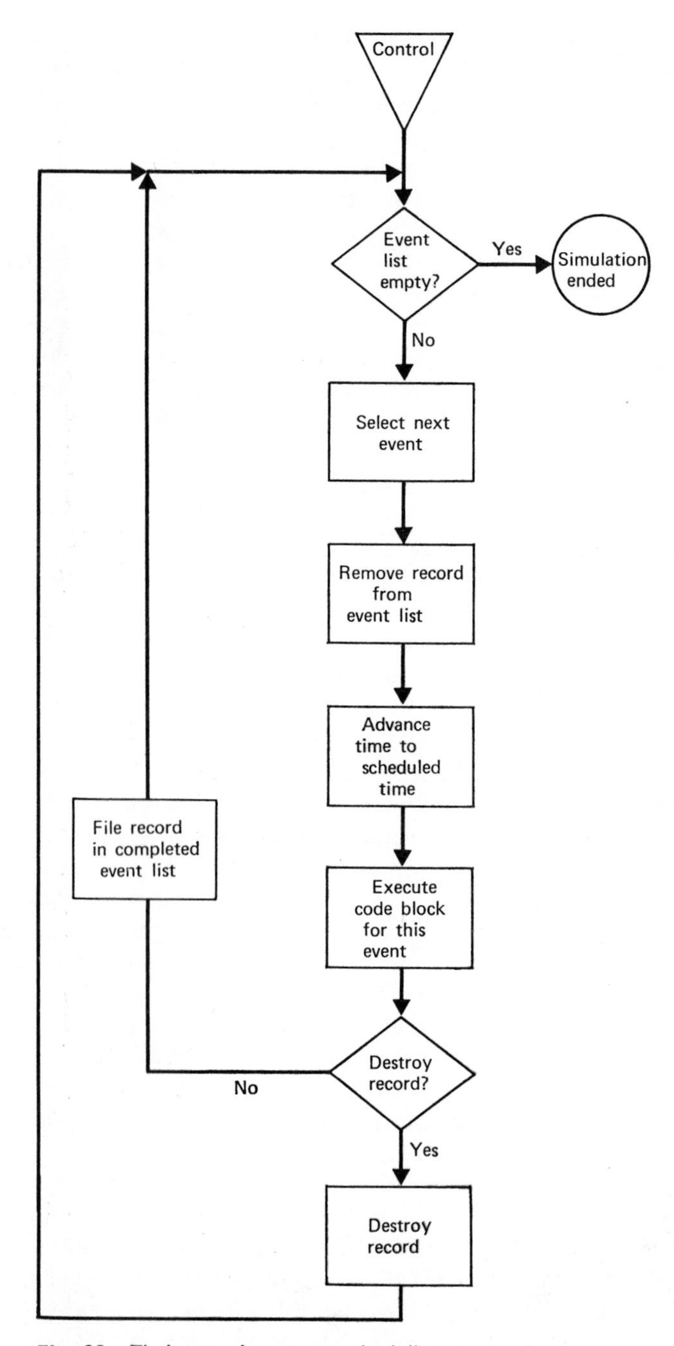

Fig. 23 Timing routine: event scheduling approach.

cycle through this waiting list until no more conditional events can be executed. Then the program advances simulated time to the time of the next scheduled unconditional event and repeats the procedure.

We must also account for entities on which work is completed. When an entity reaches a termination statement, control again passes to the simulation control program. Figure 24 illustrates one form of a simulation control program based on the process interaction approach.

One feature that distinguishes the event scheduling and process interaction approaches deserves explicit note. Each event in the event scheduling approach has one reactivation point, which occurs at the beginning of its corresponding program. A simulation program using the process interaction approach can have several reactivation points located anywhere in the program. Both types, however, may have more than one interaction (exit) point. Table 7 categorizes simulation programming languages according to the modeling approach employed.

Table 7 Simulation Programming Languages: Modeling Approaches

Event-Oriented Languages	Activity-Oriented Languages[a]	Process-Oriented Languages
GASP II [51]	AS [57]	GPSS/360 [43]
SEAL [42]	CSL [36, 44]	NSS [56]
SIMCOM [39]	ESP [66]	OPS [40]
SIMPAC [35]	FORSIM-IV	SIMPL [45]
SIMSCRIPT [53]	GSP [62]	SIMULA [64]
SIMSCRIPT I.5 [47]	SILLY [63]	SLANG [64]
SIMSCRIPT II [48]	SIMON [41]	SOL [51]
SIMSCRIPT II.5 [37]		SPL [58]

Source: [48, p. 25] with minor additions and omissions.

[a] Some of these languages are not "pure"; for example, GSP has both activity scan and event selection phases. The principal orientation is as indicated, however.

4.4 TIME FLOW

Early in Chapter 3 we noted that the next event approach employed a *variable time advance* method to update simulated time. The *variable* aspect was due to the fact that the times between successive events generally differ, and consequently the slack times skipped between events also differ. The principal exception to the use of this approach occurs in period simulations (see Section

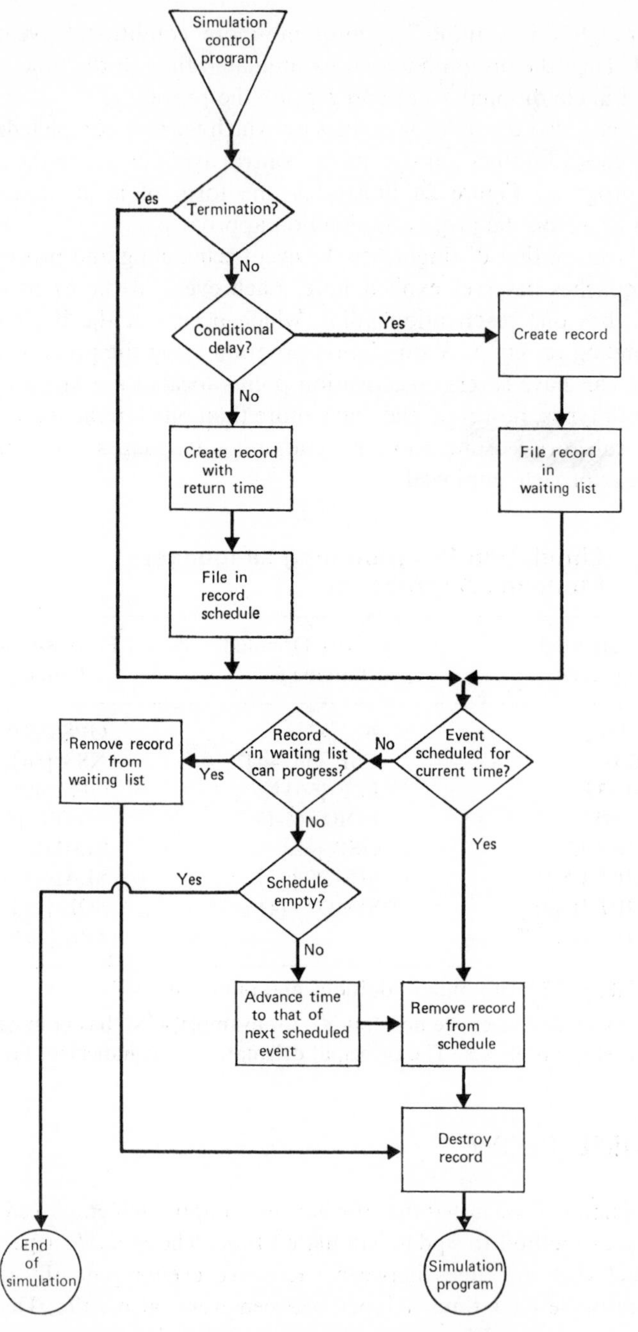

Fig. 24 Simulation control program: process interaction approach.

3.7), where activity takes place at successive time points Δt, $2\Delta t$, $3\Delta t$, ... In these cases a *fixed time advance*, where scheduling plays no role, has a distinct advantage since it eliminates the need for a timing routine.

While the two methods of time advance have easily appreciated benefits in terms of execution time in their own milieus, a question arises regarding which technique is more appropriate when a simulation contains periodically occurring events and randomly occurring events. Although Conway, Johnson, and Maxwell [38] discussed this problem over a decade ago, the fact that most simulation programming languages adopted the next event approach lessened interest in the issue. Recently, however, Morgan and Siegel [54], Nance [55], and Wickham [65] have renewed the interest of simulation methodologists in this question. The present account puts the problem into perspective, using an example from Nance [55].

Consider a machine shop where machines run continuously. However, the failure of a part in a machine causes the machine to stop. A patrolling repairman travels from machine to machine until he finds one that has failed. After putting the machine in running order he continues his rounds until he finds another failed machine. The process is then repeated. The order in which he visits the machines is fixed; the travel time between two successive machines is a constant Δt, but the time to failure of each machine is random.

Using the variable time advance method in the next event approach, we can simply schedule the repairman's arrivals, the failures of the machines, and, when necessary, the service times of failed machines. Alternatively we can use the fixed time advance approach to describe behavior. Suppose that time is advanced in increments Δt and that every machine has the attribute of failure time. As the repairman reaches a new machine, a comparison is made between current time and the machine's failure time. If the failure time is less than the current time, the repairman stays there until time is incremented beyond his arrival time plus the service time. He then advances to the next machine. The failure time of the repaired machine is then updated. Here a timing routine plays no role, nor does a list of scheduled events. The reader should note the similarity between this description and that of the activity scanning approach in Section 3.3.3.

Nance [55] describes several other algorithms for modeling time advance in this patrolling repairman problem. All produce identical simulated behavior as far as the user is concerned. Although the fixed time advance requires the least execution time for a given number of repairs [55, p. 66], it is not clear at the present time how to generalize these results to a wider class of problems containing many periodic events with different periods and many randomly occurring events. Interestingly, Morgan and Siegel [54] suggest a procedure for switching between the variable and the fixed time advance approaches, the decision rule for switching being a function of the past history of the nature of the model. The advantage of this approach would lie in the savings in execution time that an efficient time flow mechanism produces. Clearly the final word remains to be written.

4.5 RANDOM NUMBER GENERATION

Every simulation requires a mechanism for generating and assigning values to critical variables that induce or contribute indirectly to system state change. We call these critical values the *input*. Interarrival and service times play the roles of inputs in the queueing examples of Chapter 3, where we alluded to the determination of their values in the computation blocks of the flowcharts presented there. Alternatively a generating mechanism can be deterministic in the sense that the investigator specifies a way, free of stochastic elements, for producing the input numbers. However, deterministic mechanisms often do not adequately represent system input characteristics because of the investigator's imperfect knowledge of the real world. At best, he is aware of certain statistical properties of the input that enable him to specify the frequencies with which a particular input assumes different values. He would then like to assign values to the input by *sampling* from its parent population. This can be done if the selected programming language provides an algorithm for sampling from the frequency distribution of the particular population or, when this is not the case, if the investigator can write a routine for this purpose.

Chapter 8 describes algorithms for generating random variates from a wide variety of theoretical probability distributions and from empirical probability distributions. All these algorithms assume the availability of a stream of independent, uniformily distributed random numbers in the interval (0, 1). Chapter 7 describes methods of generating these numbers and discusses their corresponding programming considerations.

4.6 DATA COLLECTION, ANALYSIS, AND DISPLAY

Since the purpose of performing a computer simulation is to produce numerical results, we would like the language we select to facilitate collection and analysis of relevant data. GPSS automatically performs data collection and estimation and then prints relevant summary statistics. In SIMSCRIPT II and SIMULA the user must program this data collection, analysis, and printing.

To assist the user in the analysis phase, SIMSCRIPT II provides a convenient way for computing sums, means, variances, standard deviations, maxima, and minima, as well as other statistics.* The instruction

FOR I = 1 TO N, COMPUTE XBAR AS THE MEAN OF X(I)

sums the N X(I), divides the sum by N, and stores the resulting number in XBAR. The instruction

FOR I = I TO N, COMPUTE XVAR AS THE VARIANCE OF X(I)

computes the sample variance of X(I) and stores it in XVAR. The convenience of these instructions for statistical computation becomes apparent when a user is faced with the need for analyzing many sets of data.

In SIMULA the user may write his own *procedures* and routines to do data collection and analysis. In the version of SIMULA described in reference 64 two such procedures are already defined. One computes a histogram; the other accumulates a time integral.

Because the varied output display that different simulations require, a simulation user must often design a specialized display for his purposes. Naturally a language that simplifies this design work is attractive. SIMSCRIPT II does so by allowing free-form formatting, which permits great flexibility in display appearance. In addition to printing out a variety of statistical information automatically, GPSS allows the user to specify the creation and printing of special tabulations. However, the format is fixed. Although SIMULA in reference 64 has a procedure that prints a histogram, a user of this language can write his own output procedure with great ease.

In the absence of the GPSS, SIMSCRIPT II, and SIMULA computational conveniences, a user's programming burden is considerably increased. The motivation for using a simulation programming language thus becomes stronger.

4.7 INITIALIZATION AND TERMINATION

At the beginning of a simulation run certain *background* conditions have to be established. Some of these conditions hold throughout the run, whereas others do not. For example, suppose that the time between arrivals in the queueing model of Section 3.3 follows the exponential distribution with mean β_1. If we presume for a moment the capability of generating exponential variates in the simulation, all we require is that the mean interarrival time β_1 be assigned a numerical value at the beginning of the simulation and that this value be preserved for the period of the run. If service times also are exponentially distributed with mean service time β_2, we will also have to assign

* SIMSCRIPT II.5, a proprietary version of SIMSCRIPT II, offers ACCUMULATE and TALLY statements that simplify the computation of a histogram [37].

a value to β_2 initially and preserve it. These are permanent background conditions regarding numerical-valued parameters of the simulation.

As mentioned in Section 3.3, we also need to specify the initial length of the queue at the beginning of the simulation. Moreover, once this number is specified, we must create the concomitant records for the tasks, file these tasks in the queue, and establish the status of the server. In contrast to the mean interarrival and service times, the queue length changes during the course of the simulation. Hence we refer to the specification of queue length at the beginning of a simulation run as an *initial* condition. In addition to being a necessary part of a simulation, an initially specified condition can have a profound and often unintentional effect on the experimental results. The seriousness of this issue is discussed at length in Chapter 10.

Before transferring control to the simulation control program to begin a simulation based on the event scheduling approach, it is necessary to create events for the program to process. Otherwise when the transfer occurs, the simulation control program finds no "next" event to process and therefore terminates the simulation. Figures 23 and 24 show this. In the simple queueing problem of Section 3.3, we may, for example, schedule the arrival of the first job. Then the first act of the simulation control program is to process this arrival, thereby recursively scheduling the next arrival as in Fig. 8.

A simulation user must also specify how long he wants his experiment to run. In most queueing studies the specification is expressed either in terms of an elapsed number of simulated time units or in terms of a number of completed jobs. When this condition is met in a GPSS program, automatic termination of the run occurs and statistical results are printed. One convenient way of terminating a SIMSCRIPT program is to schedule a termination event at the desired run length time. This user-written event can contain data reduction and printing statements. It should also contain statements for terminating the simulation. A STOP statement would terminate the entire computer program. An alternative approach is to destroy any remaining scheduled events. Then the next execution of the control program passes control back to the initialization routine, where new initial conditions for subsequent simulations may be established. Figure 25 illustrates initialization and termination routines for the event scheduling approach.

4.8 ERROR MESSAGES AND DOCUMENTATION

The ease of debugging programs written in a programming language and the availability of information regarding its characteristics determine the attractiveness of the language for use. In addition to identifying coding errors, error messages or diagnostics are necessary to pinpoint trouble spots in a simulation program during execution. Since the highly interactive nature of simula-

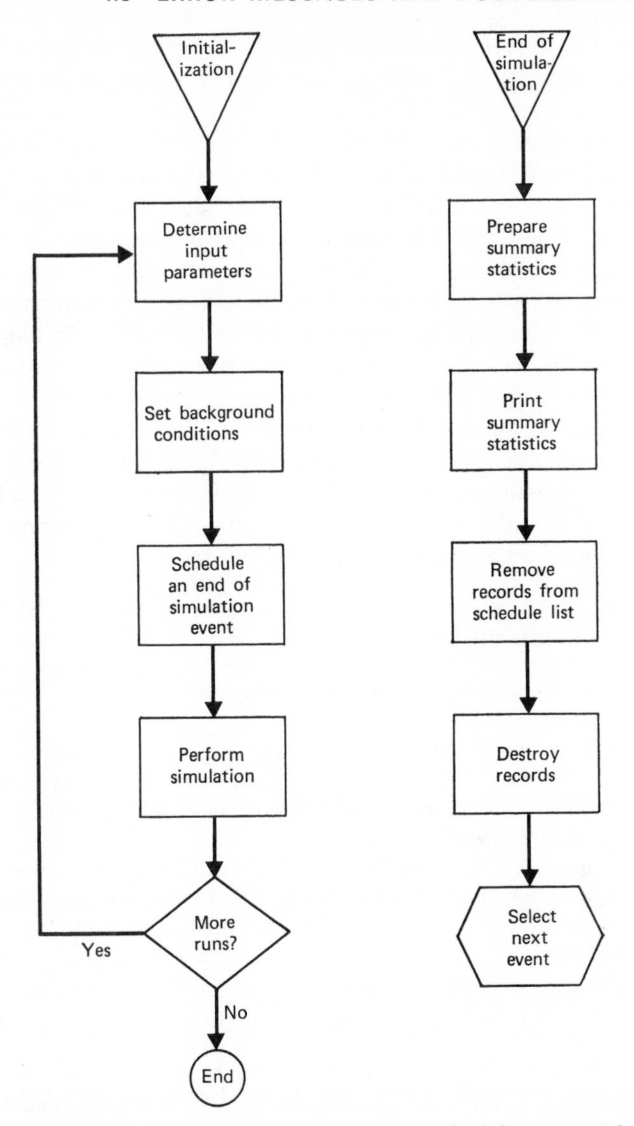

Fig. 25 Initialization and termination routines: event scheduling approach.

tion models often makes such identification especially difficult, it is desirable that error messages encountered during execution be as succinct and relevant as possible.

If one elects to write a simulation program in a general purpose language, he can hardly expect the concomitant error messages to reflect the simulation concepts he has used. This was a shortcoming of SIMSCRIPT programs which were first translated into FORTRAN statements and then compiled, assembled, and executed. The resulting error messages referred to the FORTRAN statements, not the original SIMSCRIPT ones. By contrast, we would like the error routines associated with useful simulation programming languages to provide information about errors in the context of the simulation. Both GPSS and SIMSCRIPT I.5 and II provide simulation-oriented diagnostics of this type. Table 8 lists several desirable debugging features.

Language documentation is also a crucial consideration when choosing among languages. Although many of the simulation programming languages listed in Table 7 have attractive programming features, their limited documentation or the restricted availability of their documentation retards the dissemination of information regarding these languages to the general population of simulation users. Among these languages GPSS and SIMSCRIPT have had special appeal because of the general availability of manuals and primers.* The simulation user who wishes his simulation model to have wide application is well advised to consider the issue of documentation before he becomes locked into a language with a tightly restricted set of users.

Table 8 Debugging Features

Report *compile* and *execute-time* errors by source statement-related messages.
Display complete program flow status when an an execute time error occurs. This means displaying the entry points and relevant parameters for all function and subroutine calls in effect at the time of the error.
Provide access to control information between event executions. This allows event tracing during all or selected parts of a program, and use of control information in program diagnostic routines.

Source: [48, p. 36].

4.9 SIMULATION PROGRAMMING LANGUAGES IN PERSPECTIVE

To understand how simulation programming languages relate to the main body of computer codes it is instructive to review several basic definitions and characteristics of computer programming in general. A *computer programming language* consists of a set of symbols denoting operations that a programmer wishes a computer to perform. Naturally the computer should

* See Table 7 for references.

be able to recognize these symbols, or, at least, it must contain a computer program that recognizes them.

Every computer has a *basic machine language* (BML) program, each instruction of which consists of strings of elementary symbols. Each string corresponds to a machine function such as adding two numbers, storing a number, or transferring to an address. In a binary computer the elementary symbols are zero and unity so that an instruction is a string of zeros and ones. In a decimal computer the elementary symbols are zero through nine.

Because of the elementary nature of BML instructions and the difficulty of recalling the code for each instruction in terms of its elementary symbols, most computers have an *assembly language* (AL) program in which each instruction consists of a string of mnemonic symbols; each string corresponds to a set of machine language instructions. The word *mnemonic* means "intended to assist the mind." For example, the FAP assembly language instruction CLA N means: Clear the accumulator and add N to it. Actually CLA is the mnemonic code, and N is an argument. A computer assembly program, sometimes called an *assembler*, translates a program written in AL into the corresponding instructions in BML, which are machine interpretable. Assemblers differ considerably in their powers. Some do little more than substitute BML codes for mnemonics and assign computer addresses to variable names and labels. More sophisticated assemblers can recognize complex macroinstructions and construct complicated BML programs from them.

Although the mnemonics of an AL are easier to work with than BML instructions, their direct use to construct a moderately complex computer program places a heavy burden on a programmer, both in time spent in constructing the sequence of instructions and in knowledge required to program detailed operations. Fortunately macrolanguages that use a *compiler* are available. A *compiler language* (CL) consists of macrostatements, each of which can accomplish several computer operations. A *compiler* is a program that accepts statements written in the corresponding CL and translates them into either an AL or a BML. When translation to an AL occurs an additional translation to a BML follows. In this case we say that the CL program requires compilation and assembly before execution can occur. As an example, the FORTRAN statement

```
X = A + B
```

upon compilation yields the instructions

```
CLA A
ADD B
STO X
```

These in turn are assembled into binary-coded BML instructions.

Compilation is more complex than assembly since it involves a higher level of understanding of program organization, richer input languages, and semantic as well as syntactic analysis and processing. As illustration we note that when the FORTRAN compiler encounters a DO loop it must check to determine where the loop ends in the program. This is a considerably greater ability than the assembly substitution of BML codes for AL codes.

The CL-AL-BML route is one way to direct a computer to perform desired operations. An *interpretive language* (IL) offers an alternative approach in which symbols denote commands to carry out computer operations directly rather than commands to construct an executable program through compilation and assembly. An *interpreter* (program) literally interprets the symbols of an IL and executes the commands directly. Although a CL and an IL may look alike they call for radically different treatments by the computer and for different techniques when writing programs in them.

The distinction between a *source* program and an *object* program serves to clarify the difference between ILs and CLs. A user submits a source program for execution on a computer. If the source program is written in a CL, the corresponding compiler translates the instructions into AL. Then the assembler translates the AL instructions into BML. The resulting code forms the object program, and it is this program that the computer executes.

When the source program is written in an IL, the interpreter *translates* as well as executes each IL instruction *each* time that it encounters it. Here there is no object program. If the source program calls for cycling through its instructions *n* times, then a translation takes place on each of the *n* times. By contrast translation takes place only once when a CL is used.

The relative desirability of using a CL or an IL depends on the nature of the computer investigation to be undertaken. If our work calls for writing a well-defined program and then carrying out many long production runs, it seems worth while to bear the costs of compilation and assembly *once* and then enjoy the benefits of efficient program organization and execution that a CL offers in comparison to an IL. If, however, our work is initially ill structured and we plan to devote considerable effort to developing the source program for several limited production runs, we prefer to avoid incurring the compilation and assembly costs that accompany each submission of a CL source program. Here an IL holds greater attraction.

Most compiler and interpretive languages can be classified as *problem-oriented languages* (POL). They differ from BMLs and ALs, which reflect computer hardware functions and have no problem orientation. A POL written for a particular problem area contains language statements appropriate for formulating solutions to typical problems in that area. A POL is able to express problem solutions in *computer-independent notation*, using a program that "understands" the POL to translate the problem solution expressed in source language into a BML object program or to execute it interpretively.

A *simulation programming language* is a POL containing most or all the features described in Sections 4.2 through 4.8. References 52, 61, and 62 describe the features of several different SPLs. As mentioned earlier, GPSS has become the most widely used SPL and SIMSCRIPT II is the most comprehensive.

Developers of simulation programming languages should keep in mind three critical points. First, they want to facilitate creation of a simulation program unencumbered by the limiting features of languages developed for other purposes. Second, they must recognize that most simulation studies are not well defined a priori, and consequently one could not be expected to specify a clear and unambiguous description of the problem to be programmed for analysis on the first try. Usually several modifications of a program will be made before a full scale run is undertaken. Third, developers should recognize the need for efficiency in execution, especially when several simulation runs are necessary.

With regard to the creation of simulation programs, existing simulation program languages have accomplished their purpose. The incorporation of data manipulation capabilities and timing routines attests to the greater convenience of using them than, say, FORTRAN. Although the ease of program creation unfortunately has led to a poor showing in computing efficiency relative to other languages, simulation methodologists generally agree that the benefits of simplicity outweigh the disadvantage of lower efficiency.

The remaining issues of program modification and execution, however, are not as easy to resolve. SIMSCRIPT, which is a CL, places emphasis on writing programs in modules of relatively short length compared to the total model description. For example, every event is a distinct subroutine. For an ill-structured initial concept, modular programming enables the user to recompile only the routines that have been altered and to execute the resulting object programs with the object programs for the already compiled unaltered routines. Although selective recompilation is a desirable practice to follow with any CL, the fact that SIMSCRIPT is designed in terms of events to facilitate this approach gives it a special appeal.

GPSS takes a different approach to simulation programming. There are 43 blocks in GPSS/360, each of which corresponds to a detailed subprogram, written in assembly language, for carrying out specific operations in the simulation.* At submission an assembly of these subprograms occurs. During a run control passes sequentially to each subprogram, which is interpretively

* On UNIVAC equipment a compilation using FORTRAN occurs.

executed. By providing unusually powerful macrostatements GPSS makes it easy for a novice to run a simulation program. However, the packaging of fixed subprograms gives one the feeling that redundancy must exist among them.

The benefits derived from user programming simplification and elimination of compilation make GPSS an attractive simulation tool. However, these advantages accrue at the cost of longer execution times to perform a simulation than SIMSCRIPT requires. It is important for the reader to avoid any universal judgment regarding the relative desirability of SIMSCRIPT and GPSS. Since each offers attractive features, with differing emphasis a counterexample to any universal judgment is usually easy to find.

EXERCISES

1. Write a simulation of the single server queueing problem of Section 3.3 in FORTRAN. Assume that the mean time between arrivals is 5 minutes and the mean service time is 4.5 minutes. Interarrival and service times are independent and exponentially distributed (see Section 7.1). Jobs receive service in the order of arrival. Compute sample mean queue length for a run of 1000 minutes. Be certain to specify initial conditions clearly and to write a separate timing routine.

2. Suppose that we have a list of N records and wish to find the record with minimal value in a particular field. If records are filed in no particular order, how many comparisons are needed to find the minimum?

3. Suppose that records are ordered on a particular field in order of increasing value. A new record is to be filed. How many comparisons are needed on the average to file the record? Assume a priori that the record has equal probability of belonging in any position in the list.

4. Suppose that the list in Exercise 3 has N even and that the address of record $N/2$ is available to the user. A new record is to be filed. By comparing the field in the new record with the corresponding field in record $N/2$, the user can improve on the result in Exercise 3.
   a. Show how this can be done.
   b. Compute the mean number of comparisons.
   c. Suppose that N is divisible by 4. What further improvement is possible?
   d. What is the difficulty of this method in practice?

5. Consider a simulation with both periodically and randomly occurring events. Describe how you might go about determining the relative merits

of the fixed time advance approach and the variable time advance approach described in Section 4.4. A simple example will be helpful.

6. The community dental clinic described in Exercise 10 of Chapter 3 opens every day at 8:30 A.M. and schedules one appointment for each dentist on duty at 15 minute intervals from the opening time until 4:30 P.M. When a particular appointment time is filled up with n appointments, the next request gets an appointment 15 minutes later. When this new time is filled with n appointments, the next patient is given an appointment 15 minutes later, and so on. After the nth appointment is filled at 4:30 P.M., the next request gets an appointment on the next day at 8:30 A.M.

   Assume that:

   Time is measured in days or fractions thereof.
   Patients call at any time of the day or night for first appointments.
   The earliest possible appointment time for a patient who calls after 4:00 P.M. one day and before 8:00 A.M. on the next day is 8:30 A.M.
   The earliest possible appointment time for a patient who calls between 8:00 A.M. and 4:00 P.M. is 30 minutes after his call.
   Patients in the clinic who require return visits are assigned the earliest available appointment time.
   Initially no appointment times are filled.

   Beginning with 8:30 A.M. on day zero, devise an algorithm suitable for computer use for assigning patients appointment times according to the above rules.

7. Modify the algorithm described in Exercise 6 to note that between noon and 2:00 P.M. $r < n$ dentists are always at lunch, and hence only $n - r$ appointments are scheduled for every 15 minute interval between noon and 1:45 P.M.

CHAPTER 5

GPSS/360, SIMSCRIPT II, AND SIMULA

5.1 INTRODUCTION

In this chapter we describe and illustrate the principal features of GPSS/360, SIMSCRIPT II, and SIMULA. Since each language has distinct features the reader is encouraged to familiarize himself with all of them in order to develop a conceptual understanding of their differences.

5.2 GPSS/360

GPSS is an IBM-developed interpretive simulation programming language available for use on a wide variety of computers. Since users need have no prior programming experience to work successfully with GPSS the language has a natural appeal. GPSS I, GPSS II, and GPSS III represent successively improved versions of the language. Our discussion here applies to GPSS/360 for use on the IBM/360 series of computers. The account relies heavily on the GPSS descriptions in Goeffrey Gordon's book, *System Simulation* [70], and in the IBM manual [71]. For further reading and exercises the reader may also wish to consult Thomas Schriber's book, *A GPSS Primer* [77].

GPSS uses the process interaction approach to organizing behavior in a simulation.* Certain objects called *transactions* act on the system of passive objects, which include *facilities* and *storage* units. To describe the simulation a user constructs a block diagram in which each block denotes a specific GPSS statement. Figure 26 shows 20 of the more commonly used blocks. Table 9 shows these statements with their required field information.

* See Section 3.3.5.

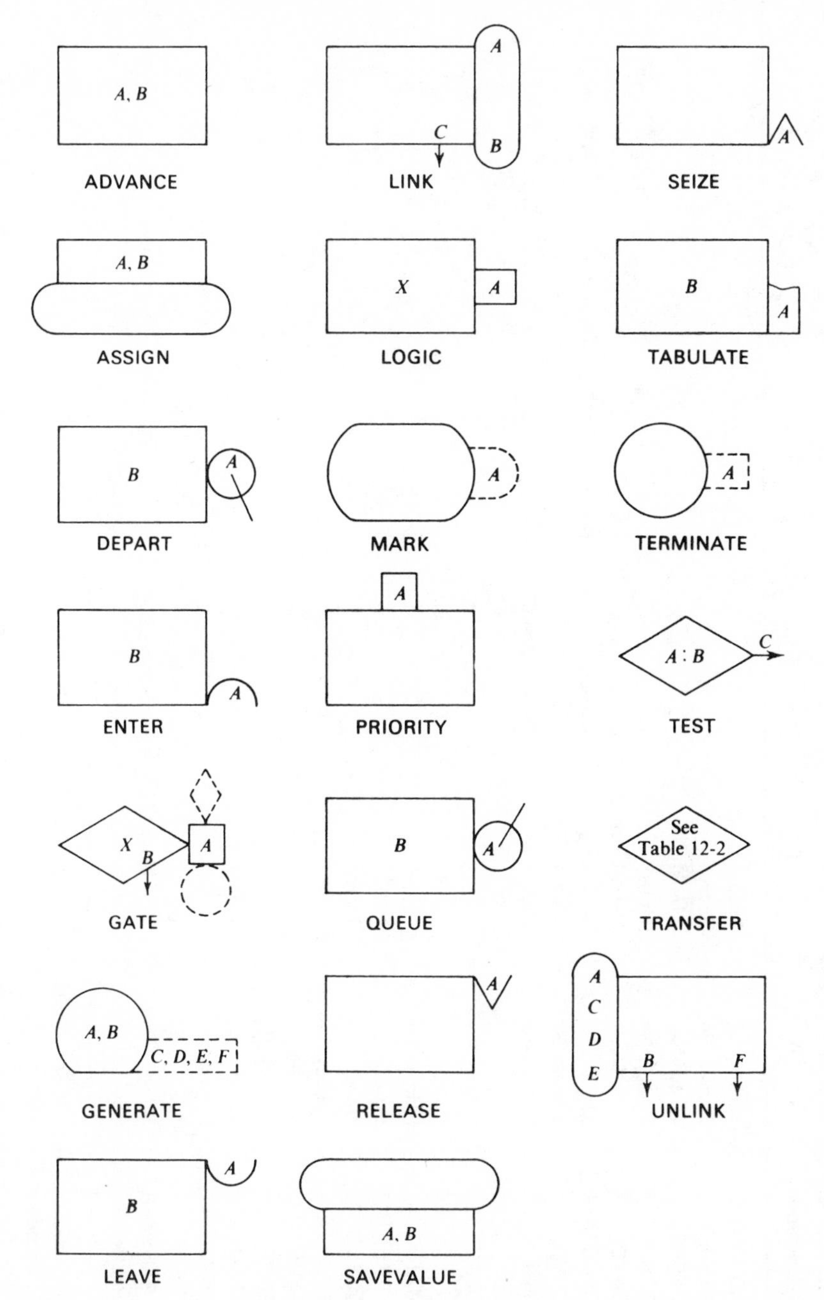

Fig. 26 GPSS block diagram symbols. *Source:* Geoffrey Gordon, *System Simulation*, © 1969, p. 192. Reprinted by permission of Prentice-Hall, Inc., Englewood Cliffs, N.J.

Table 9 GPSS Block Types

Operation	A	B	C	D	E	F
ADVANCE	Mean	Modifier				
ASSIGN	Param. No. (±)	Source				
DEPART	Queue No.	(Units)				
ENTER	Storage No.	(Units)				
GATE†	Item No.	(Next block *B*)				
GENERATE	Mean	Modifier	(Offset)	(Count)	(Priority)	(Params.)
LEAVE	Storage No.	(Units)				
LINK	Chain No.	Order	(Next block *B*)			
LOGIC {R, S, I}	Switch No.					
MARK	(Param. No.)					
PRIORITY	Priority					
QUEUE	Queue No.	(Units)				
RELEASE	Facility No.					
SAVEVALUE	S.V. No. (±)	SNA				
SEIZE	Facility No.					
TABULATE	Table No.	(Units)				
TERMINATE	(Units)					
TEST†	Arg. 1	Arg. 2	(Next block *B*)			
TRANSFER	Select. Factor	Next block *A*	Next block *B*			
UNLINK	Chain No.	Next block *A*	Count	(Param. No.)	(Arg.)	(Next block *B*)

† Columns 13 and 14. See Sec. 12-7 for codes.
( ) indicates optional field.

Source: Geoffrey Gordon, *System Simulation*, © 1969, p. 193. Reprinted by permission of Prentice-Hall, Inc., Englewood Cliffs, N.J.

Each statement corresponds to an assembly language subprogram which is interpreted and executed. The subprograms contain instructions for carrying out many operations commonly encountered in a simulation. Since GPSS statements are unusually powerful a few statements often suffice to conduct a simulation.

In GPSS active objects are called *transactions*; passive objects, *facilities* or *storage units*. For example, a job requiring service is a transaction, whereas a single server is a facility. A group of servers is a storage unit. The transaction passes through the system, encountering a facility or storage unit, depending on whether the simulation is a single or multiserver representation.

Transactions have numerical characteristics called *parameters*. Facilities and storages also have numerical attributes. Collectively they are termed *standard numerical attributes* (SNAs); a partial list is given in Table 10.

All GPSS information necessary to run a simulation is recorded on a special coding form, of which Fig. 27 shows an example. The statements listed there are for expository purposes in the rest of this section and do not constitute a GPSS program. *Variables* (which are simple arithmetic statements) can

Table 10 GPSS Standard Numerical Attributes

C1	The current value of clock time.
CHn	The number of transactions on chain *n*.
Fn	The current status of facility number *n*. This variable is 1 if the facility is busy and 0 if not.
FNn	The value of function *n*. (The function value may be computed to have a fractional part but the SNA gives only the integral part, unrounded.)
Kn	The integer *n* (the notation n may also be used).
M1	The transit time of a transaction.
Nn	The total number of transactions that have entered block *n*.
Pn	Parameter number *n* of a transaction.
Qn	The length of queue *n*.
Rn	The space remaining in storage *n*.
RNn	A computed random number having one of the values 1 through 999 with equal probability. (When the reference is made to provide the input for a function, the value is automatically scaled to the range 0 to 1.) Eight different generators can be referenced by n = 1, 2, . . . , 8.
Sn	The current occupancy of storage *n*.
Vn	The value of variable statement number *n*.
Wn	The number of transactions currently at block *n*.
Xn	The value of savevalue location *n*.

Source: Geoffrey Gordon, *System Simulation*, © 1969, p. 217. Reprinted by permission of Prentice-Hall, Inc., Englewood Cliffs, N.J.

IBM

GPSS/360 — CODING FORM

GX20-1701-1 U/M 025*
Printed in U. S. A
*No. of forms per pad may vary slightly

SHEET___OF___

NAME | DATE | PROBLEM NO. | PHONE NO. | PROGRAM TITLE

LOCATION | OPERATION | A,B,C,D,E,F →

```
 1        VARIABLE    5*RN1/50
 2        VARIABLE    S5+Q3
 1        FUNCTION    RN1,C24
0.0,0.0/0.1,0.104/0.2,0.222/0.3,0.355/0.4,0.509/0.5,0.69
0.6,0.115/0.7,1.2/0.75,1.38/0.8,1.6/0.84,1.83/0.88,2.12
0.9,2.3/0.92,2.52/0.94,2.81/0.95,2.99/0.96,3.2/.97,3.5
0.98,4.0/0.99,4.6/0.995,5.3/0.998,6.2/0.999,7/0,9997,8
```

Fig. 27 GPSS/360 coding form.

be defined as in line 1, using any of the SNAs listed in Table 10. However, operations on the SNAs are limited to addition (+), subtraction (−), multiplication (×), division (/), and modulo reduction (@). Line 2 gives an example of how variable 2 is defined as the sum of the current contents of storage unit 5 and the length of queue 3. In the body of a program V2 denotes variable 2, and whenever it is encountered in execution it assumes a value according to line 2.

The background mode in GPSS is integer. This means, for example, that variables 1 and 2 in Fig. 25 can assume integral values only. To form variable 1, RN1, which assumes values from 1 to 999 (see Table 10), is multiplied by 5, divided by 50, and then *truncated* to produce a value for V1. A user may delay this truncation until after variable 1 is employed for the particular operation for which it is intended by replacing VARIABLE in line 1 by FVARIABLE. The procedure leaves much to be desired.

The absence of a floating point capability in GPSS is an unfortunate shortcoming and can lead to serious errors. For example, the current clock time C1 is integer and any arrival times or service times are integer. In order to retain significant digits in generated times a user must establish an appropriate time unit with which to work in his simulation. We describe this point in detail in the next section.

Occasionally we wish to specify a more complex relationship than is allowed by the elementary operations mentioned above. For this purpose GPSS resorts to a *function*. If a user wishes to compute y as a function of x he must provide a table of y values corresponding to a set of specified x values. The variable x can be any of the SNAs in Table 10. As an example, lines 3 through 7 offer a table of logarithms to be used with RN1, an element of random number stream 1. Whenever FN1 is encountered in execution it is evaluated as $-\log(1 - \text{RN1})$, so that FN1 is an exponentially distributed random variable with parameter unity.* Each time RN1 is used its value is replaced by a new, uniformly distributed random number in the range (0, 1).† Since the table offers values of select points in (0, 1) GPSS uses linear interpolation to evaluate FN1, which varies from 0 to 8 in the example.

5.2.1 Transaction Creation

The statement GENERATE A,B,C,D,E,F, means, when A, B, C, D, E, and F are positive integers, that D transactions are to be generated in the simulation, each having a uniformly distributed interarrival time with range (A − B, A + B) and the first occurring at time C. Moreover each transaction has $0 \leq F \leq 100$ parameters and a selection priority of $0 \leq E \leq 127$.

* Exponentially distributed random variables are discussed in Section 8.2.3.

† See Table 10 for a definition of the values that RN1 assumes.

If F is blank then 12 parameters are assigned. A blank in E assigns a zero priority. If D is blank the number to be generated is unlimited. If C is blank the first transaction occurs as specified by A $\pm$ B. If B is blank transactions are created at time intervals of length A.

To generate other than uniform interarrival times, B may be given as FN1, indicating a user-provided tabled function from which random variates can be drawn. For a particular draw the interarrival time is the product of A (numerically valued) and the drawn value. For example, FN1 may be the cumulative exponential distribution with unit mean. Then GENERATE 3, FN1 gives an infinite number of transactions, the interarrival time of each one being drawn from the exponential distribution with mean 3 and the first arriving at time 1.

In addition to containing information needed for the execution of the user-written simulation a transaction also contains fields in which GPSS automatically stores scheduling information. This arrangement eliminates the need for the continual creation and destruction of records of scheduled events to be merged with the list of scheduled events called the *future events chain*.

5.2.2 Assignments

The statement PRIORITY A offers another way of assigning a priority to a transaction. Here $0 \leq A \leq 127$ is the priority number assigned. The statement ASSIGN A,B means, when A and B are positive integers, that parameter A is to be assigned the value B, where B can be any SNA. If *n appears for B, where n is an integer, then the statement assigns A to the parameter whose number is the contents of parameter n. The use of *n enables us to perform *indirect addressing*. The statement ASSIGN A+,B adds B to the contents of A, and the statement ASSIGN A−,B subtracts B from the contents of A. The SAVEVALUE A,B statement enables GPSS to store the current value of an SNA for later use. Here B denotes the SNA whose current value is to be saved, and A gives the *savevalue location* in which the value is to be stored. A plus or minus sign following A indicates addition or subtraction, respectively, of the contents of B to the already existing contents of A.

5.2.3 Queueing and Service

The four statements ENTER A,B; LEAVE A,B; SEIZE A; and RELEASE A concern the use of facilities and storages. In particular, ENTER A,B allows a transaction to enter storage unit A *if* it is available and augments its contents by B. When B is blank the contents SA are augmented by one. Similarly

LEAVE A,B allows a transaction to leave storage unit A and decreases its contents by B. If B is blank the contents are reduced by one. For each storage unit used in a simulation a STORAGE definition card is needed to specify the capacity of the unit, which cannot exceed $2^{31} - 1$.

We can think of a storage unit with a capacity of n as a queueing arrangement with n servers. Hence the ENTER statement causes immediate service only if at least one server is idle; otherwise it causes the transaction to wait until it is selected for service. When a transaction is delayed GPSS links it to the *current events chain*. After all events scheduled for a given time are performed GPSS scans the contents of this chain to determine which conditional events can be processed. The reader should note the power of this instruction compared to the explicit detail given in Fig. 11.

After the transaction has completed service the LEAVE statement moves the transaction out of the storage unit and puts the server in the idle mode. The SEIZE and RELEASE statements perform the same functions as the ENTER and LEAVE statements when there is one server (unit capacity). No STORAGE definition card is needed. Here the server is called a *facility*.

When a transaction is in service we need a way of indicating how long the service takes. The statement ADVANCE A,B serves this purpose by delaying the transaction for a time drawn from the uniform distribution with range A − B, A + B. To generate other than uniformly distributed delay or service times, B may be given as FN2, which indicates a user-provided tabled cumulative probability distribution function from which random variates can be drawn. For a particular draw the delay or service time is the product of A (numerically valued) and the drawn value.

5.2.4 Statistics

Statistics are gathered through the use of the QUEUE, DEPART, MARK, and TABULATE statements. To keep track of the queue length of facility A the statement QUEUE A,B adds B units to the contents queue A. Similarly DEPART A,B reduces the contents of queue A by B units when a transaction goes into service. If B is blank the addition and subtraction are by unity. The QUEUE and DEPART blocks also accumulate the product of queue length and the length of time that the queue spends in that state. We have already discussed this approach to data reduction in Section 3.3.2.

The MARK and TABULATE statements provide a way of gathering data on the transit time of a transaction through the system. MARKA causes the transaction's arrival time at the MARK block to be recorded as the contents of parameter A. If A is blank then the program records the time in a special word. TABULATE A computes the elapsed time between a transaction's arrival at the MARK block and its arrival at the TABULATE block and records this time in table A. In this way the transit time between any two blocks in the program can be determined by appropriate placement of the MARK and TABULATE blocks. Omission of a MARK block causes an elapsed time computation from a transaction's arrival in the system.*

The use of MARK and TABULATE statements necessitates a definition card called TABLE with four fields: A, B, C, and D. When A is set to M1, this statement provides specifications for computing the sample frequency distribution of transit time. Field B is the lower limit of the distribution, C is the time increment between successive ordinates, and D is the total number of ordinates to be computed.

The statement TERMINATE A removes a transaction from the system and decreases the transaction count, as specified in the START control card, by A. Field A of the START control statement contains the total number of transactions to be created during a simulation. When the A field count has been decreased to zero the run ends and the output is printed. One use of the TRANSFER A,B,C statement facilitates transfer of a transaction to either location B or C. Here a random draw from the uniform distribution on the interval (0, 1) occurs; if it is less than A, the transaction is transferred to location C. Otherwise it goes to location B. Locations B and C can have between three and five alphabetic characters, as shown in Table 8. When A is blank the statement TRANSFER, B gives an unconditional transfer to location B.

In manufacturing problems it often is necessary to know whether or not certain machinery is in working order. GPSS handles this interrogation by means of a LOGIC statement. LOGIC A, followed by S, R, or I in column 14, indicates that switch A is to be set (on), reset (off), or inverted, respectively. Logic switches can be used to indicate traffic controls, and server availability, as well as many other binary situations.

5.2.5 Logical Testing

The GATE and TEST statements provide for logical testing in GPSS programs. The statement GATE A,B, with a special condition code beginning in column 13, indicates that if entity A satisfies the condition starting in column 13 the transaction enters the GATE block. Otherwise it is transferred to location B. The conditions are as follows:

LS	n	Logic switch *n* set
LR	n	Logic switch *n* reset
U	n	Facility *n* in use

* The TABULATE statement can be used to gather other statistics (see reference 70, Section 12.2).

NU n	Facility *n* not in use
SF n	Storage *n* full
SNF n	Storage *n* not full
SE n	Storage *n* empty
SNE n	Storage *n* not empty

If B is blank the transaction waits until the condition is met and then enters the GATE block.

The TEST statement checks for the following conditions between a pair of standard numerical attributes (SNAs):

G	Greater than
GE	Greater than or equal
E	Equal
NE	Not equal
LE	Less than or equal
L	Less than

The symbol begins in column 13. For example, the statement TEST GE A,B,C means that if the contents of A exceed those of B the transaction enters the TEST block: otherwise it is transferred to location C. If C is blank then the transaction enters the TEST block when the condition becomes true. A and B assume one of the forms in Table 10.

5.2.6 Set Operations

GPSS can place a transaction on and remove it from a list by using the LINK and UNLINK statements, respectively. LINK A,B places the present transaction on list or chain A, ordered according to discipline B. The term B may be FIFO or LIFO, rules which denote orderings based on the time the entity entered the list. If Pn is used for B, transactions are ordered in ascending order of their respective values of parameter *n*. A FIFO rule resolves any ties that occur.

When a transaction encounters a LINK statement its progress through the model stops there until it is removed from the list during the course of another transaction's progress through the model. The statement UNLINK A,B,C,D,E,F removes, from chain A, C transactions for which parameter D has value E and causes these transactions to transfer to the statement whose address is B. If no transaction has value E, then the present transaction goes to the statement with address F. The count C can assume a numerical value, the contents of an SNA, or can be the word ALL; in the third case all transactions are involved. If D, E, and F are blank removal occurs from the beginning of the chain. If field F is blank the present transaction continues its progress sequentially.

5.2.7 Extended Computing Capability

Although the comprehensiveness of GPSS statements is easily appreciated the limited options available to the user for performing specialized computations are an unattractive feature. The absence of a floating point capability is an example. In an attempt to overcome such limitations GPSS/360 has a HELP statement that allows a user to write independent routines in other languages that will run in conjunction with a GPSS/360 program.

For the unsupported GPSS/360 version 1 the routine must be coded in basic assembly language (BAL). Since the average simulation user is unfamiliar with BAL, the option is not particularly comforting. For the IBM proprietary GPSS/360 version 2, the user has his choice between BAL and FORTRAN. Routines coded in BAL offer greater flexibility in terms of access to GPSS/360 control routines and statistics. FORTRAN allows access solely to the statistics. The newly augmented proprietary GPSS V permits a user to write routines in PL/1 also.

5.3 GPSS SINGLE SERVER QUEUEING PROBLEM

In this section we describe a GPSS/360 program for carrying out the steps described in Fig. 11 for the single server queueing problem with independent exponentially distributed interarrival times having a mean of 5 minutes and with independent exponentially distributed service times having a mean of 4.5 minutes.* The quantity of interest is queue length, defined as the number of jobs in the shop or, equivalently, as the number of jobs waiting for service plus the number of jobs in service. Initially there are five jobs in the queue, and the simulation is to be run for 10,000 minutes. Figure 28 shows the symbolic GPSS program.

Before describing the program several further remarks about GPSS are instructive. In GPSS each GENERATE statement creates a *transaction*. Associated with each transaction is a sequence of program statements to be executed sequentially and through time in accordance with certain conditions being met. In the present simulation three distinct types of transactions are created; one corresponds to the job flow through the facility (Fig. 11), the second to periodic data sampling (Fig. 9), and the third to establishing initial conditions. Through the fields of the GENERATE statements, the number of each type of transaction and the occurrence of transactions in time can be controlled.†

* The theoretical solution to this problem is discussed in Section 6.5 and is used in later chapters as a benchmark for evaluating simulation results.
† See Section 5.2.

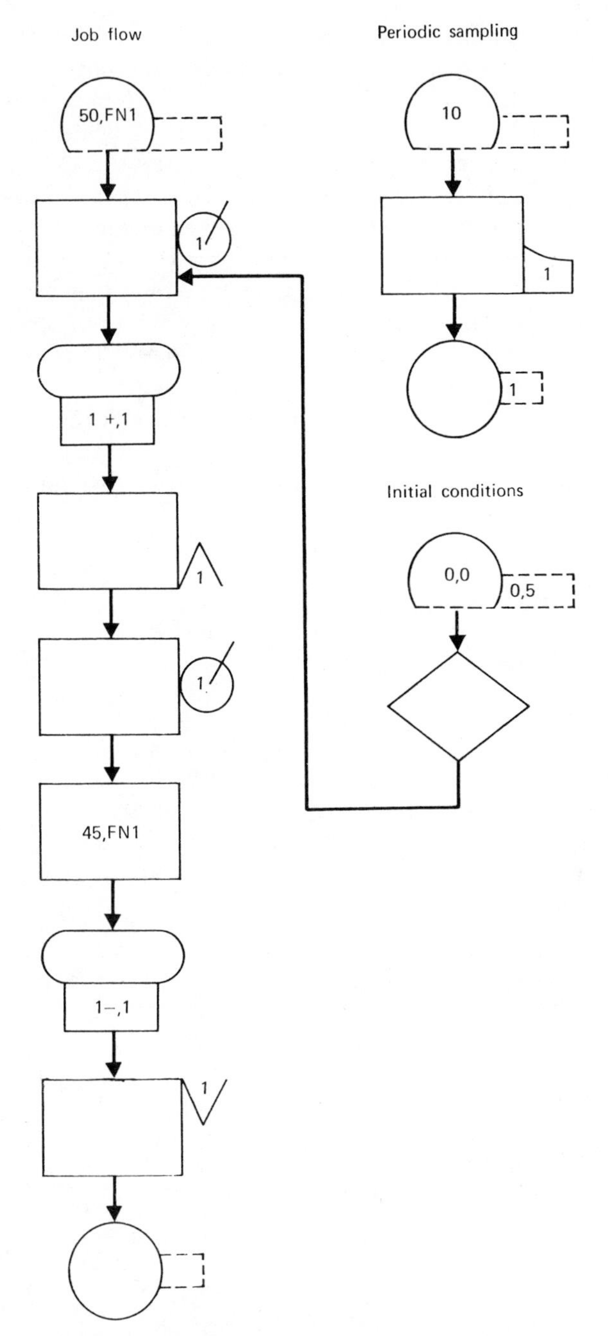

Fig. 28 GPSS block diagram of single server queueing problem.

We begin our description with line 9 of Fig. 29, which provides us with a way of generating arrivals in the simulation. Because time can assume only integral values in GPSS we can accommodate the 4.5 minute mean service rate (which arises later in the ADVANCE statement) by defining the GPSS time unit as 0.1 minute. Hence the mean arrival rate is 50 × 0.1 minutes. The function FN1 in field B indicates that the table in lines 4 through 7 should be used to generate a number which is then multiplied by the mean of 50 to determine the elapsed time to the next arrival. The table is of the exponential distribution with unit mean. Line 3 indicates that the argument of function 1 is RN1, the current pseudorandom number of stream 1. After computing RN1, GPSS linearly interpolates in the table to find FN1. The record of the arriving job is also created at this point.

Line 10 adds 1 to the contents of queue 1, and line 11 adds 1 to the queue length which is stored in X1. Here queue length is the total number of jobs in the system. In the absence of specifying a queue discipline GPSS takes

```
Line
 1  *          SIMULATE
 2  *          SINGLE SERVER QUEUEING PROBLEM
 3  1          FUNCTION    RN1,C24
 4  0.0,0.0/0.1,0.104/0.2,0.222/0.3,0.355/0.4,0.509/0.5,0.69
 5  0.6,0.915/0.7,1.2/0.75,1.38/0.8,1.6/0.84,1.83/0.88,2.12
 6  0.9,2.3/0.92,2.52/0.94,2.81/0.95,2.99/0.96,3.2/0.97,3.5
 7  0.98,4.0/0.99,4.6/0.995,5.3/0.998,6.2/0.999,7/0.9997,8
 8             TABLE       X1,0,1,100
 9             GENERATE    50,FN1
10  WAIT       QUEUE       1
11             SAVEVALUE   1+,1
12             SEIZE       1
13             DEPART      1
14             ADVANCE     45,FN1
15             SAVEVALUE   1-,1
16             RELEASE     1
17             TERMINATE
18             GENERATE    10
19             TABULATE    1
20             TERMINATE   1
21             GENERATE    0,0,0,5
22             TRANSFER    ,WAIT
23             START       10000
24  *
25             END
```

Fig. 29 GPSS coding of single server queueing problem.

jobs for service in the order of their arrival. Hence a transaction takes over the server in line 12 when all preceding jobs have been served.

Line 14 generates the service time for the job from the exponential distribution with mean 45×0.1 minutes and then causes this service time to elapse before progressing through the remainder of the program. This particular program is re-entered at line 15 when the service time has elapsed, and the timing routine finds notice of this re-entry as the next event to be executed.

Once the server completes a job, queue length, as computed in SAVEVALUE 1, is diminished by 1. Line 15 effects this change in a way analogous to line 11. Line 16 puts the server in the idle mode, and the TERMINATE statement in line 17 makes the transaction inactive.

In this particular simulation we wish to observe queue length at 1 minute intervals and then use these observations to create a queue histogram. Lines 18, 19, 20, and 8 allow us to do this. The START 10000 statement in line 23 specifies that 10,000 counted transactions are to be created before completion of the simulation. The GENERATE 10 statement in line 18 creates a transaction every 10×0.1 minutes or every minute. In line 19 TABULATE 1 instructs GPSS to add an observation according to the table specification in line 8 to the histogram being computed. The TABLE X1,0,1,100 statement indicates that the observation is stored in SAVEVALUE 1 and that the histogram has 100 intervals each of 1 unit width, the upper limit of the first interval being zero. From our earlier discussion the reader may recall that queue length is stored in this savevalue. Note that the collection of data for further analysis can also be accomplished in this step (see Fig. 9) by use of a MSAVEVALUE statement, which would store the appropriate vector.*

The TERMINATE 1 statement in line 20 causes the record of this transaction to become inactive and the transaction count to be decreased by 1. Since this TERMINATE statement in line 17 involves no decrement, the program continues until 10,000 transactions as in lines 18, 19, and 20 have been performed. Then the simulation ends. The effect is to cause the simulation to run for $10{,}000 \times 10 \times 0.1 = 10{,}000$ simulated minutes.

Lines 21 and 22 provide a way of establishing the initial queue size at time zero. The GENERATE 0,0,0,5 causes five jobs to arrive at time zero so that initially five jobs are waiting for service. To facilitate the movement of each of these jobs through the system we use the TRANSFER statement in line 22, which unconditionally transfers the program to the appropriate program logic. Note that this approach to establishing initial conditions differs from the framework given in Fig. 25. However, the concept is similar.

Once the simulation has ended, GPSS automatically computes the

* See reference 71.

statistics shown in Table 11. With the exception of the histogram and the four lines preceding it, the computation is automatic. The histogram and the preceding four lines are computed in response to our instructions regarding statistics for queue length in lines 18 and 19.

Notice that the entry under AVERAGE CONTENTS in Table 11 is 5.960 whereas the entry under MEAN ARGUMENT is 6.825. Both refer to sample mean queue length. The first of these entries is automatically computed by GPSS whereas the second is generated at user command. The apparent conflict is eliminated when we note that GPSS regards queue length as the number of jobs waiting for service but our definition defines queue length as the number of jobs in the system. Consequently we expect the sample mean computed at our request to exceed the GPSS computed sample mean, which it does. We also note that some difference in the sample means is attributable to the fact that GPSS has computed its statistic using the state time approach of Section 3.3.2 but the user command statistic is based on the periodic sampling approach of that section.

5.4 SIMSCRIPT II

SIMSCRIPT II represents the latest in a sequence of discrete event digital simulation languages that use the SIMSCRIPT title. In the early 1960s Markowitz, Karr, and Hausner [75] developed the original SIMSCRIPT at the Rand Corporation. A translator converts a SIMSCRIPT program into an equivalent FORTRAN program, which is then compiled and assembled. This translation naturally adds to the computer time necessary to obtain an executable program. SIMSCRIPT I.5, described in [72], eliminates this translation step by providing equivalence between SIMSCRIPT I.5 statements and assembly language instructions.

SIMSCRIPT II represents a considerable extension of the original features and capabilities. Developed at Rand by Kiviat, Villanueva, and Markowitz, the language is divided into five levels [73, p. v]:

Level 1 A simple teaching language designed to introduce programming concepts to nonprogrammers.

Level 2 A language roughly comparable in power to FORTRAN, but departing greatly from it in specific features.

Level 3 A language roughly comparable in power to ALGOL or PL/I, but again with many specific differences.

Level 4 That part of SIMSCRIPT II that contains the entity-attribute-set features of SIMSCRIPT. These features have been updated and augmented to provide a more powerful list processing capability.

This level also contains a number of new data types and programming features.

Level 5 The simulation-oriented part of SIMSCRIPT II, containing statements for time advance, event processing, generation of statistical variates, and accumulation and analysis of simulation-generated data.

SIMSCRIPT II has its own compiler, which relates a SIMSCRIPT II statement to a sequence of assembly language instructions so that no intermediate translation is necessary. In addition to the unsupported version an improved compiler, which reduces compilation time and contains a number of SIMSCRIPT II instructions not implemented in the original compiler, was developed by Simulation Associates, Inc., and marketed as SIMSCRIPT II Plus. Consolidated Analysis Centers, Inc., now markets this newer compiler, with further improvements, as SIMSCRIPT II.5 [67]. Since a user may wish to employ instructions not implemented in the original compiler, it is advisable for him to determine whether or not he has access to the improved compiler.

The principal building blocks of SIMSCRIPT II are *entities*, *attributes*, and *sets*. Entities denote objects of a system; attributes, characteristics of the entities; and sets, entity groupings. Every SIMSCRIPT II program contains a *preamble* that defines the system's entities, attributes, and sets. For each entity we have a statement*

EVERY entity name HAS A attribute list.

For example, we define a task for which we wish to retain service time as†

EVERY TASK HAS A SERVICE.TIME.

Here SERVICE.TIME is an arbitrarily named attribute that happens to correspond literally to the attribute we wish to define. If each task requires a special server, we write

EVERY TASK HAS A SERVICE.TIME AND SERVER.

In the executable program we can then use SERVICE.TIME(TASK) and SERVER(TASK) to refer to the service time and server, respectively, of a particular task.

* The words in capital letters denote SIMSCRIPT II code. In addition to reference 73, the reader will find the reference manual [74] helpful.

† The period is a legal SIMSCRIPT II character. As used here, it facilitates code reading. Entities, attributes, and variables can assume names of any length in SIMSCRIPT II, provided that the first six letters are distinct.

Table 11 GPSS/360 Sample Output

RELATIVE CLOCK 100000 ABSOLUTE CLOCK 100000

BLOCK COUNTS

BLOCK	CURRENT	TOTAL	BLOCK	CURRENT	TOTAL	BLOCK	CURRENT	TOTAL	BLOCK	CURRENT	TOTAL	BLOCK	CURRENT	TOTAL
1	0	1934	11	0	10000									
2	0	1939	12	0	10000									
3	1	1939	13	0	5									
4	0	1938	14	0	5									
5	0	1938												
6	1	1938												
7	0	1937												
8	0	1937												
9	0	1937												
10	0	10000												

FACILITY	AVERAGE UTILIZATION	NUMBER ENTRIES	AVERAGE TIME/TRAN	SEIZING TRANS. NO.	PREEMPTING TRANS. NO.
1	.865	1938	44.682	27	

CONTENTS OF FULLWORD SAVEVALUES (NON–ZERO)

SAVEVALUE	NR.	VALUE	NR.	VALUE	NR.	VALUE	NR.	VALUE	NR.	VALUE
	1	2								

QUEUE	MAXIMUM CONTENTS	AVERAGE CONTENTS	TOTAL ENTRIES	ZERO ENTRIES	PERCENT ZEROS	AVERAGE TIME/TRANS	$ AVERAGE TIME/TRANS	TABLE NUMBER	CURRENT CONTENTS
1	36	5.960	1939	239	12.3	307.397	350.614		1

$ AVERAGE TIME/TRANS = AVERAGE TIME/TRANS EXCLUDING ZERO ENTRIES

Table 11 (continued)

TABLE 1

ENTRIES IN TABLE	MEAN ARGUMENT		STANDARD DEVIATION		SUM OF ARGUMENTS	
10000	6.825		6.769		68251.000	NON-WEIGHTED
UPPER LIMIT	OBSERVED FREQUENCY	PER CENT OF TOTAL	CUMULATIVE PERCENTAGE	CUMULATIVE REMAINDER	MULTIPLE OF MEAN	DEVIATION FROM MEAN
0	1327	13.26	13.2	86.7	− .000	−1.008
1	1019	10.18	23.4	76.5	.146	− .860
2	889	8.88	32.3	67.6	.293	−.712
3	776	7.75	40.1	59.8	.439	−.565
4	738	7.37	47.4	52.5	.586	−.417
5	646	6.45	53.9	46.0	.732	−.269
6	674	6.73	60.6	39.3	.879	−.121
7	574	5.73	66.4	33.5	1.025	.025
8	420	4.19	70.6	29.3	1.172	.173
9	377	3.76	74.3	25.6	1.318	.321
10	276	2.75	77.1	22.8	1.465	.468
11	258	2.57	79.7	20.2	1.611	.616
12	192	1.91	81.6	18.3	1.758	.764
13	207	2.06	83.7	16.2	1.904	.912
14	203	2.02	85.7	14.2	2.051	1.059
15	218	2.17	87.9	12.0	2.197	1.207
16	144	1.43	89.3	10.6	2.344	1.355
17	204	2.03	91.4	8.5	2.490	1.503
18	157	1.56	92.9	7.0	2.637	1.650
19	111	1.10	94.0	5.9	2.783	1.798
20	89	.88	94.9	5.0	2.930	1.946
21	66	.65	95.6	4.3	3.076	2.093
22	55	.54	96.1	3.8	3.223	2.241
23	63	.62	96.8	3.1	3.369	2.389
24	50	.49	97.3	2.6	3.516	2.537
25	34	.33	97.6	2.3	3.662	2.684
26	39	.38	98.0	1.9	3.809	2.832
27	45	.44	98.5	1.4	3.955	2.980
28	24	.23	98.7	1.2	4.102	3.127
29	43	.42	99.1	.8	4.249	3.275
30	18	.17	99.3	.6	4.395	3.423
31	9	.08	99.4	.5	4.542	3.571
32	10	.09	99.5	.4	4.688	3.718
33	12	.11	99.6	.3	4.835	3.866
34	14	.13	99.8	.1	4.981	4.014
35	7	.06	99.8	.1	5.128	4.162
36	6	.05	99.9	.0	5.274	4.309
37	6	.05	100.0	.0	5.421	4.457

REMAINING FREQUENCIES ARE ALL ZERO

As a second example, in a multiserver queueing problem it is desirable to be able to determine the status of servers. We can then write

EVERY SERVER HAS A STATUS

and use STATUS(SERVER) to store the server's state.

As the student may surmise, this SIMSCRIPT II facility is an expeditious and flexible way of defining special purpose arrays that contain the attribute values of entities. The language also permits more conventional definition of variables and arrays by use of the statement*

DEFINE variable name AS A

$$\begin{Bmatrix}\text{REAL}\\ \text{INTEGER}\end{Bmatrix}\begin{Bmatrix}\text{i–DIMENSIONAL}\\ \text{i–DIM}\end{Bmatrix}\begin{Bmatrix}\text{VARIABLE}\\ \text{ARRAY}\end{Bmatrix}.$$

When defining scalars, the dimension modifier is suppressed and VARIABLE is used in the right braces.

The statement

TEMPORARY ENTITIES

before a sequence of EVERY statements means that the entities defined in these statements are created and destroyed during the simulation. Upon creation, SIMSCRIPT II searches for a block of space long enough to store the corresponding entity record and assigns the space to the entity. Upon destruction the record is erased. Hence SIMSCRIPT II dynamically assigns space to temporary entities. The statements

CREATE A TASK

DESTROY THIS TASK

have self-evident meanings.

The statement

PERMANENT ENTITIES

before a sequence of EVERY statements indicates that these statements define entities that remain in existence for the length of the simulation. Once N.entity name, the number of entities called "name," is specified, a CREATE statement is used to allocate an entire block of space to that entity.

* The braces mean that a choice must be made between the enclosed arguments.

For example, if N.SERVER is 10, then

CREATE EACH SERVER

reserves a block of space for ten servers, and, recalling our earlier EVERY statement, this space includes a STATUS block for each server.

After the TEMPORARY ENTITIES statement one often finds a statement

EVENT NOTICES INCLUDE event names.

This statement defines temporary entities called *event notices*, which are created whenever an event is scheduled. For example,

EVENT NOTICES INCLUDE ARRIVAL AND DEPARTURE

defines the two basic events of the single server queueing model first discussed in Chapter 3. SIMSCRIPT II destroys these event notices upon execution of the corresponding events when, for example, the first statement in the event routine is

EVENT event name.

If, however, the first statement is

EVENT event name SAVING THE EVENT NOTICE

no destruction takes place. SIMSCRIPT II automatically defines a creation time attribute for every event notice in a simulation. This system variable is called TIME.A and can be referenced, for example, by TIME.A(ARRIVAL). TIME.V is a real system variable that gives the current value of simulated time in decimal days.

The preamble also includes the definitions of set relationships. The statement

THE SYSTEM OWNS A list

identifies the sets that are to exist in the simulation program. For example,

THE SYSTEM OWNS A QUEUE

defines a list called QUEUE. The statement

EVERY TASK MAY BELONG TO THE QUEUE

implies that every temporary entity created during the simulation may, at one time or another, be linked to the list called QUEUE. The definition of a list automatically creates a set of associated variables that the running simulation constantly updates. The quantities N.name, F.name, and L.name denote the number of entities in the list "name," the address of the first entity, and the address of the last entity, respectively.

The preamble statement

$$\text{DEFINE set name AS A SET RANKED BY } \begin{Bmatrix} \text{HIGH} \\ \text{LOW} \end{Bmatrix} \text{ attribute}$$

allows a user to order the events in a list according to the value assumed by a specific attribute. In the absence of such a statement, entities are ranked in the order of entry into the list.

In the executable segment of a SIMSCRIPT II program the statement*

$$\begin{Bmatrix} \text{CAUSE} \\ \text{SCHEDULE} \\ \text{RESCHEDULE} \end{Bmatrix} \begin{Bmatrix} \text{A} \\ \\ \text{AN} \end{Bmatrix} \text{ event name IN variable } \begin{Bmatrix} \text{MINUTES} \\ \text{HOURS} \\ \text{DAYS} \end{Bmatrix}$$

does several operations explicitly delineated in our event scheduling flowcharts. It determines the time of the next event of a specified type, creates an event notice, and files it in the list of scheduled events. SIMSCRIPT II measures time in days or decimal fractions of days. If a time is computed in minutes (hours) then MINUTES (HOURS) should be chosen in the right braces of the above statement to guarantee the correct conversion to days for scheduling purposes.

When appropriately used with a SCHEDULE statement, the SAVING THE EVENT NOTICE phrase mentioned earlier can lead to a more efficiently running simulation. Suppose, for example, that we have

EVENT ARRIVAL

SCHEDULE AN ARRIVAL IN 5 MINUTES.

The SCHEDULE statement is used here to keep the simulation running. Every time an ARRIVAL event is executed, an ARRIVAL event notice is destroyed and a new one is created. However, the statements

EVENT ARRIVAL SAVING THE EVENT NOTICE

SCHEDULE THIS ARRIVAL IN 5 MINUTES

enable SIMSCRIPT II to update and use the already existing ARRIVAL notice for the scheduling of the next notice. Since no destruction or recreation takes place, computer time is saved. The key word here is THIS.

* In the left braces CAUSE, SCHEDULE, AND RESCHEDULE are synonyms.

SIMSCRIPT II provides a variety of statements for filing and removing records from lists. Examples are as follows:

FILE entity name IN list

$$\text{REMOVE THE} \begin{Bmatrix} \text{FIRST} \\ \text{LAST} \end{Bmatrix} \text{entity name FROM list}$$

FOR EVERY entity name WITH attribute of entity name

$$= \begin{Bmatrix} \text{variable} \\ \text{number} \end{Bmatrix}, \text{ FIND THE FIRST CASE}$$

IF FOUND, REMOVE THIS entity name FROM list.

SIMSCRIPT II also provides a collection of computational statements, including the following:

arithmetic operations

$$\text{LET A} = \text{X} \begin{Bmatrix} + \\ - \\ \dagger \\ / \end{Bmatrix} \text{Y}$$

ADD A TO B

SUBTRACT A FROM B,

logical control phrases†

$$\text{FOR I} = 1 \text{ TO N,} \begin{Bmatrix} \text{DO} \\ \vdots \\ \text{COMPUTE} \end{Bmatrix},$$

⋮

LOOP

and *logical testing*

$$\text{IF A} \geq \text{B, LET A} = \text{X} + \text{Y} \begin{Bmatrix} \text{ELSE} \\ \text{OTHERWISE} \\ \text{ALWAYS} \end{Bmatrix}.$$

Notice that logical control phrases and logica testing require key words to demarcate the extent of their ranges.

† The DO and COMPUTE in braces are examples. Many other statements are also possible.

For convenience in statistical computation, SIMSCRIPT II provides

$$\text{COMPUTE} \left\{ v \left\{ \begin{array}{c} \text{AS} \\ \\ = \end{array} \right\} \left[\text{THE} \right] \left\{ \begin{array}{l} \text{AVERAGE} \\ \text{AVG} \\ \text{MEAN} \\ \text{SUM} \\ \text{NUMBER} \\ \text{NUM} \\ \text{VARIANCE} \\ \text{VAR} \\ \text{STD.DEV} \\ \text{STD} \\ \text{SUM.OF.SQUARES} \\ \text{SSQ} \\ \text{MEAN.SQUARE} \\ \text{MSQ} \\ \text{MINIMUM } (e) \\ \text{MIN } (e) \\ \text{MAXIMUM } (e) \\ \text{MAX } (e) \\ \text{MINIMUM} \\ \text{MIN} \\ \text{MAXIMUM} \\ \text{MAX} \end{array} \right\} \right\} \text{OF } e.$$

The brackets indicate that THE is optional. These compute statements must be controlled by a logical control phrase. However, the computation of the final statistic may occur after the LOOP statement is encountered. For example,

```
FOR I = 1 TO N, DO
  COMPUTE XBAR AS THE MEAN OF X(I)
  COMPUTE XVAR AS THE VARIANCE OF X(I)
LOOP
```

causes division by N for the mean and by N − 1 for the variance to occur after the LOOP statement is encountered.

If, for example, a statement

```
DEFINE X AS A REAL 2-DIM ARRAY
```

occurs in the preamble, then a statement

```
RESERVE X AS N BY M
```

must occur in the executable program before the use of the X array. Naturally the dimensions N and M must be assigned values before the program encounters the RESERVE statement.

SIMSCRIPT II provides a flexible input-output capability for tightly specified or free-form formatted data. The statement

READ Z

causes the program to read the next number it finds on the input data cards and to store this number in Z. When the next READ statement occurs, the program reads the next number on the card or, if none remains, reads the first number on the next data card. This contrasts with the FORTRAN procedure, which requires a new data card for each READ encountered.

To begin a simulation we use

START SIMULATION.

This causes transfer of control to the SIMSCRIPT II timing routine for the duration of the simulation. Upon the termination of all aspects of the simulation, control returns to the statement following START SIMULATION.

5.5 SIMSCRIPT II SINGLE SERVER QUEUEING PROBLEM

Although a full account of programming in the SIMSCRIPT II programming language cannot be given here, we can present an example of a simulation of the simple queueing problem in Section 3.3.1. This exercise acquaints the student with many SIMSCRIPT II features that he will use repeatedly in practice. A detailed description follows the program given in Fig. 30.

Every SIMSCRIPT II program begins with a *preamble*, in which background conditions that are to prevail during execution are defined. The key words in line 2 under "Declarations" are SYSTEM, OWNS, and HAS. This statement declares that the system under study has a set called QUEUE and an entity called SERVER. Lines 3 and 4 define the temporary entity TASK that is created whenever a job has to wait (see Fig. 8) and notes that this entity may be filed in the set called QUEUE. Line 5 identifies the events in the program, and line 6 establishes the order of execution if several events are scheduled to occur at the same time. The remaining statements in the preamble are definitions.

The MAIN routine initializes the program, starts the simulation, and terminates program execution. Line 4 reads user-specified numbers into the system variables SEED.V(1) and SEED.V(2), which hold the current values of the pseudorandom number seeds for streams 1 and 2, respectively. Before

Declarations

```
Line
 1  PREAMBLE
 2  THE SYSTEM OWNS A QUEUE AND HAS A SERVER
 3  TEMPORARY ENTITIES
 4        EVERY TASK MAY BELONG TO THE QUEUE
 5  EVENT NOTICES INCLUDE ARRIVAL, DEPARTURE,
    COLLECTION AND END.OF.SIMULATION
 6  PRIORITY ORDER IS ARRIVAL, DEPARTURE, COLLECTION
    AND END.OF.SIMULATION
 7  NORMALLY MODE IS INTEGER
 8  DEFINE IDLE TO MEAN 0
 9  DEFINE BUSY TO MEAN 1
10  DEFINE S AS A REAL, 1-DIMENSIONAL ARRAY
11  DEFINE X AS A 1-DIMENSIONAL ARRAY
12  DEFINE N, MEAN.INTERARRIVAL, MEAN.SERVICE, P,
    QBAR, QVAR, T, AND T.1 AS REAL VARIABLES
13  DEFINE INITIAL.Q, MAX.Q, SEED.START.1, SEED.2 AS
    INTEGER VARIABLES
14  END
```

Initialization (Fig. 2)

```
 1  MAIN
 2  READ MEAN.INTERARRIVAL, MEAN.SERVICE, INITIAL.Q, N
 3  RESERVE X AS N AND S AS 100
 4  READ SEED.V(1),SEED.V(2)
 5  LET SEED.START. = SEED.V(1)
 6  LET SEED.2 = SEED.V(2)
 7  IF INITIAL.Q>0,
 8     LET SERVER = BUSY
 9     SCHEDULE A DEPARTURE IN
       EXPONENTIAL.F(MEAN.SERVICE, 2) MINUTES
10          FOR I = 1 TO INITIAL.Q-1, DO
11               CREATE A TASK
12               FILE THIS TASK IN THE QUEUE
13          LOOP
14  ELSE
15  SCHEDULE A COLLECTION IN 1 MINUTE
16  SCHEDULE AN END.OF.SIMULATION IN N MINUTES
17  SCHEDULE AN ARRIVAL NOW
18  START SIMULATION
19  END
```

Fig. 30 SIMSCRIPT II coding of single server queueing problem.

Arrival Event (Fig. 8)

```
Line
 1  EVENT ARRIVAL SAVING THE EVENT NOTICE
 2  RESCHEDULE THIS ARRIVAL IN
    EXPONENTIAL.F(MEAN.INTERARRIVAL, 1) MINUTES
 3     LET T = TIME.V*1440
 4     ADD T-T.1 TO S(N.QUEUE+SERVER+1)
 5     LET T.1 = T
 6     LET MAX.Q = MAX.F(MAX.Q,N.QUEUE+SERVER+1)
 7  IF SERVER = BUSY,
 8     CREATE A TASK
 9     FILE THIS TASK IN QUEUE
10     RETURN
11  ELSE
12  LET SERVER = BUSY
13  SCHEDULE A DEPARTURE IN
    EXPONENTIAL.F(MEAN.SERVICE, 2) MINUTES
14  RETURN
15  END
```

Departure Event (Fig. 8)

```
 1  EVENT DEPARTURE SAVING THE EVENT NOTICE
 2  LET T = TIME.V*1440
 3  ADD T-T.1 TO S(N.QUEUE+SERVER+1)
 4  LET T.1 = T
 5  IF QUEUE IS EMPTY,
 6     LET SERVER = IDLE
 7     RETURN
 8  ELSE
 9  REMOVE THE FIRST TASK FROM THE QUEUE
10  DESTROY THIS TASK
11  SCHEDULE THIS DEPARTURE IN
    EXPONENTIAL.F(MEAN.SERVICE, 2) MINUTES
12  RETURN
13  END
```

Collection Event (Fig. 9)

```
 1  EVENT COLLECTION SAVING THE EVENT NOTICE
 2  RESCHEDULE THIS COLLECTION IN 1 MINUTE
 3  LET T = TIME.V*1440
 4  LET X(T) = N.QUEUE + SERVER
 5  RETURN
 6  END
```

Fig. 30 (continued)

End of Simulation Event (Fig. 25)

```
Line
 1  EVENT END.OF.SIMULATION
 2 FOR I = 1 TO MAX.Q+1, DO
 3      LET S(I) = S(I)/T.1
 4      COMPUTE QBAR AS THE SUM OF (I-1)*S(I)
 5  LOOP
 6  FOR I = 1 TO MAX.Q + 1, COMPUTE QVAR AS THE SUM OF
    S(I)*(I-1-QBAR)*(I-1-QBAR)
 7  NOW REPORT
 8  CALL ANALYSIS (N)
 9  FOR I = 1 TO EVENTS.V,
10      FOR EACH ITEM IN EV.S(I), DO
11           REMOVE THE ITEM FROM EV.S(I)
12           DESTROY THIS ARRIVAL CALLED ITEM
13      LOOP
14  RETURN
15  END
```

Output Report

```
 1  ROUTINE REPORT
 2  DEFINE N.PRIME AS A REAL VARIABLE
 3  START NEW PAGE
 4  PRINT 1 LINE THUS
 5  SIMULATION RESULTS
 6  SKIP 3 LINES
 7  PRINT 7 LINES WITH SEED.START.1,  SEED.V(1), SEED.2,
    SEED.V(2) THUS
 8  STREAM 1 STARTING RANDOM NUMBER = ************
 9
10        STREAM 1 LAST RANDOM NUMBER = ************
11
12  STREAM 2 STARTING RANDOM NUMBER = ************
13
14        STREAM 2 LAST RANDOM NUMBER = ************
15  SKIP 3 LINES
16  PRINT 1 LINE WITH INITIAL.Q THUS
17  INITIAL QUEUE LENGTH=***
18  SKIP 1 LINE
19  PRINT 1 LINE WITH MEAN.INTERARRIVAL AND
    MEAN.SERVICE THUS
20  MEAN INTERARRIVAL TIME = **.**
    MEAN SERVICE TIME = **.**
21  SKIP 3 LINES
```

Fig. 30 (continued)

```
Line
22  PRINT 2 LINES WITH QBAR AND SQRT.F(QVAR) THUS
23  AVERAGE QUEUE LENGTH IS ****.** TASKS
24  STD.DEV. IS ****.**
25  SKIP 3 LINES
26  BEGIN REPORT
27  BEGIN HEADING
28  PRINT 2 LINES THUS
29  QUEUE LENGTH HISTOGRAM
30  NO.       FREQ.
31  END
32  FOR I = 1 TO MAX.Q + 1, PRINT 1 LINE WITH I-1 AND
    S(I) AS FOLLOWS
33  ***       *.***
34  END
35  SKIP 3 LINES
36  PRINT 2 LINES THUS
37  ANALYSIS OF SERVER
38  IDLE      BUSY
39  PRINT 1 LINE WITH S(1) AND 1-S(1)
40  *,***   *.***
41  RETURN
42  END
```

Fig. 30 (continued)

this step SIMSCRIPT II has loaded specified seeds automatically.* Therefore our input of values for SEED.V(1) and SEED.V(2) is not necessary but, as the following discussion shows, is desirable from the viewpoint of experimental design.

The retention of beginning and ending pseudorandom numbers in a simulation facilitates the running of successive independent replications of the same experiment. For example, the last pseudorandom numbers of one replication can be used as the pseudorandom number seeds for a second replication, thereby ensuring no overlap between the sequences of random numbers used in the two replications. This nonoverlap theoretically makes these replications independent. The last pseudorandom numbers in the second replication can then serve as the seeds for the third replication. Recording the first and last numbers also allows us to keep track of the order in which replications are run.

Another reason for assigning initial pseudorandom numbers is a desire

* These seeds are listed in Section 7.6, where we discuss the SIMSCRIPT II pseudorandom number generator.

to control sampling variation. If, for example, we wish to run a simulation again for a different combination of input parameters, statistical theory suggests that we use the same streams for both experiments in order to minimize the difference in experimental results attributable to random variation. Recording of seeds enables us to ensure this statistical property. We have provided the output report in the REPORT routine with the capability for printing beginning and ending pseudorandom numbers, so that the suggestions regarding independent replications and pair comparisons can be effected.

Returning to the MAIN routine, lines 7 through 13 prime the system. The remaining three scheduling statements are clear. The expression EXPONENTIAL.F(MEAN.SERVICE, 2) denotes the generation of an exponentially distributed variate and, as used in line 9, means that the number so generated is to be measured in minutes, for scheduling purposes, with MEAN.SERVICE using pseudorandom number stream 2. The START SIMULATION command transfers control to the timing routine, which immediately executes the event ARRIVAL already scheduled. Notice that an ARRIVAL is necessary to get the simulation going.

The ARRIVAL event program follows the flowchart in Fig. 8. The declaration of the ARRIVAL event in line 1 carries the words SAVING THE EVENT NOTICE. This preserves the record of the arrival for scheduling the next arrival in line 2. Two SIMSCRIPT II system variables play explicit roles in this event. The variable TIME.V contains the current value of simulated time in decimal days. Since our measurements in this simulation are in minutes, we multiply TIME.V by 1440 minutes in a day to convert time units to minutes. This procedure could have been avoided by using UNITS instead of MINUTES in line 2. The variable N.QUEUE denotes the number of tasks in the set QUEUE.

Lines 3 through 6 relate to the accumulation of relevant statistics. Lines 7 through 10 are needed for arrivals that must wait. Line 12 puts the server in the busy mode, and line 13 schedules the arrival's departure. The RETURN commands in lines 10 and 14 cause return to the timing routine for determination and execution of the next event. This return distinguishes event routines in SIMSCRIPT II from general subroutines which return to their explicit calling routines.

The DEPARTURE and COLLECTION event programs follow the corresponding flowcharts in Figs. 8 and 9, respectively. The END.OF. SIMULATION event program, which follows the corresponding flowchart in Fig. 25, computes summary statistics in lines 2 through 6, which are then printed out in the REPORT routine called for by the command NOW REPORT in line 7. A study of the REPORT routine makes apparent the convenience of the SIMSCRIPT II free-form format.

The ANALYSIS routine called in line 8 can contain instructions for performing a more comprehensive statistical analysis of results than the meager computations of lines 2 through 6 permit. We discuss the contents of such an ANALYSIS routine in Chapter 10. Lines 9 through 13 provide a way of deleting from the list of scheduled events those whose scheduled execution time exceeds N. The return from the END.OF.SIMULATION event to the timing routine then finds no more scheduled events and terminates the simulation, returning control to the MAIN routine, which executes its next command in line 19. Failure to delete scheduled events causes the simulation to continue. An alternative way to terminate the simulation is to use a STOP statement. However, in a more general purpose simulation, such a statement precludes return to the MAIN routine, which the user may, for example, employ to establish initial conditions for a subsequent simulation run.

Table 12 shows an example of the output from the REPORT routine. In addition to the sample mean queue length and the sample variance of queue length, the histogram provides the user with the sample probability of having a particular number of jobs in the system. The report routine is programmed to print only the nonzero sample probabilities which occur, in this case, for 0 through 32.

Although the descriptive information that Table 12 conveys is helpful, no attempt is made there to assess the accuracy of the sample descriptors with regard to statistical variation in the simulation experiment. For example, the sample mean queue length, 7.34, is a point estimate of the true underlying mean queue length. However, Table 12 does not tell us how representative an estimate it is. Table 13 provides such an evaluation for mean queue length. This table contains results computed in the ANALYSIS routine (Appendix B) called for by the END.OF.SIMULATION event. In addition to the point estimate 7.34, Table 13 gives an approximate 90 per cent confidence interval (5.20, 9.48) for mean queue length. The theoretically correct answer is 9.

Chapter 10 describes how this interval is estimated. The most crucial information needed to compute the interval is an estimate of the variance of the sample mean. In Table 13, this is 1.44. Had an analyst assumed that observations were independent, he would estimate the variance of the sample mean as the ratio of the sample population variance, 37.41, and the sample size, 10,000. This estimate, $37.41/10{,}000 = 0.003741$, would underestimate the variance of the sample mean by a factor of $1.44/0.003741 = 384.92$. Naturally this would have also led to a gross underestimate of the confidence interval.

The equivalent degrees of freedom and the computed critical t value are data that the ANALYSIS routine computes and uses in the construction of

Table 12 Single Server Queueing Problem: SIMSCRIPT II Report Routine Statistics

```
SIMULATION RESULTS

STREAM 1 STARTING RANDOM NUMBER =     2116429302
    STREAM 1 LAST RANDOM NUMBER =      215867681
STREAM 2 STARTING RANDOM NUMBER =      683743814
    STREAM 2 LAST RANDOM NUMBER =      181518447

INITIAL QUEUE LENGTH=  0
MEAN INTERARRIVAL TIME =   5.00 MEAN SERVICE TIME =   4.50
AVERAGE QUEUE LENGTH IS    7.34 TASKS
STD.DEV.  IS       6.12

QUEUE LENGTH HISTOGRAM
NO.        FREQ.
  0          .090
  1          .080
  2          .072
  3          .070
  4          .069
  5          .070
  6          .067
  7          .070
  8          .062
  9          .058
 10          .060
 11          .039
 12          .026
```

the confidence interval to compensate for the substitution of the estimated variance of the sample mean for its true variance. The remaining output are data computed in the process of constructing the estimated variance of the sample mean. In particular, an *autoregressive* model is fitted to the data to account for the dependence among observations. In addition to facilitating estimation of the variance of the sample mean, the autoregressive structure enables us to estimate other measures such as the autocorrelation function and the spectrum. Chapter 6 discusses the significance of these descriptors and shows how the remaining output in Table 13 can be put to use.

The reader may have already noted that, although the GPSS and SIMSCRIPT simulations are for the same problem with identical parameters and simulated times, the results for corresponding quantities in Tables 11 and 12 differ markedly. These discrepancies are due principally to the different sequences of pseudorandom numbers used in the two simulations. Variation

Table 12 (continued)

QUEUE LENGTH HISTOGRAM

NO.	FREQ.
13	.024
14	.018
15	.015
16	.013
17	.015
18	.011
19	.014
20	.011
21	.011
22	.010
23	.004
24	.002
25	.003
26	.004
27	.005
28	.002
29	.002
30	.002
31	.001
32	.000

ANALYSIS OF SERVER

IDLE	BUSY
.090	.910

in results would also occur, for example, among several SIMSCRIPT simulation runs of this problem if different seeds were used to prime the system each time. It is this variation in output that a simulation user must explain before he can assert the degree to which his results are representative of true system behavior. Chapter 10 is devoted to methods of accounting for such variation.

5.6 SIMULA

SIMULA was designed and implemented by Dahl and Nygaard [68] at the Norwegian Computing Center under a contract with the UNIVAC division of the Sperry Rand Corporation. It is presently available on UNIVAC 1107 and 1108 computers [78] and on CDC 6000 series computers. An updated

Table 13 Single Server Queueing Problem: SIMSCRIPT II Analysis Routine Statistics

```
STATISTICAL RESULTS OF SIMULATION EXPERIMENT USING AUTOREGRESSIVE APPROACH
                          SAMPLE MEAN= 7.34110E 00
            .050 LOWER CONFIDENCE POINT= 5.20319E 00
            .050 UPPER CONFIDENCE POINT= 9.47901E 00
             SAMPLE POPULATION VARIANCE= 3.74146E 01
          SAMPLE VARIANCE OF SAMPLE MEAN= 1.43888E 00
                      FINAL SAMPLE SIZE= 10000
          EQUIVALENT DEGREES OF FREEDOM=   12
             COMPUTED CRITICAL T VALUES= 1.78
                   AUTOREGRESSIVE ORDER= 1
             SAMPLE RESIDUAL VARIANCE= 3.87161E-01

              SAMPLE AUTOREGRESSIVE COEFFICIENTS
B( 0)= 1.0000E 00      B( 1)= -9.9481E-01
  TEST STATISTIC =     19.88   D.F. =   24   CRITICAL 0.025 VALUE =  41.92

                  SAMPLE AUTOCORRELATIONS
1.0000E-00  9.9481E-01  9.8982E-01  9.8485E-01  9.7994E-01  9.7504E-01  9.7013E-01  9.6516E-01
9.6012E-01  9.5519E-01  9.5024E-01  9.4554E-01  9.4083E-01  9.3618E-01  9.3158E-01  9.2684E-01
9.2209E-01  9.1748E-01  9.1291E-01  9.0843E-01  9.0378E-01  8.9918E-01  8.9441E-01  8.8966E-01
8.8499E-01  8.8040E-01

AUTOREGRESSIVE ORDER DETERMINED AS IN FISHMAN, G.S., CONCEPTS AND METHODS
IN DISCRETE EVENT DIGITAL COMPUTER SIMULATION
```

version, SIMULA-67, is under development [69]. Unlike SIMSCRIPT, which shares a common syntactic structure with FORTRAN, SIMULA is an ALGOL-based simulation language. It augments the ALGOL syntax with concepts appropriate to simulation work.

Whereas an ALGOL program specifies a data structure and a sequence of operations on data local to the program, SIMULA extends this notion to include a collection of such programs, called *processes*, that can conceptually operate in parallel. It provides a means for describing and dynamically generating processes in addition to referencing already existing ones. It enables data local to one process to be used by others and provides a timing routine, called the *sequencing set*, that orders processes according to the time at which activity occurs.

In SIMULA a set of data declarations, together with a sequence of statements that describe the behavior of a type of component of a system, defines an *activity*. When numerical values are assigned to the elements of the data structure the result is called a *process*. One activity can produce many processes. Here a *transaction* of GPSS would be one type of process. Unlike the situation with GPSS, where transactions play *active* roles and storages and facilities play *passive* roles, all components of a simulation written in SIMULA can be active or passive. It is therefore necessary to write an activity procedure for each type of component. SIMULA arranges for the interaction among the resulting processes, thereby reducing the modeling burden that a user faces.

SIMULA also makes available to a user several sampling and data analytic procedures. One such procedure defines the pseudorandom number generator (described in Section 7.6). Others allow a user to generate independent Bernoulli trials, continuous and discrete uniform deviates, normal variates, gamma variates, and Poisson variates. A procedure for sampling from empirical distributions is also included. For data analysis SIMULA provides a procedure for constructing a histogram and a procedure for accumulating time integrals. In addition, as in SIMSCRIPT II, a user can write a procedure to meet his specialized needs.

5.7 SIMULA EXAMPLE

To illustrate SIMULA we use an example described in McNeley [76] and programmed by Mr. Joseph Faulkner of UNIVAC. Consider a store or supermarket with two clerks and many customers. Clerk 1 serves customers with fewer than seven items, and clerk 2 serves the remaining customers. To describe the problem we use underlined lower-case letters to denote *key* words, routines, and variables known to SIMULA. There are two activities in this model; one, customer, describes a customer in lines 4 through 14 of Fig. 31. The other, clerk, describes a clerk in lines 15 through 24. Lines 25 through 33 establish the initial conditions and create the stream of customers.

In line 2 SIMULA defines a set "queue," an integer "ncus," and an array C. The H1 and H2 arrays in line 3 are needed to construct a histogram. Line 6 initializes H1 and H2. In line 27 SIMULA creates clerk 1 and stores his address in C(1), that is C(1) is a pointer. Clerk 1 is then scheduled to become

```
Line
 1  simula begin comment store;
 2  set array queue (1:2); element array c(1:2); integer ncus,seed,i;
 3  array h1(1:26),h2(1:25);
 4  activity customer(n); integer n;
 5  begin real stime; integer i;
 6     stime := time;
 7     if n leq 6 then i := 1 else i := 2;
 8     if empty(queue(i)) then activate c(i) at time;
 9     wait(queue(i));
10     histo(h1,h2,time - stime,1);
11     hold(0.25);
12     activate c(i) at time;
13     ncus := ncus + 1; remove(first(queue(i)));
14     end;
15  activity clerk(i); integer i;
16  begin
17  L1: if empty(queue(i)) then go to L2;
18        inspect first(queue(i)) when customer do
19            begin hold(max(0.25,0.10 * n));
20               activate first(queue(i));
21               passivate;
22            go to L1; end;
23  L2: passivate;
24           go to L1; end;
25  seed := 1234567;
26  for i := 1 step 1 until 25 do h2(i) := 0.50 + (i - 1) * 0.50;
27  c(1) := new clerk(1); activate c(1) at time;
28  c(2) := new clerk(2); activate c(2) at time;
29  L1: activate new customer(randint(1,25,seed)) at time;
30  hold(negexp(1./3.0,seed));
31  if ncus leq 1000 then go to L1;
32  hprint (h1,h2);
33  end simula;
```

Source: [76].

Fig. 31 SIMULA coding of supermarket with 2 clerks and 1000 customers.

active at the current simulated time denoted as "time." SIMULA performs analogous steps for clerk 2 in line 28.

At the same simulated time line 28 creates and activates a customer with N items (see line 4), where N is drawn from a discrete uniform distribution on the integers (1, 25). After a time delay drawn from the exponential distribution with mean $\frac{1}{3}$ minute, in line 30, the program continues to loop through lines 29 and 30 until 1000 customers are created and activated. The augmentation of ncus takes place in line 13.

To see how the simulation works we recall that clerk 1 is the first activated process. If queue 1 is empty in line 17 then clerk 1 becomes passive in line 23. This is the case on the first activation of clerk 1. Note that the reactivation point is line 24. Alternatively, if the first process in queue 1 is a customer, then lines 19 through 21 are executed. Line 19 assumes that 0.1 minute is required to check an item but at least 0.25 minute is needed to check any customer. After the service time has elapsed, line 20 activates the customer and line 21 makes clerk 1 passive. Note that the reactivation point is line 22. Analogous steps occur for clerk 2.

Upon activating the first customer, "stime" assumes the value of current simulated time in line 5. Line 7 determines which clerk will serve the customer; if no one is waiting for this clerk he is activated in line 8. Remember that for the first customer the selected clerk becomes active at line 22. Other customers may have to wait, and they are assigned to an appropriate queue and are made inactive in line 9. When a customer is activated (line 20) execution begins at the reactivation point, line 10. There the customer's waiting time, time − "stime", is entered into the histogram. The expression in line 10 causes H1(I) to be incremented by 1, where I is the smallest integer such that time − "stime" < H2(I). (See line 26 for the construction of H2.) A delay of 0.25 minute occurs at line 11 to enable a customer to pick up his change. Line 12 activates the appropriate clerk at line 22. Line 13 augments the customer count by 1 and removes the customer from the queue, and the subsequent end statement destroys this process.

EXERCISES

1. Modify the SIMSCRIPT II program given in Section 5.5 to reflect the testing procedure described in Exercise 1*a* of Chapter 3, and run the modified program using the same input parameters and the ANALYSIS program in Appendix B. Assume that the probability of passing the test is 0.95, and use Bernoulli trials (Section 6.1) to determine which jobs pass the test and which do not. Assume that the test time has the exponential distribution with mean 2.

2. Modify the GPSS/360 program given in Section 5.3 to reflect the testing procedure described in Exercise 1b of Chapter 3, and run the modified program using the same input parameters. Assume that the probability of passing the test is 0.95, and use Bernoulli trials (Section 6.1) to determine which jobs pass the tests and which do not. Assume that test time has the exponential distribution with mean 2.

3. Modify the SIMSCRIPT II program given in Section 5.5 to permit job selection based on minimum service time (see Exercise 2a and reference 49 in Chapter 3). Run the program, using the same input parameters.

4. Modify the GPSS/360 program given in Section 5.3 to permit job selection based on minimum service time (see Exercise 2b in Chapter 3 and reference 71). Run the program, using the same input parameters.

5. Program Fig. 12 in SIMULA, using the mean interarrival and service characteristics given in the SIMSCRIPT II example in Section 5.5. Include a histogram of queue length in the output.

6. Incorporate the testing feature described in Exercise 1b of Chapter 3 into the SIMULA program in Exercise 5.

7. Write a SIMSCRIPT II program for Exercise 8 in Chapter 3. Assume that interarrival times and service times are independent and exponentially distributed with mean times 3.8 and 16 minutes, respectively. The probability that a customer leaves because no barber is available is 0.3. Assume that there are four barbers and two manicurists and that the probability that a customer wants a manicure is 0.3. If his haircut is completed before his manicure has started, he leaves without receiving a manicure.

   In the output provide an analysis of quarter-hourly queue length and hourly revenues, using the SIMSCRIPT II ANALYSIS routine in Appendix B. In other words, two time series analyses are to be performed. Run the simulation for 6000 minutes, and begin the simulation in the empty and idle condition with an arrival at the beginning of the run.*

8. Write a GPSS/360 program for Exercise 9 in Chapter 3, using the specifications in Exercise 7 and producing all the output requested there. To perform an analysis on the queue length and revenue data it is necessary first to arrange for the output of these data and to use the output as input to the FORTRAN ANALYS routine in Appendix B.†

9. Write a SIMULA program for Exercise 9 in Chapter 3, using the specifications in Exercise 7 and producing all the output requested there. To perform an analysis on the queue length and revenue data it is necessary

first to arrange for the output of these data and to use the output as input to the FORTRAN ANALYS routine in Appendix B.†

10. Run the program in Exercise 7 again with the same input, but with one additional barber.* Compare your results with those of Exercise 7 and explain the differences. If barbers receive $4.00 hourly, do you think there is merit in adding a fifth barber?

11. Run the program in Exercise 8 again with the same input, but with one additional barber.† Compare your results with those of Exercise 8 and explain the differences. If barbers receive $4.00 hourly, do you think there is merit in adding a fifth barber?

12. Run the program in Exercise 9 again with the same input, but with one additional barber.† Compare your results with those of Exercise 9 and explain the differences. If barbers receive $4.00 hourly, do you think there is merit in adding a fifth barber?

* If each student uses different pseudorandom number seeds the teacher can collect and compare the results to give an example of the differences due to random variation. The instructor may wish to discuss the statistical effect on dependence of a *quarter*-hourly analysis on queue length in contrast to an hourly analysis on revenue. The ANALYSIS output is to be used in the exercises in Chapter 6.

† The ANALYSIS output is to be used in the exercises in Chapter 6.

CHAPTER 6

STATISTICAL DEFINITIONS AND CONCEPTS

In addition to model building and programming considerations, many diverse statistical problems face the potential simulation user. They include:

1. random number generation,
2. random variate generation from specified distributions,
3. input parameter determination based on sample data.
4. simulation output analysis, and
5. comprehensive experimental design for performing a series of related experiments.

Although the science of statistics undeniably can contribute to the solutions of the probabilistic and inferential problems in simulation, a question arises regarding how statistically sophisticated a simulation user must be to resolve these problems. Just as we cannot expect every simulation user to be familiar with the nature of every programming problem that the list processing languages described earlier are designed to solve, so we can hardly expect him to be the compleat statistician. Moreover, a simulation user is interested in getting answers from his simulation. Generally, he has only incidental interest in the methodological statistical problems of simulation. If a statistical methodology is simple and can improve his results, he *may* use it. For a long time this attitude prevailed with regard to the choice of programming languages, even though the easier model building capabilities of list processing languages were acknowledged.

In the rest of this book we describe a variety of the statistical problems that a simulation user is knowingly or unknowingly likely to encounter during the course of a simulation study. We also offer methods of studying and often solving these problems. Some problems are naturally harder than others, and the corresponding suggested solutions vary accordingly. It is hoped that this problem articulation will encourage a reader to use the suggestions provided here to enrich the statistical content of his simulation.

Since many statistical problems in simulation involve series of random events we begin our discussion of statistics in simulation in this chapter with a brief introduction to the theory of stationary stochastic processes, which provides a particularly helpful framework for studying these series. The exposition relies principally on description with only an occasional effort at proof when it seems beneficial for the reader. As a prelude to stationary processes we review the notions of independence, dependence, and correlation. Readers familiar with these concepts may wish to skip to Section 6.2.

6.1 STATISTICAL ASSOCIATION

Independence

To facilitate a clear exposition of new ideas we first review several basic definitions in probability theory. The first notion to be defined is *statistical independence.* Consider two continuous random variables X_1 and X_2 with probability density functions (p.d.f.s) $f(x_1)$ and $f(x_2)$, respectively. Their joint p.d.f. is

$$f(x_1, x_2) = f(x_1 \mid x_2)f(x_2) = f(x_2 \mid x_1)f(x_1) \tag{1.1}$$

where $f(x_1 \mid x_2)$ is the conditional p.d.f. of X_1, given X_2, and $f(x_2 \mid x_1)$ is the conditional p.d.f. of X_2, given X_1. The p.d.f.s $f(x_1)$ and $f(x_2)$ are often called the *marginal* p.d.f.s of X_1 and X_2, respectively, since they may be obtained by the integrations

$$\begin{aligned} f(x_1) &= \int f(x_1, x_2)\, dx_2 \\ f(x_2) &= \int f(x_1, x_2)\, dx_1. \end{aligned} \tag{1.2}$$

We say that the events X_1 and X_2 are *independent* if

$$\begin{aligned} f(x_1 \mid x_2) &= f(x_1) \\ f(x_2 \mid x_1) &= f(x_2). \end{aligned} \tag{1.3}$$

This means that if X_1 occurs before X_2 then X_1 is in no way statistically dependent on X_2. Similarly if X_2 occurs first it is in no way statistically dependent on X_1. For n random variables $X_1, \ldots, X_n$ to be *mutually independent*, we require that their *joint* p.d.f. be

$$f(x_1, \ldots, x_n) = \prod_{s=1}^{n} f(x_s). \tag{1.4}$$

The convenience of the independence assumption follows from the fact that, if we are given the marginal p.d.f.s of the n random variables, we may obtain a functional form for their joint p.d.f. simply by multiplication as in (1.4). The form may not be recognizable as a commonly encountered p.d.f., but it is derivable. Moreover, there is no necessity for specifying the conditional p.d.f.s.

Dependence

When we consider dependent events, (1.4) no longer holds since, for example,

$$f(x_1 \mid x_2) \neq f(x_1)$$
$$f(x_2 \mid x_1) \neq f(x_2). \tag{1.5}$$

Here the conditional p.d.f. of X_1, given X_2, is not identical to the marginal p.d.f. of X_1, nor is the conditional p.d.f. of X_2, given X_1, identical to the marginal p.d.f. of X_2. If we wish to describe the jointly dependent behavior of X_1 and X_2 using (1.1), we require either $f(x_1 \mid x_2)$ and $f(x_2)$ or $f(x_2 \mid x_1)$ and $f(x_1)$. For the joint behavior of $n > 2$ random variables, we require additional conditional p.d.f.s.

Generally we do not know the numerical values of the parameters of these conditional and marginal p.d.f.s, nor do we know the functional or mathematical forms of the p.d.f.s. To find these forms and to estimate their parameters require an unusually time-consuming effort. The only joint p.d.f. for which a broad and well-organized theory of statistical inference currently exists is the *multivariate normal distribution*.

Correlation

Because of these difficulties in specifying dependence between random variables, attention often focuses on a subset of the dependence properties.

Consider again the random variables $X_1, \ldots, X_n$. The *mean* of X_s is defined as

$$\mu_s \equiv E(X_s) = \int x_s f(x_s)\, dx_s \tag{1.6}$$

and its *variance* as

$$R_{ss} \equiv E(X_s - \mu_s)^2 = \int (x_s - \mu_s)^2 f(x_s)\, dx_s \qquad s = 1, \ldots, n. \tag{1.7}$$

The mean and variance are assumed to be finite. The *covariance* between X_s and X_t is

$$\begin{aligned} R_{st} &\equiv E[(X_s - \mu_s)(X_t - \mu_t)] \\ &= \int\int (x_s - \mu_s)(x_t - \mu_t) f(x_s, x_t)\, dx_s\, dx_t \\ &= \int (x_s - \mu_s) f(x_s)\left[\int (x_t - \mu_t) f(x_t \mid x_s)\, dx_t\right] dx_s \\ &= \int (x_t - \mu_t) f(x_t)\left[\int (x_s - \mu_s) f(x_s \mid x_t)\, dx_s\right] dx_t \\ &\qquad s, t = 1, \ldots, n. \end{aligned} \tag{1.8}$$

Notice that the mean μ_s can assume any (real) value, the variance R_{ss} is always nonnegative [since both $(x_s - \mu_s)^2$ and $f(x_s)$ are both nonnegative], and $R_{st} = R_{ts}$. Moreover, for X_s to be a random variable, the variance R_{ss} must be positive.

If

$$\int (x_s - \mu_s) f(x_s \mid x_t)\, dx_s = 0, \tag{1.9}$$

then

$$R_{st} = R_{ts} = 0 \qquad s \neq t, \tag{1.10}$$

and we say that X_s and X_t are *uncorrelated.* If X_s and X_t are independent, so that

$$f(x_s \mid x_t) = f(x_s), \tag{1.11}$$

then

$$\begin{aligned} \int (x_s - \mu_s) f(x_s \mid x_t)\, dx_s &= \int (x_s - \mu_s) f(x_s)\, dx_s \\ &= \int x_s f(x_s)\, dx_s - \mu_s \int f(x_s)\, dx_s = \mu_s - \mu_s = 0 \end{aligned} \tag{1.12}$$

so that X_s and X_t are uncorrelated. Notice that independence implies the absence of correlation, but lack of correlation does not necessarily imply independence.

If $R_{st} > 0$ for $s \neq t$ we say that X_s and X_t are *positively correlated.* This generally means that on average a deviation in X_s above (below) the mean μ_s occurs when a deviation in X_t above (below) the mean μ_t occurs. If $R_{st} < 0$ for $s \neq t$ we say that X_s and X_t are *negatively correlated.* This generally means that on average a deviation in X_s above (below) the mean μ_s occurs when a deviation in X_t below (above) the mean μ_t occurs.

Figure 32 illustrates the notion of correlation. In Fig. 32*a*, no dependence exists and the points whose coordinates are X_1 and X_2 form a random pattern. Figure 32*b* shows positive correlation, where small (large) values of X_1 generally occur with small (large) values of X_2 and the pattern of points looks somewhat like a straight line with positive slope. Figure 32*c* shows negative correlation, where small (large) values of X_1 occur with large (small) values of X_2, the pattern being roughly linear with negative slope.

By manipulating the expression

$$\iint \left[\frac{x_s - \mu_s}{\sqrt{R_{ss}}} \pm \frac{x_t - \mu_t}{\sqrt{R_{tt}}} \right]^2 f(x_s, x_t)\, dx_s\, dx_t \geq 0 \tag{1.13}$$

we may easily show that

$$-\sqrt{R_{ss}R_{tt}} \leq R_{st} \leq \sqrt{R_{ss}R_{tt}}. \tag{1.14}$$

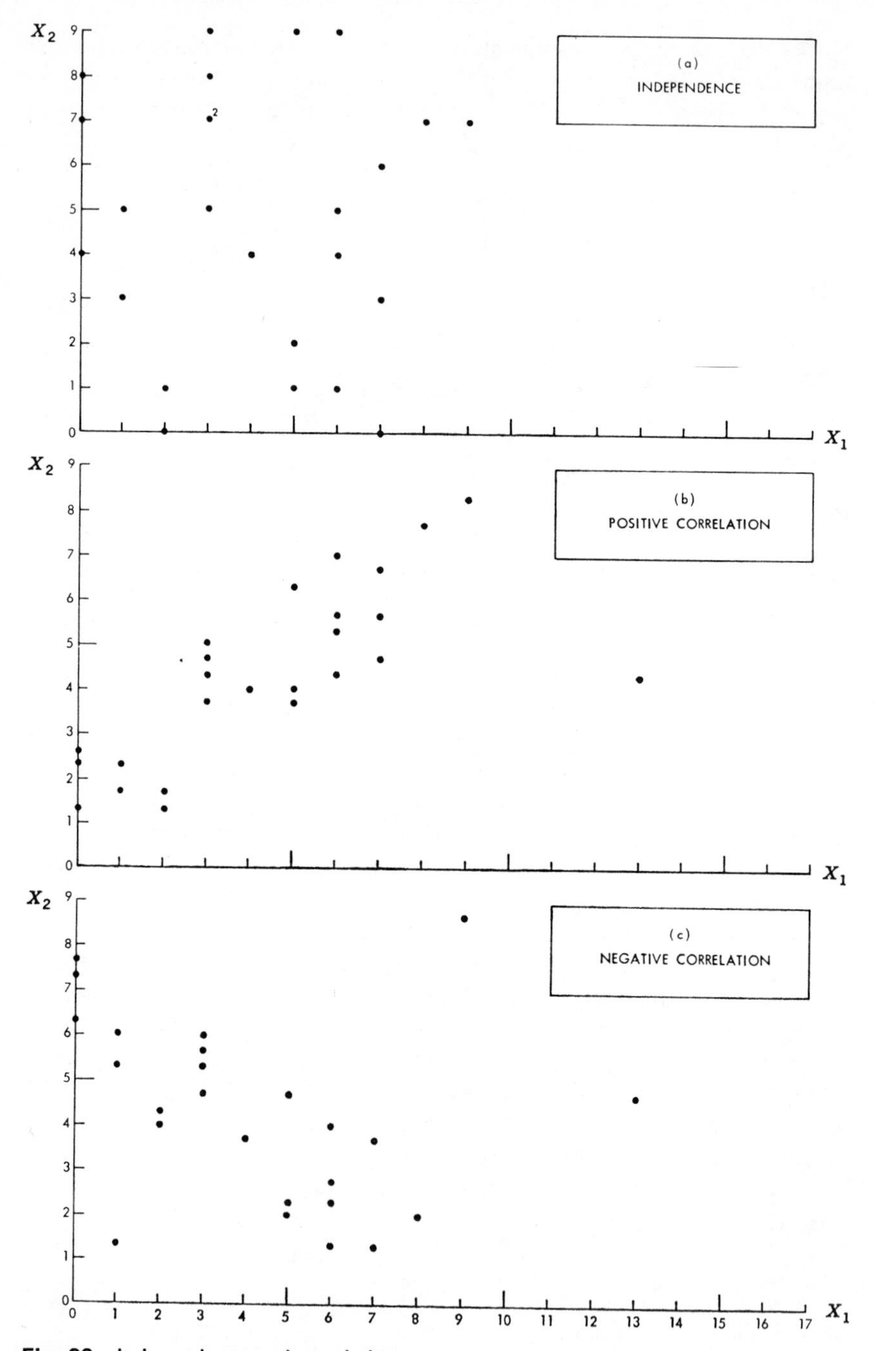

Fig. 32 Independence and correlation.

To measure the extent of correlation we define the *correlation coefficient*:

$$\rho_{st} \equiv \frac{R_{st}}{\sqrt{R_{ss}R_{tt}}}, \tag{1.15}$$

and, using (1.14), we obtain

$$-1 \le \rho_{st} \le 1. \tag{1.16}$$

The larger $|\rho_{st}|$ is, the greater is the correlation between X_s and X_t.

The following interpretation provides an intuitive understanding of correlation. Suppose that we represent X_s by

$$X_s = \alpha X_t + \varepsilon \tag{1.17}$$

where α is a constant and ε is a random variable that accounts for the random behavior in X_s beyond that represented by X_t. For example, ε would account for the influence of X_t^2, X_t^3, . . . , $\log X_t$, e^{X_t} *as well as* the random variation in X_s that is independent of any function of X_t. Rewriting (1.17), we have

$$X_s - \alpha X_t = \varepsilon, \tag{1.18}$$

and taking expectations leads to

$$\mu_s - \alpha\mu_t = \mu_\varepsilon \tag{1.19}$$

where μ_θ denotes the mean of θ. Subtracting (1.19) from (1.18), squaring both sides, and taking expectations gives us

$$R_{ss} + \alpha^2 R_{tt} - 2\alpha R_{st} = \sigma_\varepsilon^{\,2} \tag{1.20}$$

where $\sigma_\varepsilon^{\,2}$ is the variance of ε. Let us now choose α so as to minimize $\sigma_\varepsilon^{\,2}$. This occurs when

$$2\alpha R_{tt} - 2R_{st} = 0 \tag{1.21}$$

$$\alpha = \frac{R_{st}}{R_{tt}}. \tag{1.22}$$

Substituting into (1.20) yields

$$R_{ss} - \frac{R_{st}^{\,2}}{R_{tt}} = \min \sigma_\varepsilon^{\,2}$$

$$\frac{R_{st}^{\,2}}{R_{tt}} = R_{ss} - \min \sigma_\varepsilon^{\,2} \tag{1.23}$$

$$\rho_{st}^{\,2} = \frac{R_{ss} - \min \sigma_\varepsilon^{\,2}}{R_{ss}}.$$

The square of ρ_{st} is called the *coefficient of determination*, and it measures the proportion of the variance of X_s that is accounted for by the variable X_t. If $\min \sigma_\varepsilon^2 = 0$, then ε is a constant and we see that (1.17) implies a perfect *linear association* between X_s and X_t. That is, knowledge of X_t gives us X_s up to the two quantities α and ε. Also

$$\rho_{st}^2 = 1 \tag{1.24}$$

so that

$$\rho_{st} = \pm 1. \tag{1.25}$$

If $\min \sigma_\varepsilon^2 = R_{ss}$, then $\alpha = 0$, so that

$$X_t = \varepsilon$$
$$\rho_{st}^2 = \rho_{st} = 0. \tag{1.26}$$

This means that X_s accounts for none of the variance in X_t and no linear association exists. Hence the correlation is zero.

When working with independent observations an investigator generally estimates means and variances and uses these statistics to gain his first insight into the nature of his data. The estimates are simple to compute, and he knows that, if the underlying probability law is normal, a knowledge of means and variances suffices to describe the law completely.

If the observations are dependent, the investigator not only estimates the means and variances but also the covariances to provide a first insight. Moreover, if the observations jointly have a multivariate normal distribution, a knowledge of the means, variances, and covariances completely describes the joint probability law governing the mechanism that generates the observations. It is important to note that, when the multivariate normal distribution applies and the covariances are zero, the observations are not only uncorrelated but also independent.

Suppose that we are given the variances and covariances for n random variables and we array them in a *covariance matrix*:

$$R_n = \begin{bmatrix} R_{11} & R_{12} & R_{13} & \cdots & R_{1n} \\ R_{21} & R_{22} & & & \\ R_{31} & & \ddots & & \vdots \\ \vdots & & & & \\ R_{n1} & & \cdots & & R_{nn} \end{bmatrix}. \tag{1.27}$$

From this matrix we can derive all the relevant correlation coefficients to learn about the correlation between observations. For example, the correlation between the first and the nth random variables is

$$\rho_{1n} = \frac{R_{1n}}{\sqrt{R_{11}R_{nn}}}. \tag{1.28}$$

Since it is most convenient to express the statistical functional relationship among variables in linear terms whenever possible, examination of the covariance structure is a natural first step in data analysis. The same emphasis applies in analyzing the results of simulation experiments, where study of the covariance structure of a set of observations enables us to develop useful and meaningful inferences about the processes under investigation.

Table 14 summarizes a number of facts regarding independence, dependence, and correlation.

Table 14 Independence, Dependence, and Correlation

Independence[a]	$f(x_1 \mid x_2) = f(x_1)$,	$f(x_2 \mid x_1) = f(x_2)$
Dependence[b]	$f(x_1 \mid x_2) \neq f(x_1)$,	$f(x_2 \mid x_1) \neq f(x_1)$
Correlation[c]		
Definition	$\rho_{st} \equiv R_{st}/\sqrt{R_{ss}R_{tt}}$	
No	$\rho_{st} = 0$	
Perfect	$\rho_{st} = \pm 1$	
Positive	$\rho_{st} > 0$	
Negative	$\rho_{st} < 0$	
Restriction	$-1 \leq \rho_{st} \leq 1$	

[a] Independence implies no correlation.
[b] Dependence does not imply correlation.
[c] Correlation implies dependence. No correlation does not imply independence.

6.2 STOCHASTIC SEQUENCES*

Consider an ordered series of events $X_1, \ldots, X_M$, where M is any positive integer. The quantity X_s may be the waiting time for the sth job to receive service in a job shop, so that the sequence is a series of M waiting times. The arrangement of the order of the waiting times may vary. For example, the index s may denote the order in which jobs arrive, the order in which they receive service, or the order in which they emerge from the service facility. Figure 33 shows a sequence of waiting times arranged in the three orders.

Alternatively, the quantity X_s may denote the state of an activity at time s. For example, X_s may be the number of jobs waiting for service at time s. Here the ordering of the events is determined by the time at observation. For convenience of analysis the time between successive observations is usually fixed at unity by appropriately scaling the time axis.

* See references 79–81, 83, 85, and 87 for more complete discussions.

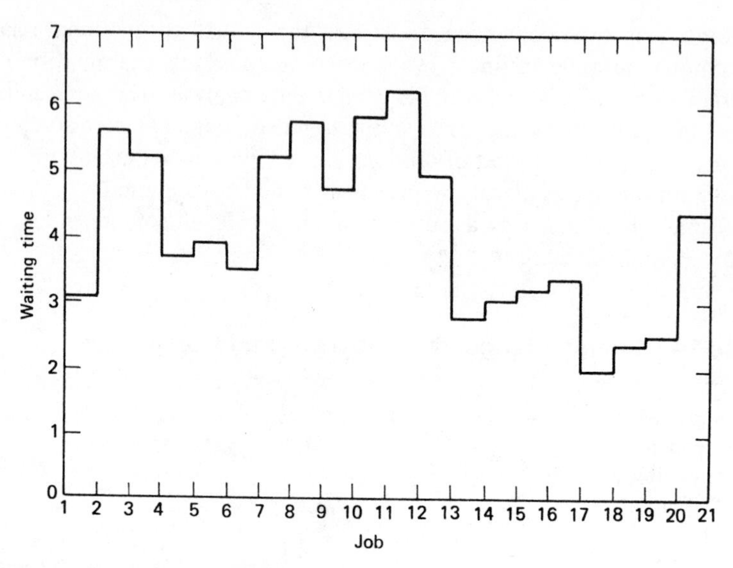

Fig. 33a Waiting times in order of arrival.

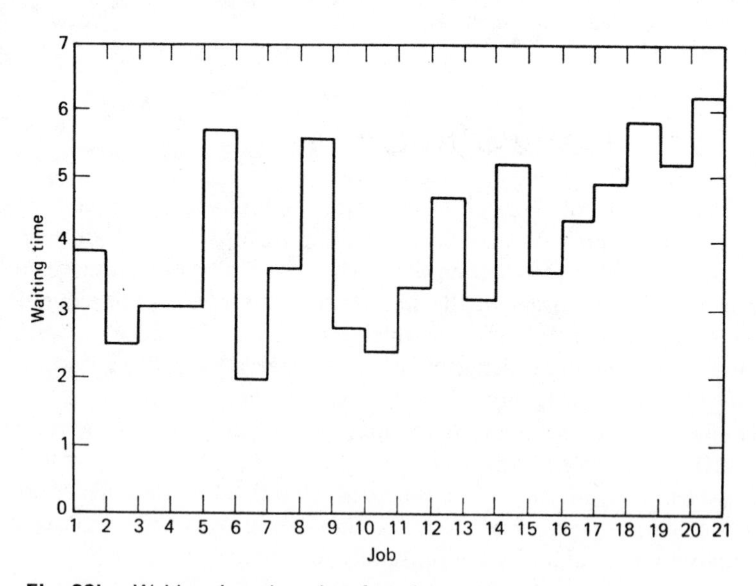

Fig. 33b Waiting times in order of service.

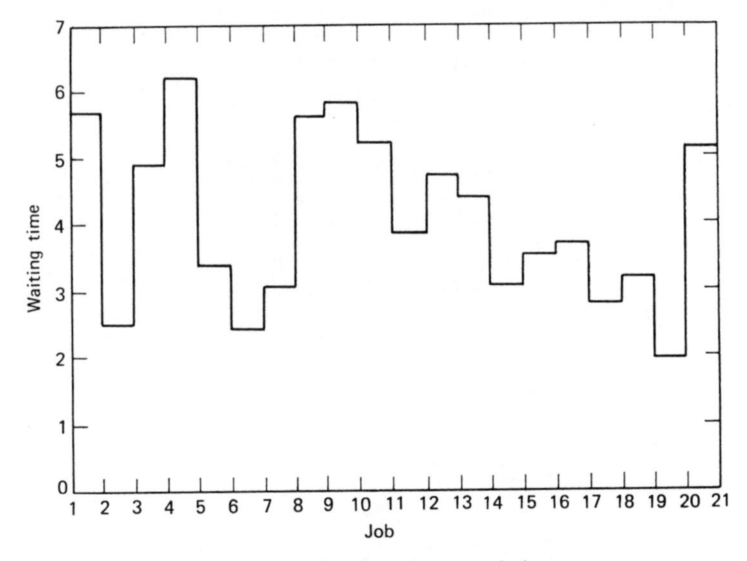

Fig. 33c Waiting times in order of service completion.

When the index s indicates time the variable X_s may sometimes denote an average over the interval between observations. Suppose that an underlying process $X(t)$ is developing through time. Then the average value of the process at time s is

$$X_s = \int_{s-1}^{s} dX(t) \tag{2.1}$$

where the *Stieltjes* integral in (2.1) is used to indicate that $X(t)$ may be developing discontinuously in time. It may occasionally be desirable to collect a sequence of averages such as (2.1) rather than to collect observations on the state of the system at a single moment in time. This is especially true when the state of the system varies widely over short time intervals since averaging smooths out some of the wide fluctuations.

If the outcome of each of the M events is subject to random behavior, we say that $X_1, \ldots, X_M$ is a *stochastic sequence*. If the outcomes are statistically dependent, then we must speak of the joint behavior of the M events to describe the events completely. Just as a random variable X can assume one of several values in accordance with its p.d.f., so the series of events $X_1, \ldots, X_M$ can assume one of several sets of M values, the probability of occurrence of any set being determined by the joint p.d.f. Each set is called a *realization* of the sequence.

Consider a stochastic sequence X_1, X_2, X_3 (i.e., $M = 3$), where X_s for $s = 1, 2, 3$ can assume the value 0 or 1. The $(2)^3 = 8$ possible realizations are shown in Table 15.

Table 15 Example of a Stochastic Sequence

		Probability	
Q	Realization	(a)	(b)
1	1 1 1	$\frac{1}{8}$	$\frac{1}{32}$
2	0 1 1	$\frac{1}{8}$	$\frac{3}{32}$
3	1 0 1	$\frac{1}{8}$	$\frac{9}{32}$
4	1 1 0	$\frac{1}{8}$	$\frac{3}{32}$
5	0 0 1	$\frac{1}{8}$	$\frac{3}{32}$
6	0 1 0	$\frac{1}{8}$	$\frac{9}{32}$
7	1 0 0	$\frac{1}{8}$	$\frac{3}{32}$
8	0 0 0	$\frac{1}{8}$	$\frac{1}{32}$
		1.0	1.0

Let us now investigate the joint probability of X_1, X_2, and X_3 or, equivalently, the probability of occurrence of any of the eight realizations. Let Q denote the number of the realization as given in Table 15. Suppose that events in the sequence are independent and that 0 and 1 are equally likely occurrences on any trial, that is,

$$\begin{aligned} &\Pr(Q) = \Pr\{X_1, X_2, X_3\} = \Pr(X_1)\Pr(X_2)\Pr(X_3) \\ &\Pr(X_s = 0) = \Pr(X_s = 1) = \tfrac{1}{2} \qquad s = 1, \ldots, 3. \end{aligned} \tag{2.2}$$

The realization probabilities given by (2.2) are shown under (a) in Table 15. We see that all realizations are equally likely.

Suppose now that X_1, X_2, and X_3 are not independent, so that

$$\Pr(Q) = \Pr(X_1, X_2, X_3) = \Pr(X_1)\Pr(X_2 \mid X_1)\Pr(X_3 \mid X_1, X_2) \tag{2.3}$$

since the events occur in order of the indices. Let

$$\left.\begin{aligned} &\Pr(X_1 = 0) = \Pr(X_1 = 1) = \tfrac{1}{2} \\ &\Pr(X_s = 1 \mid X_{s-1} = 0) = \Pr(X_s = 0 \mid X_{s-1} = 1) = \tfrac{3}{4} \\ &\Pr(X_s = 1 \mid X_{s-1} = 1) = \Pr(X_s = 0 \mid X_{s-1} = 0) = \tfrac{1}{4} \\ &\Pr(X_3 \mid X_1, X_2) = \Pr(X_3 \mid X_2) \end{aligned}\right\} \qquad s = 2, 3. \tag{2.4}$$

The realization probabilities are given under (b) in Table 15, based on (2.3). Notice that realizations 3 and 6 are most likely and 1 and 8 are least likely.

More generally, if X_s were to assume k different values and a sequence consisted of M points, there would be k^M possible realizations. If X assumes an infinity of values, then an infinite number of realizations can occur. For an ongoing process the number of events M becomes arbitrarily large. For example, each job passing through a job shop has a waiting time and the number of observations of queue length increases linearly with time. Then as an underlying model we may take $M = \infty$ and define a stochastic sequence as $\{X_s; s = 1, \ldots, \infty\}$ or, more concisely, as $\{X_s\}$.

Although we cannot observe every event because of time limitations and cost, we can observe a subset, say n events, $X_s, X_{s+1}, \ldots, X_{s+n-1}$, which we call a *finite realization* or, more commonly, a *time series*. The term time series has come into common usage since many series of events are collected through time. We use the term here but emphasize that any finite ordered series of stochastic events, for example, a series of waiting times, constitutes a time series.

Let us again draw an analogy between a random variable Y and a stochastic sequence $\{X_s\}$. Suppose that we collect one observation on Y. The corresponding action for $\{X_s\}$ is to collect one time series of an arbitrary number of events n. In other words, one observation on $\{X_s\}$ consists of a series of n ordered events. Collecting k observations on Y corresponds to collecting k time series on $\{X_s\}$.

To generate k time series on a stochastic sequence of interest in a simulation experiment, we may run the experiment k times and collect k time series. When this is done we say we are *replicating* the experiment, the kth time series corresponding to the kth replication. Figure 34 shows time series for queue size for two replications of the same simulation experiment, the only difference in initial conditions being in the pseudorandom number seed. Table 16 summarizes several definitions already discussed.

Just as our ability to make statistical inferences about the probability law governing a single event improves as the sample size k increases, so our ability to make inferences about the joint probability law governing n events in $\{X_s\}$ improves as the number of replications and hence the number of time series increases.

Table 16 Definitions in Stochastic Sequences

Stochastic sequence	$\{X_s; s = 1, \ldots, \infty\}$, X_s is a random variable
Realization	A single sequence of numbers assumed by $\{X_s\}$
Time series	A finite sample of a realization
Replication	One of several time series obtained on each run of an experiment

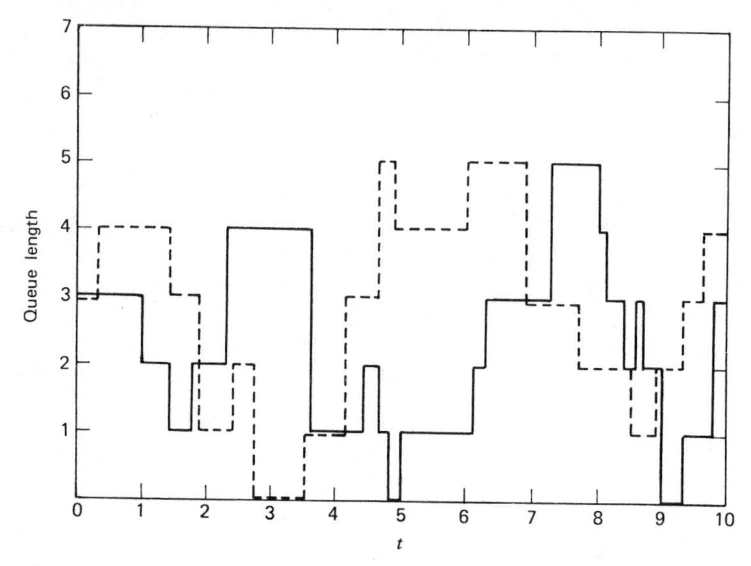

Fig. 34 Two replications of queue length.

Since each time series has n observations in it, it is natural to ask what effect increasing n has on making statistical inferences on $\{X_s\}$. It turns out that, if $\{X_s\}$ has certain *ergodic* properties, the accuracy of statistical inferences about $\{X_s\}$ improves as n increases. Moreover only one time series is needed to realize the improved accuracy. This means that we have two ways of learning about $\{X_s\}$; one is replicating the experiment a number of times, and the other is increasing the number of observations in one time series (i.e., one replication). In Chapter 10 we discuss the benefits and costs of the two alternatives in a simulation environment. In Section 6.4 we describe how to ensure that the sequence of interest has the required ergodic properties.

6.3 STATIONARITY

Suppose that every series $X_s, X_{s+1}, \ldots, X_{s+n}$ for $s = 1, 2, \ldots, \infty$ has the same joint p.d.f. For example, $X_1, X_2, \ldots, X_n$; $X_{11}, X_{12}, \ldots, X_{10+n}$; and $X_{21}, X_{22}, \ldots, X_{20+n}$ all have the same joint p.d.f. Then we say that X_s is a *strictly stationary sequence*. The property implies that the joint probability law governing $\{X_s\}$ is invariant with regard to location in the set of indices. If the index is time we say that the probability law governing the sequence is independent of historical time. An alternative name for this property is *strong stationarity*.

Let us now restrict attention to the covariance properties of a stochastic sequence. We have the formal definitions of the mean:

$$\mu_s = E(X_s) = \int x_s f(x_s)\, dx_s \tag{3.1}$$

and the covariances:

$$\begin{aligned} R_{st} &= E[(X_s - \mu_s)(X_t - \mu_t)] \\ &= \int\int (x_s - \mu_s)(x_t - \mu_t) f(x_s, x_t)\, dx_s\, dx_t \qquad s, t = 1, 2, \ldots, \infty. \end{aligned} \tag{3.2}$$

To emphasize that these covariances describe behavior in a single stochastic sequence, we denote R_{st} as the *autocovariance* between the sth and tth events in the sequence. The sequence of means $\{\mu_s; s = 1, 2, \ldots, \infty\}$ forms the *mean function*, and the sequence of autocovariances $\{R_{st}; s, t = 1, 2, \ldots, \infty\}$ forms the *autocovariance function*. We denote these functions by μ and R, respectively.

With the exception of the restriction that s and t be ordering indices, the definitions of the mean and autocovariance functions follow the development of Section 6.1. What we would like now are mild restrictions on $\{\mu_s\}$ and $\{R_{st}\}$ that enable us to perform statistical analyses more easily than is possible with the unrestricted model. Suppose that

$$\begin{aligned} \mu_s &= \mu < \infty \\ R_{st} &= R_{s-t} < \infty \qquad s, t = 1, 2, \ldots, \infty. \end{aligned} \tag{3.3}$$

That is, the mean is a finite constant μ for all events in the sequence, and the covariance between any two events in the sequence is finite and a function only of their separation in the series. A stochastic sequence with these properties is termed *covariance stationary*. Some writers prefer to call such sequences *wide sense*, *mean square*, or *second-order* stationary.

A strictly stationary sequence is covariance stationary, but the converse does not necessarily hold. Any covariance stationary sequence in which the elements of any arbitrarily chosen finite set of events jointly have a multivariate normal distribution is called a covariance stationary *normal process*.

Table 17 lists the definitions and several properties of stationarity for convenient reference. The autocorrelation function, denoted by ρ, is a convenient descriptor of certain stochastic properties, as the next section shows. The positive definite property plays a role when we consider autoregressive properties in Section 6.6.

6.4 THE AUTOCORRELATION AND SPECTRAL DENSITY FUNCTIONS

The autocorrelation function may assume many forms, each one being a consequence of one or more statistical relationships between events in the

Table 17 Definitions and Properties of Stationarity

Definitions	
Strict stationarity	$f(x_s, \ldots, x_{s+n}) = f(x_{s+t}, \ldots, x_{s+t+n})$
Covariance stationarity	$\mu = E(X_s) < \infty,\ R_{s-t} = E[(X_s - \mu)(X_t - \mu)] < \infty$
Autocovariance	$R_\tau = E[(X_s - \mu)(X_{s+\tau} - \mu)]$
Autocovariance function	$\{R_\tau;\ \tau = 0, \pm 1, \ldots, \pm\infty\}$
Autocorrelation	$\rho_\tau = R_\tau / R_0$
Autocorrelation function	$\{\rho_\tau;\ \tau = 0, \pm 1, \ldots, \pm\infty\}$

Properties		
Bounds	$-R_0 \le R_\tau \le R_0$	$-1 \le \rho_\tau \le 1$
Symmetry	$R_\tau = R_{-\tau}$	$\rho_\tau = \rho_{-\tau}$
Positive definite	$\sum_{s,t=1}^{n} a_s a_t R_{s-t} > 0$	
Positive semidefinite	$\sum_{s,t=1}^{n} a_s a_t R_{s-t} \ge 0$	

sequence (i.e., more than one linear relationship can result in the same autocorrelation structure). These relationships are of interest to us, but rather than describe them now we consider it more instructive to study several common forms of the autocorrelation function to obtain useful insights. In Section 6.6 we show how these forms are generated by linear functions of the events in a sequence.

Suppose that

$$\rho_\tau = \begin{cases} 1 & \tau = 0 \\ 0 & \tau \neq 0. \end{cases} \tag{4.1}$$

Then events are uncorrelated, and we say that a sequence with this autocorrelation function is an *uncorrelated sequence*.

Now consider the form

$$\rho_\tau = \alpha^{|\tau|} \qquad -1 < \alpha < 1 \tag{4.2}$$

as shown in Fig. 35*a*. Notice in the figure that the correlation between events monotonically decreases as the separation between them increases. For many situations one would intuitively expect the correlation to be greater between

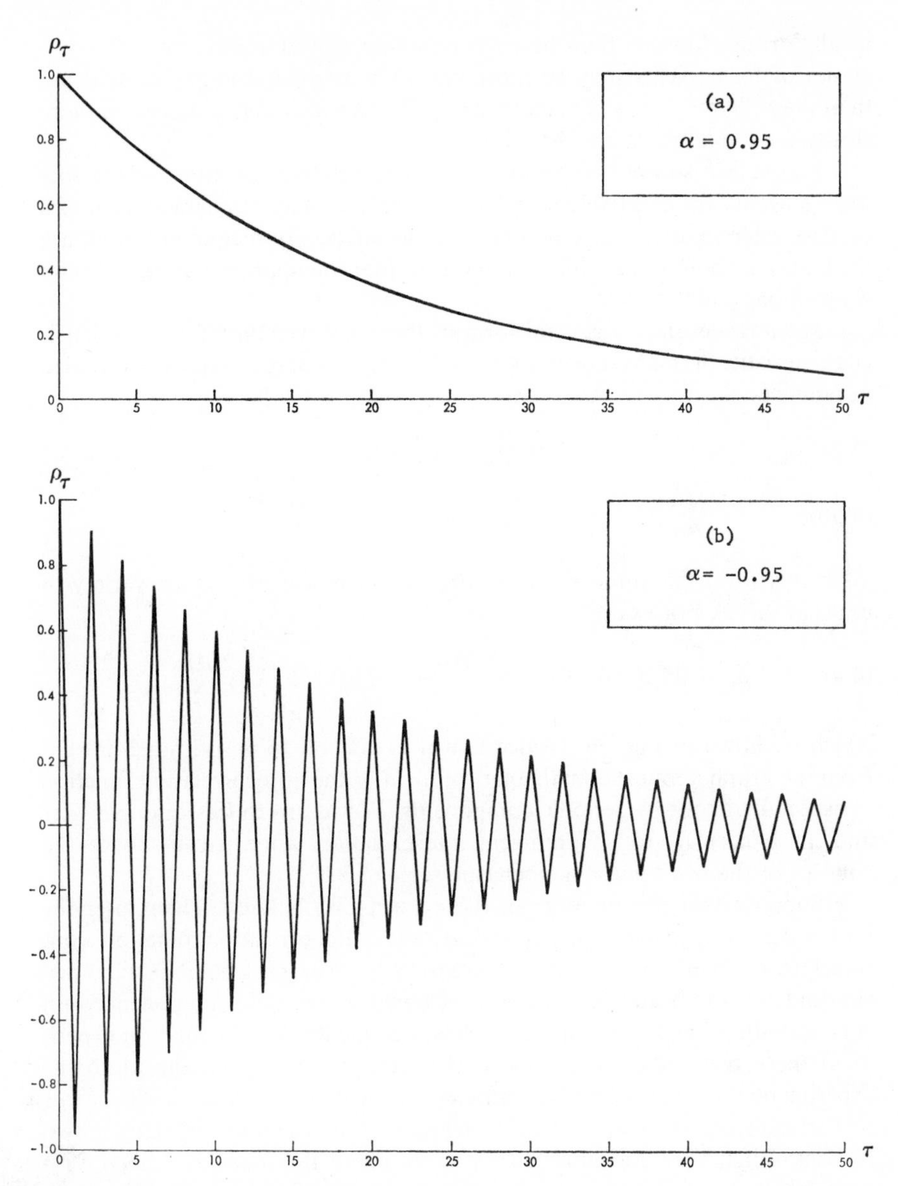

Fig. 35 Autocorrelation function, $\rho_\tau = \alpha^{|\tau|}$.

locally grouped events than between remotely spaced events, since the outcomes of local events may be more plausibly attributed to the same cause than may the outcomes of remote events. Notice also that ρ decreases more slowly for the larger α.

Figure 35*b* shows two oscillating autocorrelation functions where successive events are negatively correlated. This form may be characteristic of a process wherein attempts to dampen out deviations from equilibrium induce fluctuations about the equilibrium level, so that the approach to equilibrium is not monotonic.

We now consider a general form of the autocorrelation function that is characteristic of many covariance stationary stochastic sequences studied in practice:

$$\rho_\tau = \sum_{i=1}^{p} \beta_i \alpha_i^{|\tau|} \cos \lambda_i \tau \tag{4.3a}$$

$$\sum_{i=1}^{p} \beta_i = 1 \qquad -1 < \alpha_i \leq 1, \quad i = 1, \ldots, p. \tag{4.3b}$$

With this form the autocorrelation function may assume a wide variety of unusual forms. For example, consider

$$\rho_\tau = 0.25(-0.95)^{|\tau|} \cos \frac{3\Pi\tau}{2} + 0.75(0.75)^{|\tau|} \cos \frac{6\Pi\tau}{7}. \tag{4.4}$$

which is shown in Fig. 36. Notice that it is difficult to derive much insight from the graph about the mathematical form of the autocorrelation function ρ, especially about its periodic components. We return to this example later and, in particular, to its graphical interpretation when we introduce the concept of the *spectrum* of a stochastic sequence.

Suppose that one or more of the α_is in (4.3a) is unity. Then, after the remaining damped elements disappear, the autocorrelation function continues to oscillate undamped for increasing $|\tau|$. The presence of undamped sinusoids in ρ indicates the existence of regularly periodic components with corresponding frequencies in the stochastic sequence. In a simulation experiment these regularities may appear as a consequence of building into the experiment rules that contribute an element of regularly recurring behavior to the sequence of interest. Their presence is undesirable for at least two reasons. First, they add unnecessary variation to the sequence, thereby increasing the length of a time series needed to estimate the mean with a given accuracy. Second, they create statistical problems, especially for the estimation of the autocovariance function. One may show that, if undamped periodic components are present in ρ, it is not possible to estimate the autocovariance function consistently from a single time series [79, 85]. A *consistent* estimator is one whose accuracy improves as the number of observations in

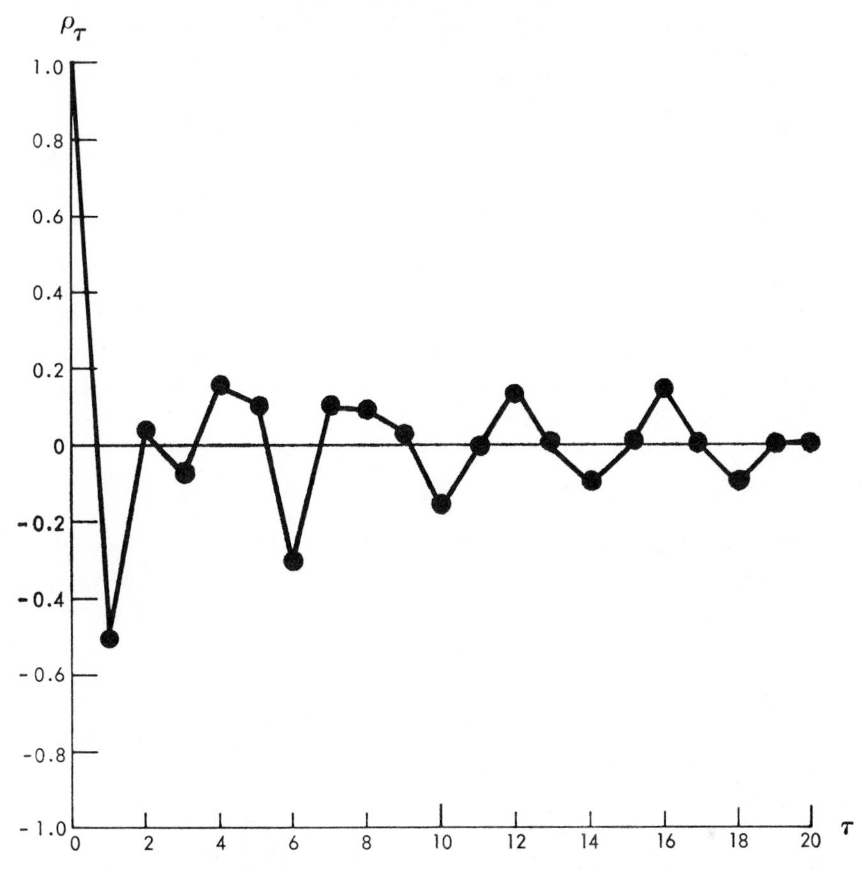

Fig. 36 $\rho_\tau = 0.25(-0.95)^{|\tau|} \cos(3\Pi\tau/2) + 0.75(0.75)^{|\tau|} \cos(6\Pi\tau/7)$.

the time series increases. Then we say that the sequence is not *ergodic* with regard to the autocovariance function.

The undesirable properties introduced by periodic components should encourage the investigator to remove all regularly periodic contributions to the sequence of interest before analyzing the corresponding sequence. This admonition may be more easily given than obeyed, but certainly, whenever possible, the removal of periodic components is beneficial.

Hereafter we limit our attention to sequences with no regularly periodic components but with autocorrelation functions that can be reasonably well approximated by (4.3) for $|\alpha_i| < 1$. This implies that

(4.5a) $$\lim_{\tau \to \infty} \rho_\tau = 0.$$

In order to estimate the mean and autocovariance function consistently from a single time series we require for the mean

$$\lim_{T\to\infty} T^{-1} \sum_{s=-T+1}^{T-1} \rho_s = 0 \tag{4.5b}$$

and for the autocovariance function

$$\lim_{T\to\infty} T^{-1} \sum_{s=-T+1}^{T-1} \rho_s^{\,2} = 0. \tag{4.5c}$$

Expression (4.5a) ensures these convergences, provided that ρ_τ approaches zero rapidly enough. Then the sequence X is *ergodic* in the mean and autocovariance function. If a sequence is a normal process with the autocovariance function satisfying (4.5b) and (4.5c), it is an ergodic sequence.

As statistical descriptors the autocovariance and autocorrelation functions are direct extensions of classical statistical theory. They replace the covariance and correlation coefficient when the ordering index becomes a relevant consideration for analysis. Because of the definition of covariance stationarity there exists an alternative, but completely equivalent, way of studying covariation between events in a sequence. This alternative approach deserves careful consideration, for it often leads to a simpler interpretation of behavior and to more desirable sampling properties for estimators.

If $\{X_s\}$ is a covariance stationary sequence with autocovariance function R and autocorrelation function ρ and (4.5a) holds, then we may represent R_τ and ρ_τ by

$$R_\tau = \int_{-\Pi}^{\Pi} g(\lambda) e^{i\lambda\tau}\, d\lambda \tag{4.6}$$

$$\rho_\tau = \int_{-\Pi}^{\Pi} f(\lambda) e^{i\lambda\tau}\, d\lambda, \tag{4.7}$$

respectively, where

$$g(\lambda) = g(-\lambda) \tag{4.8}$$

$$g(\lambda) = R_0 f(\lambda). \tag{4.9}$$

The functions g and f are called the *spectrum* and the *spectral density function*, respectively. The principal appeal of the spectrum occurs for $\tau = 0$ when

$$R_0 = \int_{-\pi}^{\pi} g(\lambda)\, d\lambda = 2 \int_{0}^{\pi} g(\lambda)\, d\lambda. \tag{4.10}$$

Here we may think of the variance of the sequence as being made up of additive uncorrelated contributions of variance, each attributable to a small, nonoverlapping band of frequencies about λ, as λ ranges from zero to π. Then the spectral density function tells us the *relative* contribution to the variance attributable to each small band of frequencies. Expressions (4.6) and (4.7) are the essence of the *Wiener-Khintchine* theorem, which forms the basis for modern harmonic analysis.

It is sometimes more convenient to talk about the average period of a fluctuation than about its frequency. This is especially true when we are studying observations collected through time. The period corresponding to frequency λ is $2\pi/\lambda$ and is measured in the same time units as the original index for the sequence.

As already mentioned, the *time-domain* functions R and ρ and the *frequency-domain* functions g and f contain the same information in a mathematical sense about the covariance properties of $\{X_s\}$. As Figs. 35 and 36 have shown, a graph of the autocorrelation function ρ enables us to determine the correlation between events at different separations. Suppose that an event occurs in a simulation at time t and influences the value of X_t. Often we wish to know how the event influences behavior τ periods later. Looking at the correlation between X_t and $X_{t+\tau}$ provides a measure of this influence. If $\rho_\tau \simeq 0$ then the influence has been practically dissipated at a time separation of τ periods. This information is contained in the graph, and we need not know the precise mathematical form of ρ that generated the graph.

But what if we are faced with a graph like that in Fig. 36? The damped oscillations between positive and negative values indicate the presence of damped periodicity around the mean and arouse our curiosity. This concern is especially important during exploratory studies, when we are most ignorant of what is going on in the process. Figure 36 does not allow us to identify the periodicities readily. However, a graph of the spectrum or spectral density function (s.d.f.) does provide *identification* of prominent periodicities. Figure 37 shows the s.d.f. corresponding to the autocorrelation function in Fig. 36. Notice that there are two distinct variance concentrations near frequencies $\pi/2$ and $43\pi/50$. If the ordering index of the sequence were time then prominent fluctuations would have periods of about $2\pi/(\pi/2) = 4$ and $2\pi/(43\pi/50) = 2.3$ time units.

In designing simulation experiments many simulated activities are brought together, in either a series or parallel arrangement. The outputs of some of these activities serve as inputs to others. To improve our understanding of these interactions, we may wish to examine the correlation properties of some of these interim outputs. We may also want to compare the correlation properties of these outputs when changes are made in the logical rules that control the simulation.

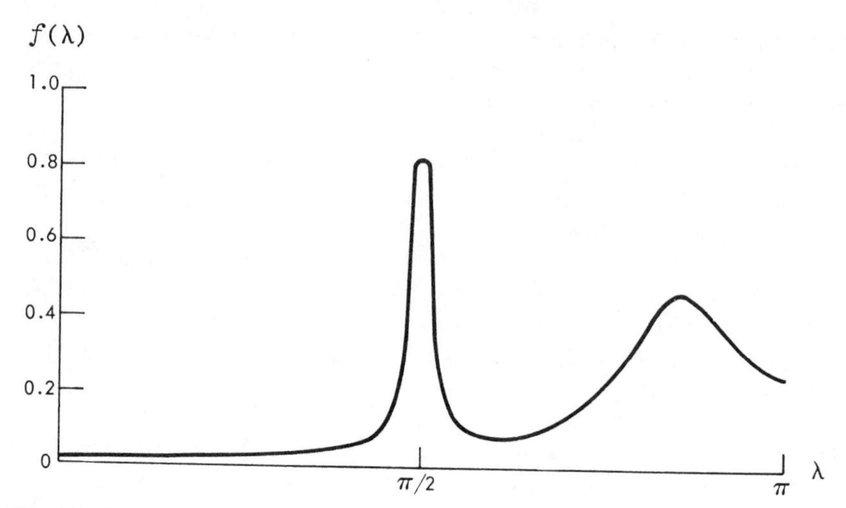

Fig. 37 Spectral density function corresponding to Fig. 36.

As an example, consider a job shop with a single server and Poisson job arrivals. Suppose that service for each job requires exactly 20 hours and each job receives service in the order of its arrival. Figure 38*a* shows a *correlogram* (sample autocorrelation function) of queue length samples at hourly intervals. The extent of correlation between events is apparent, but the scalloped appearance, though it arouses interest, is not unusually revealing. Figure 38*b* shows the corresponding sample spectrum, which reveals distinctive peaks at multiples of $2\pi/20$. These imply the presence of a cyclic, though not regularly periodic, component in the queue length process with fundamental frequency $2\pi/20$ corresponding to a period of 20 hours. A little reflection explains this phenomenon. Whenever the queue is not empty a job leaves the queue for service every 20 hours. This generates a declining step behavior in the queue on which the random arrivals are superimposed.

The efflux from this service facility will also be periodic whenever jobs are queueing. This output may be intended to serve as the input to another activity whose resulting behavior may not be clearly anticipated. Then it would also be instructive to examine the output of the second activity.

The information gained from the sample spectrum in Fig. 38*b* could have been obtained with a bit of clear reasoning before performing the experiment. However, this is not always the case, and therefore we must study the data to obtain useful insights. As we see, the sample spectrum of the data presents a helpful picture.

* This example is discussed more completely in reference 84.

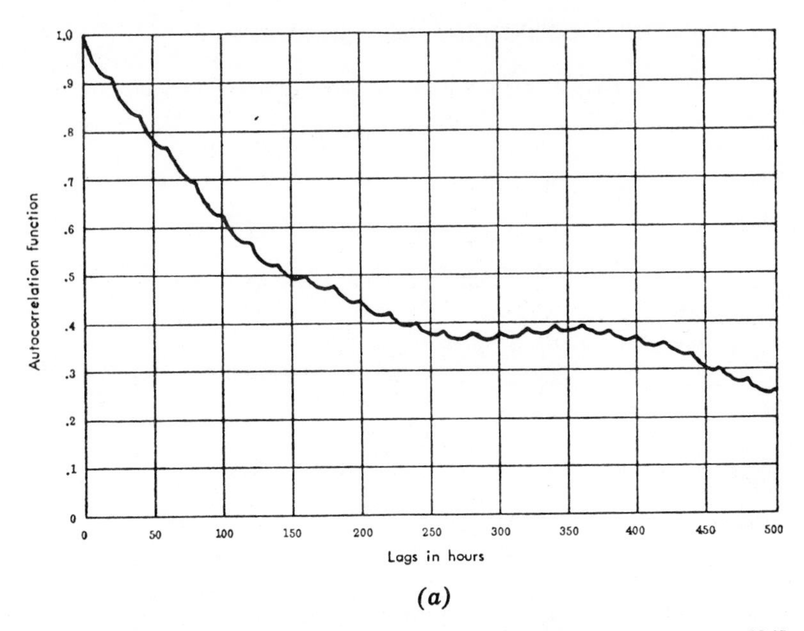

(a)

Fig. 38a Estimated queue length correlogram, constant service time. *Source:* [84].

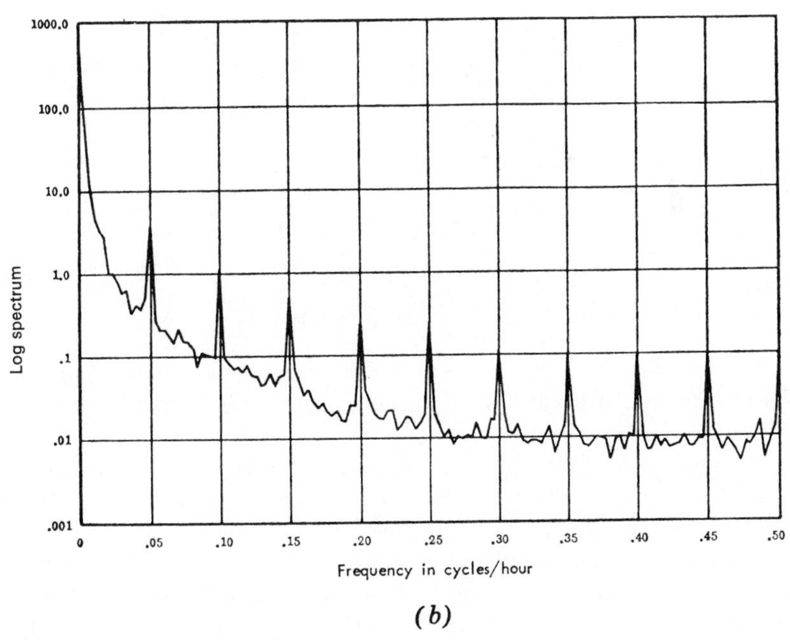

(b)

Fig. 38b Estimated queue length spectrum, constant service time. *Source:* [84].

To derive analytical expressions for $g(\lambda)$ and $f(\lambda)$ we use the *inverse Fourier transforms*:

$$g(\lambda) = (2\pi)^{-1} \sum_{s=-\infty}^{\infty} R_s e^{-i\lambda s} \tag{4.11}$$

$$f(\lambda) = (2\pi)^{-1} \sum_{s=-\infty}^{\infty} \rho_s e^{-i\lambda s} \qquad -\pi \leq \lambda \leq \pi. \tag{4.12}$$

Let us consider the uncorrelated sequence with

$$\rho_\tau = \begin{cases} 1 & \tau = 0 \\ 0 & \tau \neq 0. \end{cases} \tag{4.13}$$

Clearly

$$f(\lambda) = (2\pi)^{-1} \tag{4.14}$$

so that the sequence has a uniform spectrum at all frequencies $-\pi \leq \lambda \leq \pi$. This means that all frequencies contribute equally to the variance of the sequence.

For Fig. 37, the corresponding s.d.f. is obtained by applying (4.12) to (4.4):

$$\begin{aligned} f(\lambda) = \frac{1}{4\pi}\Bigg\{0.0244 & \left[\frac{1}{1.9025 + 1.9 \cos[(3\pi/2) - \lambda]}\right. \\ & + \left.\frac{1}{1.9025 + 1.9 \cos[(3\pi/2) + \lambda]}\right] \\ & + 0.3281\left[\frac{1}{1.5625 - 1.5 \cos[(6\pi/7) - \lambda]}\right. \\ & + \left.\frac{1}{1.5625 - 1.5 \cos[(6\pi/7) + \lambda]}\right]\Bigg\}. \end{aligned} \tag{4.15}$$

Observe, however, that we do not rely on the mathematical expression (4.15) to identify the prominent frequencies. These are discernible from the graph of the s.d.f. alone.

To summarize briefly, the mean provides information about the level of the process, a graph of the autocorrelation function shows the extent of correlation between events in the sequence, and a graph of the spectral density function reveals prominent periodicities, if they exist. We now begin to see that we can learn a considerable amount about a sequence from its covariance structure in both the time and frequency domains. Table 18 lists properties of these domains that prove useful in the rest of the book.

Table 18 Properties of the Time and Frequency Domains

Time Domain	Frequency Domain
$R_s = \int_{-\pi}^{\pi} g(\lambda)e^{i\lambda s}d$	$g(\lambda) = (2\pi)^{-1} \sum_{s=-\infty}^{\infty} R_\tau e^{-i\lambda s}$
$s = 0, \pm 1, \ldots, \pm\infty$	$-\pi \leq \lambda \leq \pi$
R is positive semidefinite	$g(\lambda) \geq 0$
$R_s = R_{-s}$	$g(\lambda) = g(-\lambda)$
$R_s \leq R_0 < \infty$	$g(\lambda) < \infty$
$\rho_s = R_s/R_0$	$f(\lambda) = g(\lambda)/R_0$
ρ measures extent of correlation	f identifies prominent periodicities

6.5 A QUEUEING PROBLEM

In later sections we describe methods of statistical inference designed to estimate parameters of interest in a simulation with a specified degree of accuracy. To evaluate the usefulness of these methods it is instructive to compare the known analytical solution to a problem of interest and the simulated solution to the same problem. Since queueing problems often crop up in simulation work, we use a well-known elementary queueing problem for comparative purposes. We present the theoretical solution here for reference.

Consider a single server queueing problem in which the times between arrivals are independent and exponentially distributed, with mean inter-arrival time $1/\omega_1$. Jobs receive service in the order of their arrivals, the service times being independent and exponentially distributed with mean service rate $1/\omega_2$. The activity level is defined as

$$\beta \equiv \frac{\omega_1}{\omega_2} \tag{5.1}$$

and from theory we know the mean queue length to be*

$$\mu = \frac{\beta}{1 - \beta} \tag{5.2}$$

* See, for example, reference 81.

For $\beta < 1$ and the variance of queue length to be

$$R_0 = \frac{\beta}{(1-\beta)^2}. \tag{5.3}$$

The autocorrelation function is*

$$\rho(s) = \frac{\lambda(\omega_2 - \omega_1)^3}{\Pi} \int_0^{2\Pi} \frac{\sin^2 \theta e^{-ws}}{w^3} d\theta \tag{5.4}$$

$$w \equiv \omega_1 + \omega_2 - 2(\omega_1\omega_2)^{1/2} \cos \theta$$

and the spectrum is

$$g(\lambda) = \frac{R_0(\omega_2 - \omega_1)^3}{\pi^2} \int_0^{2\pi} \frac{\sin^2 \theta \, d\theta}{w^2(w^2 + \lambda^2)}. \tag{5.5}$$

As the reader may have noted, Sections 5.3 and 5.5 have already presented simulation programs of this problem in GPSS/360 and SIMSCRIPT II, respectively.

6.6 AUTOREGRESSIVE PROCESSES

One convenient way of describing certain stochastic sequences is in terms of their *autoregressive-moving average* representation. Since a detailed treatment of this subject could easily fill an entire book, we concentrate here exclusively on the autoregressive representation,† which can play several roles in simulation work. As described in Chapter 8, the autoregressive form provides one way of generating an autocorrelated time series. Below and again in Chapter 10 we describe how this form can assist us in the analysis of simulation output. At the outset we note that not all stochastic sequences have autoregressive representations. However, the convenience that they offer encourages us to use such forms even when theory does not completely justify them.

Suppose that $\{X_t\}$ allows the representation

$$\sum_{s=0}^{p} b_s(X_{t-s} - \mu) = \varepsilon_t \qquad b_0 \equiv 1 \tag{6.1}$$

$$E(X_t) = \mu, \qquad R_{s-t} = E(X_sX_t) - \mu^2$$

$$E(\varepsilon_t) = 0, \qquad E(\varepsilon_s\varepsilon_t) = \begin{cases} \sigma^2 & s = t \\ 0 & s \neq t \end{cases} \tag{6.2}$$

$$E(\varepsilon_s X_t) = 0 \qquad s > t.$$

* See reference 86.

† See references 79 and 85.

It is easily seen that

(6.3)

$$E\left[(X_{t-r} - \mu) \sum_{s=0}^{p} b_s(X_{t-s} - \mu)\right] = E[(X_{t-r} - \mu)\varepsilon_t]$$

$$= \sum_{s=0}^{p} b_s R_{s-r} = \begin{cases} \sigma^2 & r = 0 \\ 0 & r > 0. \end{cases}$$

These are known as the *Yule-Walker* equations. If the roots of the characteristic equation

(6.4)
$$\sum_{s=0}^{p} b_s z^s = 0$$

lie outside the unit circle then $\{X_t\}$ is a covariance stationary sequence with autocorrelation function and spectrum expressible as*

(6.5)

$$R_s = \sigma^2 \sum_{j=1}^{p} \frac{\alpha_j^{|s|+p-1}}{(1 - \alpha_j)^2 \prod_{i \neq j}[(1 - \alpha_i\alpha_j)(\alpha_j - \alpha_i)]} \qquad s = 0, \pm 1, \ldots$$

(6.6)

$$g(\lambda) = \frac{\sigma^2}{2\pi} \sum_{j=1}^{p} \frac{\alpha_j^{p-1}}{(1 - 2\alpha_j \cos\lambda + \alpha_j^2) \prod_{i \neq j}[(1 - \alpha_i\alpha_j)(\alpha_j - \alpha_i)]} \qquad |\lambda| \le \pi,$$

respectively, where $\alpha_1, \ldots, \alpha_p$ are the reciprocals of the roots of (6.4). We have already encountered a form of (6.5) in (4.3a). This linear combination of complex exponential quantities can produce a wide variety of autocorrelated behavior. For $p = 1$ we have the *first-order* scheme

(6.7)
$$X_t + b_1 X_{t-1} = \varepsilon_t,$$

which gives for $b_1 = -\alpha$

(6.8)

$$R_s = \frac{\sigma^2\alpha^{|s|}}{1 - \alpha^2}, \qquad \rho_s = \alpha^{|s|}$$

$$f(\lambda) = \frac{1}{2\pi(1 - 2\alpha\cos\lambda + \alpha^2)} \qquad |\lambda| \le \pi.$$

This representation for ρ_s was displayed in Fig. 35.

* We assume that the roots are distinct.

The attractiveness of the autoregressive approach for describing second-order properties comes from (6.1) and (6.3). First, from (6.3) we note that

(6.9a) $$R_r = -\sum_{s=1}^{p} b_s R_{s-r}$$

(6.9b) $$\rho_r = -\sum_{s=1}^{p} b_s \rho_{s-r} \qquad r > 0,$$

so that the autocovariances and autocorrelations can be generated *recursively*. In general the sample autocorrelation function or correlogram is computed from

(6.10) $$\hat{\rho}_s = \frac{\hat{R}_s}{\hat{R}_0},$$

where

(6.11) $$\hat{R}_s = \frac{1}{n}\sum_{t=1}^{n-s}(X_t - \bar{X})(X_{t+s} - \bar{X}) \qquad s \geq 0$$
$$\bar{X} = \frac{1}{n}\sum_{t=1}^{n} X_t.$$

Here each lag s requires the computation of an additional autocovariance R_s. This can easily become a time-consuming computation. By contrast, if one can compute estimates of $\{b_s\}$, say $\{\hat{b}_s\}$, and has $\hat{\rho}_s$ for $s = 0, \ldots, p$, he can use (6.9b) with sample parameters replacing corresponding ones to generate a correlogram.

To illustrate the use of (6.9b) we collected $n = 4000$ observations at unit intervals on the simulated queue length process of the single server queueing problem described in Section 6.5 and programmed in Section 5.5. We set $\omega_1 = 4.5$ arrivals and $\omega_2 = 5$ jobs serviced per hour. By use of the SIMSCRIPT II ANALYSIS routine in Appendix B, the $\{b_s\}$ and σ^2 were estimated by a technique to be discussed in Chapter 10. Substituting into (6.9b) and *priming* the expression with the first p sample autocorrelations, we produced the sample autocorrelation function (a.c.f.) shown in Fig. 39. The theoretical a.c.f. obtained by using (5.4) is also shown, along with the sample a.c.f. based on (6.10) exclusively.

Notice that the correlogram based on (6.9b), with substitutions, closely resembles the correlogram based on (6.10) for small τ. However, Fig. 39 shows that (6.9b) leads to the correct asymptotic behavior whereas (6.10) does not. The poor sampling properties of (6.10) account for this poor showing for large τ. Section (10.14) discusses these sampling properties in more detail. Given that (6.9b) does as well or better than (6.10) in estimating the

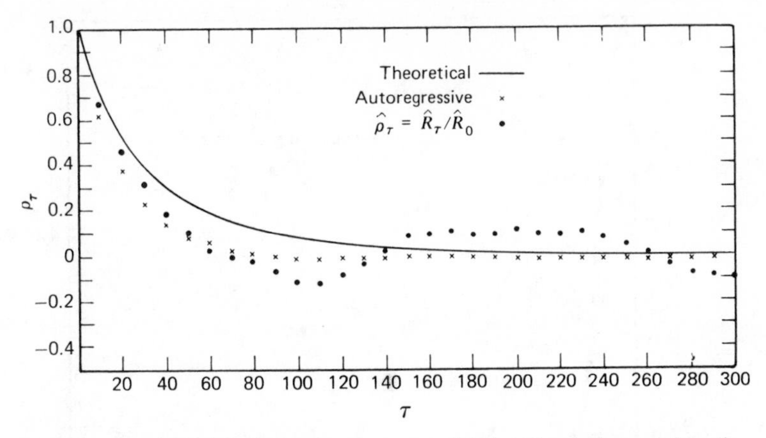

Fig. 39 Theoretical and sample autocorrelation functions of queue length.

theoretical autocorrelation function and that a self-contained computer program, requiring the computation of far fewer quantities, can be used to produce a correlogram, the appeal of the autoregressive approach is clear.

As mentioned before, the second beneficial feature of autoregressive schemes for describing second-order properties has to do with (6.1). We note that squaring both sides of (6.1) and taking expectations produce

$$\sigma^2 = \sum_{r,s=0}^{p} b_r b_s R_{r-s}. \tag{6.12}$$

Replacing R_{r-s} by its Fourier transform (4.6) gives

$$\sum_{r,s=0}^{p} b_r b_s \int_{-\pi}^{\pi} g(\lambda) e^{i\lambda(r-s)}\, d\lambda = \int_{-\pi}^{\pi} g(\lambda) |b_s e^{-i\lambda s}|^2\, d\lambda = \sigma^2 \tag{6.13}$$

which is satisfied by

$$g(\lambda) = \frac{\sigma^2}{2\pi |\sum_{s=0}^{p} b_s e^{-i\lambda s}|^2} \qquad |\lambda| \le \pi. \tag{6.14}$$

A computationally simpler form is

$$g(\lambda) = \frac{\sigma^2}{2\pi \sum_{r,s=0}^{p} b_r b_s \cos \lambda(r-s)}. \tag{6.15}$$

To estimate the spectrum we can substitute estimates of the b_s and σ^2 into (6.15). Figure 40 shows the corresponding sample spectrum obtained by using (6.15), the theoretical spectrum resulting from (5.5), and an alternative spectrum estimated from *spectrum analysis*, a statistical technique that we

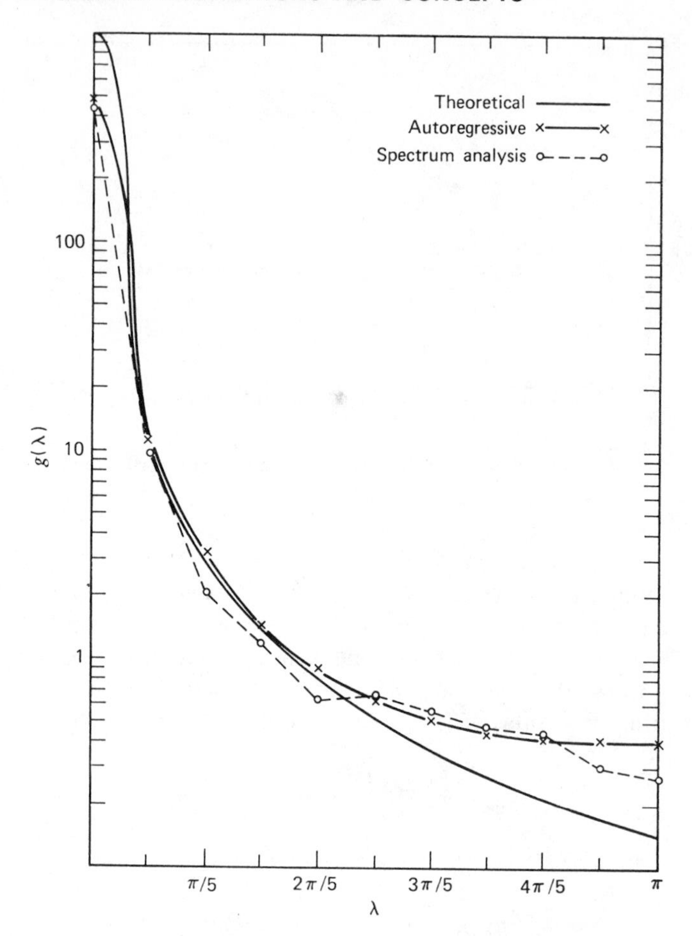

Fig. 40 Theoretical and sample spectra of queue length.

discuss briefly in Chapter 10. Notice that the autoregressive approach does better at estimating the true spectrum at low frequencies than the spectrum analysis approach does. For high frequencies, both do equally well. Since the spectrum analysis approach is considerably more complex and time-consuming in practice, the appeal of the autoregressive approach is again apparent. The poor showing of both estimation techniques for high frequencies should not be a concern in general since the logarithmic scale indicates that the variance attributable to these frequencies is relatively small.

In closing this section we wish to emphasize that most time series analysts would prefer to estimate the correlogram and spectrum by means of techniques that do not require the assumption that an autoregressive representation exists for a particular stochastic sequence. For instance, in our queueing example the queue length process does not have such a representation and hence the autoregressive analysis must be regarded as approximate. However, the reader with a limited statistics background needs to learn considerably less in order to use the approach described here than is required to apply the sophisticated tools of a comprehensive time series analysis.

In Section 5.5 we showed a sample output using the SIMSCRIPT II autoregressive analysis program given in Appendix B. We leave the computation of the correlogram and sample spectrum using the results in Section 5.5 as an exercise for the reader.

EXERCISES

1. Consider the autocovariance function

$$R_s = \frac{\sigma^2}{(1-\alpha_1\alpha_2)(\alpha_1-\alpha_2)}\left(\frac{\alpha_1^{|s|+1}}{1-\alpha_1{}^2} - \frac{\alpha_2^{|s|+1}}{1-\alpha_2{}^2}\right).$$

   a. Find the corresponding spectrum.
   b. Suppose that α_1 and α_2 are complex quantities; $\alpha_1 = qe^{i\theta}$, $\alpha_2 = qe^{-i\theta}$, $|q| < 1$. Find the frequency $0 < \lambda < \pi$ where a concentration of variance occurs in the spectrum.

2. Let $\{X_t\}$ have the spectral density function in (6.8).
   a. Plot the s.d.f. roughly for $\alpha = 0.5$ and 0.9.
   b. Plot the s.d.f. roughly for $\alpha = -0.5$ and -0.9.
   c. What shortcoming do you observe from the second graph in assuming that a particular process, known to have a concentration of variance at one frequency, has the autocorrelation function (6.8)?

3. Let $\{X_t\}$ have the autocorrelation function (4.2).
   a. Show that $\{X_t\}$ is ergodic in the mean.
   b. Show that $\{X_t\}$ is ergodic in the autocovariance function.
   c. Does this ergodicity continue to hold when $\alpha = 1$?

4. Use the ANALYSIS outputs of Exercises 7 and 10 in Chapter 5.
   a. Plot queue length correlograms using (6.9b).
   b. Plot queue length sample spectra using (6.15).

5. Perform the steps in Exercise 4 for the output of Exercises 8 and 11 in Chapter 5.

6. Perform the steps in Exercise 4 for the output of Exercise 9 in Chapter 5.

7. Clearly, adding a barber in Exercise 10 of Chapter 5 should increase revenues. However, one additional reason for the new barber is to make it possible to give each barber a 15 minute coffee break every 3 hours. These breaks do not overlap, so that there are always four barbers working.
   a. Modify the program in Exercise 10 of Chapter 5 to reflect the coffee breaks, and run it again with the same input.
   b. Plot a correlogram for the queue length output, and compare it to the queue length correlogram in Exercise 10 of Chapter 5.
   c. Plot a sample spectrum for the queue length output and compare it to the queue length sample spectrum in Exercise 10 of Chapter 5.

8. Perform the steps in Exercise 7 on the program in Exercise 11 of Chapter 5.

9. Perform the steps in Exercise 7 on the program in Exercise 12 of Chapter 5.

CHAPTER 7

RANDOM NUMBER GENERATION

7.1 THE IMPORTANCE OF UNIFORM DEVIATES

A simulation based on random behavior naturally requires a mechanism for generating sequences of events where each sequence obeys a probability law governing a particular component of the random behavior in question. The probability law may take many forms. One commonly encountered in simulation work assumes that events in the sequence are independent and identically distributed; for example, with the exponential, gamma, Weibull, or normal distribution if the events are continuous random variables, and with the Poisson binomial or negative binomial distribution if the events are discrete random variables.

The event generating mechanism must therefore have the capability fo producing random variates from a variety of continuous and discrete probability distributions. It is a remarkable fact of probability theory that variates can be generated for a wide variety of theoretical and empirical distributions, *provided* only that a sequence of independent random variates, each with uniform distribution on the interval (0, 1), can be generated. A random variate with these distributional properties is often referred to as a *uniform deviate*. This assertion is easily proved for the case of a continuous random variable X with density function f_X and cumulative probability distribution

$$F_X(x) = \int_{-\infty}^{x} f_X(\theta)\, d\theta. \tag{1.1}$$

Consider the new random variable

$$U = F_X(X) = \int_{-\infty}^{X} f_X(\theta)\, d\theta. \tag{1.2}$$

From elementary probability theory we know that the p.d.f. of U is

$$f_U(u) = |J| f_X(x) \tag{1.3}$$

$$u = F_X(x) = \int_{-\infty}^{x} f_X(\theta)\, d\theta \qquad 0 \leq u \leq 1 \tag{1.4}$$

where the quantity J, called the *Jacobian* of the transformation, is given as $J \equiv \partial x/\partial u$. In the present case,

$$J = \frac{1}{f_X(x)} \tag{1.5}$$

$$f_U(u) = \begin{cases} 1 & 0 \leq u \leq 1 \\ 0 & \text{elsewhere,} \end{cases} \tag{1.6}$$

which is the p.d.f. of a uniformly distributed random variable on (0, 1).

As an example, suppose that X has the exponential distribution with p.d.f.

$$f_X(x) = \begin{cases} (1/\beta)e^{-x/\beta} & 0 \leq x \leq \infty \\ 0 & x < 0. \end{cases} \tag{1.7}$$

Then

$$F_X(x) = 1 - e^{-x/\beta} \tag{1.8}$$

and, using (1.2), we have

$$U = 1 - e^{-X/\beta} \tag{1.9}$$

$$X = -\beta \log(1 - U). \tag{1.10}$$

That is, one generates a uniformly distributed random variable U and uses (1.10) to transform it into an exponentially distributed random variable X with parameter β.

This approach is known as the *inverse transformation* method, since it requires knowledge of the inverse transformation

$$X = \Phi[F_X(X)] = \Phi(U) \tag{1.11}$$

to generate X. It is particularly convenient for generating exponential, gamma, Weibull, and beta variates, as we show in Chapter 8.

To show how uniform deviates facilitate the generation of discrete random variates we consider the events $X = 0$ with probability p and $X = 1$ with probability $1 - p$. Here X has the *Bernoulli* distribution. Let us generate a uniformly distributed deviate U and assign a value to X according to the rule

$$X = \begin{cases} 0 & 0 < U \leq p \\ 1 & p < U < 1. \end{cases} \tag{1.12}$$

Since $\text{Prob}(0 < U \leq p) = p$ and $\text{Prob}(p < U < 1) = 1 - p$ it is clear that we can generate an event X with the prescribed probability law. It turns out that we can also use combinations of events generated in this way to create geometric, binomial, and negative binomial variates. Poisson variates are usually generated by exploiting certain relationships between exponential and Poisson variates.

This brief introduction makes clear the central role that uniform deviates play in simulation experiments. The first step in preparing the input for a simulation, therefore, is to find an algorithm that produces uniform deviates. In reality we must take one step back and find a source of *random numbers* that can be converted into the desired uniform deviates. Suppose that in a sampling experiment we draw a number x which assumes any value in $(0, N)$ with equal probability. Then x is a random number, and since x/N assumes any value in $(0, 1)$ with equal probability it has the uniform distribution on $(0, 1)$. However, the independence of the variates in such a sequence is an additional issue. If drawings are independent then x/N is a uniform deviate.

7.2 CONSIDERATIONS IN RANDOM NUMBER GENERATION*

Three considerations play influential roles in determining whether or not a particular source provides random numbers that are adequate for experimental purposes. The numbers must first pass a battery of statistical tests designed to reveal departures from *independence* and *uniformity*. For a truly random sequence $Z_1, Z_2, Z_3, \ldots (0 \leq Z_i \leq N)$, the elements of any subsequence of these numbers are jointly independent and each has a uniform distribution over $(0, N)$. Failure of a sequence to have this property can lead to severely misleading results in simulation work. Section 7.7 describes procedures for testing for a variety of departures.

The second significant property requires that the random number contain enough digits so that generation of numbers on the interval $(0, 1)$ is sufficiently *dense*. Suppose that $\{Z_i\}$ consists of integers, N being the largest integral value that any Z_i can assume. Then the probability of assuming a particular value is $1/(N + 1)$. If $N = 9$ the uniform variate can assume

* See reference 105 and 108 for more complete descriptions of random number generation.

values $0(\cdot 1)1$ but cannot produce values $m + 0.001(0.001)0.009$ for $m = 0(0.01)0.99$. Such wide gaps between assumable values on the unit interval are undesirable, for a nonlinear transformation of a supposed continuous uniform variate, which is actually a discrete uniform variate, intended to produce a variate with a desired distribution may not in fact produce such a variate. Exercise 1 illustrates this point.

More generally, n decimal digits allow us to produce up to 10^n random numbers. For $n = 2$, we have 100 possible random numbers; for $n = 4$, 10,000; and for $n = 6$, 1,000,000. This concern over the number of digits arises because of the finite length of words in computer memories. Since most medium-size to large computers can store at least 31 binary digits, it is possible, at least in theory, to produce a fairly dense sampling on (0, 1) of the 2^{31} available numbers. If double precision random numbers are used, at least 2^{62} numbers can be sampled in theory, but at an increased consumption of computer time and memory space.

The third significant property concerns the *efficiency* with which a particular source produces random numbers. The faster an algorithm produces a number, the more desirable that algorithm is. A minimal utilization of storage is also attractive, especially with the new generation of computers whose user charges often depend on space utilization as well as computing time.

Since the three desirable properties seldom, if ever, characterize any one method of producing random numbers, some compromise must be made. It is generally agreed that the presence of sufficient independence and uniformity to preserve the integrity of the particular experiment under consideration should be the prevailing criterion in determining adequacy.

7.3 A TABLE OF RANDOM NUMBERS

One method of generating random numbers on a digital computer is to prepare and store a table of random numbers, either in the internal magnetic core memory of the machine or on one of its fast-access peripheral devices. Assuming that the numbers in the table have passed a variety of tests designed to reveal departures from independence and uniformity, one would expect this method to relieve the user of any concern about the statistical appropriateness of the numbers. The Rand Corporation has prepared, tested, and listed in [115] a set of a million random digits that can be used to form a table of random numbers.

MacLaren and Marsaglia [111] describe a procedure for using a table based on the Rand source for the IBM/7094 computer. Each number is a random integer in the set $\{0, 1, \ldots, 2^{27} - 1\}$, so that theoretically the probability of drawing a particular integer in the set is $1/2^{27}$. They describe a procedure for storing the numbers on tape and then reading them into a core block of 1024 words while a similarly filled block is being used to provide random numbers. The writer is unaware of any published procedure for using a table of random numbers on the more recent generation of computers such as the IBM/360 and 370 systems. However, the larger storage capacities and the faster access from disk should increase the desirability of using this method of obtaining random numbers, provided the numbers have the desired statistical properties.

One shortcoming of using a table of random numbers is the possibility of exhausting the table. To handle this contingency, MacLaren and Marsaglia describe two algorithms for transforming the tabled random numbers into *new* random numbers. They conjecture that the association between the old and new numbers is not a substantive issue. If the new random numbers are used after the old numbers have been exhausted and in the same corresponding order, one would indeed expect little, if any, joint influence on the experiment.

With regard to statistical acceptability, MacLaren and Marsaglia report that the tabled numbers and the two algorithms using them generally give better results than six other proposed generators. The purpose of our presentation of this tabled approach is to inform the reader of its existence. The potential user is encouraged to compare statistical test results from this method with those from any alternative generators that he may consider.

7.4 PSEUDORANDOM NUMBER GENERATION

Since the last two sections have emphasized the desirability of independence and uniformity, it may surprise the reader to learn that the most commonly employed random number generation algorithm produces a nonrandom sequence of numbers, each number being completely determined by its predecessor and, consequently, all numbers being determined by the initial number. Even more surprising is the fact that, if the parameters of the generator are chosen in an informed way, the numbers appear sufficiently independent and uniform for many practical purposes. To emphasize their inherently nonrandom character, we call these numbers *pseudorandom numbers*.

The generator has the form

$$Z_i \equiv aZ_{i-1} + c \pmod{m} \qquad i = 1, \ldots, n, \tag{4.1}$$

where Z_0 is the initially specified number or *seed*, and the *congruence* symbol $\equiv$ and modulo notation (mod m) mean that

$$(4.2) \qquad Z_i = aZ_{i-1} + c - \left[\frac{aZ_{i-1} + c}{m}\right] \times m,$$

the brackets indicating the largest integer in the quantity inside them. As an example, let $a = 3$, $c = 0$, $Z_0 = 4$, and $m = 5$. Then we have the number generation in Table 19.

Table 19 Number Generation Example

$Z_i \equiv 3Z_{i-1} \pmod{5} \qquad Z_0 = 4$

i	Z_i
0	4
1	12 − [12/5] × 5 = 2
2	6 − [6/5] × 5 = 1
3	3 − [3/5] × 5 = 3
4	9 − [9/5] × 5 = 4

It should be clear from (4.1) that no Z_i, $i = 1, \ldots, n$, can exceed m. Then numbers on the unit interval (0, 1) are formed by

$$(4.3) \qquad U_i = \frac{Z_i}{m}.$$

One may solve the first-order difference equation

$$(4.4) \qquad Z_i - aZ_{i-1} \equiv c \pmod{m}$$

to obtain the expression

$$(4.5) \qquad Z_n \equiv a^n Z_0 + \frac{a^n - 1}{a - 1} c \pmod{m}$$

for the nth number. The reader may easily verify this result by using (4.5) to compute the result in Table 19.

Generators of the form (4.1) and (4.5) are called *linear congruential generators*. When $c = 0$ we call them *multiplicative congruential*; otherwise, *mixed congruential*. They have the desirable feature of requiring little core storage, and when a, c, and m are chosen in accordance with the characteristics of the computer being used congruential generators can produce numbers very efficiently. Moreover a sequence generated in this manner can easily be reproduced, provided that the seed is saved. Another benefit of the congruential generator is that generation can be interrupted and restarted without loss of order in the sequence simply by saving the last number. This is particularly helpful when one wishes to perform several replications of a simulation. Knowledge of the last number generated makes it possible to avoid using the same stream of numbers in successive replications.

Lehmer [109] apparently first suggested the congruential generator with $c = 0$. This multiplicative congruential generator performed favorably on a variety of statistical tests for selected a and m [99]. Later the mixed congruential generator was introduced to improve computational efficiency and presumably to increase the range of numbers that could be generated; also the number theoretic considerations of these generators were easier to understand. Unfortunately the mixed generators have not fared exceptionally well statistically. Recently there has been a renewed emphasis on multiplicative generators with parameters a and m chosen to overcome certain number theoretic shortcomings inherent in earlier choices.

In practice pseudorandom numbers are processed in a computer as integers. Let γ be the number base of the computer, γ being 10 or 2 for decimal or binary machines, respectively. If we set $m = \gamma^\beta$, where β is an integer, then the modulo reduction

$$(aZ_{i-1} + c)/m$$

is accomplished by shifting the decimal point of $aZ_{i-1} + c$ to the left by β positions and retaining only the digits that remain to the left of the new decimal point location. The mapping of Z_i onto (0, 1) via Z_i/m is also accomplished by shifting the decimal point β positions to the left. Since a shift consumes considerably less time than a division on most computers it is desirable to select m to facilitate this shift substitution for a division.

Since $Z_i < m$ it is clear that the sequence $\{Z_i;\ i = 1, \ldots, n > m\}$ can contain at most m distinct numbers. Moreover, as soon as any number is repeated, the entire sequence is repeated. We wish therefore to choose m as large as possible to ensure a sufficiently large number of distinct numbers, and we also wish to select the remaining parameters a, c, and Z_0 so that as many as possible of the m numbers occur in a cycle. If p is the total number of numbers that can occur among the m possible ones, we say that the generator has *full period* when $p = m$.

7.4.1 Mixed Congruential Generators

It can be shown that the generator defined in (4.1) has full period, p, provided that the following conditions hold [100]:

1. c is *relatively prime* to m,

2. $a \equiv 1 \pmod{q}$ for each *prime factor* q of m, and
3. $a \equiv 1 \pmod{4}$ if 4 is a *factor* of m.

Condition 1 means that the greatest common divisor of c and m is unity. This condition is easily met. Condition 2 means that

$$a - q\left[\frac{a}{q}\right] = 1.$$

If $k = [a/q]$ we may write a as

$$a = 1 + qk \tag{4.6}$$

where q is a prime factor of m. Condition 3 means that we set

$$a = 1 + 4k \tag{4.7}$$

if $m/4$ is an integer.

If we are using a decimal machine and $m = 10^\beta$ then we may choose p equal to 2 or 5 since $m = 2^\beta \times 5^\beta$. Clearly, we must choose c not divisible by 2 or 5. Moreover, we require $a \equiv 1 \pmod{2}$, $a \equiv 1 \pmod{5}$, and, for $\beta \geq 2$, $a \equiv 1 \pmod{4}$. Let $k_1 = [a/2]$, $k_2 = [a/5]$, and $k_3 = [a/4]$. We require that

$$a = 1 + 2k_1 = 1 + 5k_2 = 1 + 4k_3$$

so that $k_2 = 2k_1/5$ and $k_3 = k_1/2$. Let $k_1 = 10k$, $k_2 = 4k$, and $k_3 = 5k$, where k is an integer. Then

$$a = 1 + 20k \tag{4.8}$$

satisfies the requirement. If we select $m = 2^\beta$ for a binary machine, then c must be odd and we require $a \equiv 1 \pmod{2}$ and $a \equiv 1 \pmod{4}$. Since the latter congruence subsumes the former we have

$$a = 1 + 4k. \tag{4.9}$$

In early work on pseudorandom number generation emphasis was placed on selecting k to permit an efficient computation of aZ_{i-1}. In the decimal case a choice of $k = 5 \times 10^\delta$, so that $a = 1 + 10^{\delta+2}$, enables the user to compute aZ_{i-1} by a shift of the decimal point $\delta + 2$ places to the right and an addition of the original number Z_{i-1}, thus avoiding a multiplication. On a binary machine $k = 2^\delta$ leads to $a = 1 + 2^{\delta+2}$, which permits analogous steps. The choice of δ remains to be made.

Expressions (4.8) and (4.9) still leave wide latitude in the choice of δ and consequently k and a. One approach to selection is to choose a so as to produce a desirable statistical property for the resulting pseudorandom sequence. Greenberger has shown [96] that the correlation between U_1 and U_{i+1} is*

$$\text{corr}(U_i, U_{i+1}) \sim \frac{1}{a} - \frac{6c}{am}\left(1 - \frac{c}{m}\right) + K \qquad (4.10)$$

$$-\frac{a}{m} < K < \frac{a}{m}.$$

One way to choose a is to minimize the upper bound of $\text{corr}(U_i, U_{i+1})$ with respect to a.† This gives

$$a = \sqrt{m - (6c/m)[1 - (c/m)]} \simeq m^{1/2} \qquad (4.11)$$

since m is usually much larger than c. For $a = m^{1/2}$, we have

$$-\frac{6c}{m^{3/2}}\left(1 - \frac{c}{m}\right) < \text{corr}(U_i, U_{i+1}) < \frac{2}{m^{1/2}}. \qquad (4.12)$$

This result consoles us with regard to the lower bound ($K = -a/m$), which is of order m smaller than the upper bound ($m^{1/2}$). For binary computers with $m \sim 2^\alpha$, this gives lower and upper bounds of orders $2^{-3\alpha/2}$ and $2^{1-\alpha/2}$, respectively. These results suggest that we select $k = 2^{\alpha/2-2}$.

7.4.2 Multiplicative Congruent al Generators

Historically, as mentioned earlier, multiplicative congruential generators preceded mixed ones. Since c plays no role, multiplicative generators have a natural appeal. Because their periods depend on Z_0 and a, we want to select these quantities to guarantee a maximum possible period length. We may ensure that the period length of a multiplicative generator is a maximum provided that Z_0 is relatively prime to m (i.e., the largest common divisor of Z_0 and m is unity) and a meets certain congruence conditions [100]. As examples, for a decimal computer with $m = 10^\alpha$, $\alpha \geq 3$,

$$a \equiv \pm 3 \pmod{200} \qquad (4.13)$$

so that

$$a = \pm 3 + 200k \qquad (4.14)$$

* In the same article steps are also described for deriving correlations between U_i and U_j for $i - j > 1$.

† The second derivative is easily shown to be positive, guaranteeing a maximum.

satisfies the conditions for a maximum period of $5 \times 10^{\alpha-2}$, $\alpha \geq 4$, as does

$$a \equiv \pm 3 \pmod 8 \tag{4.15}$$

$$a = \pm 3 + 8k$$

for a binary computer with $m = 2^{\alpha}$ for a maximum period* of $2^{\alpha-2}$.

As in the mixed congruential case the above choices of m allow us to replace a division by a shift. Also, for a decimal machine a choice of $k = 5 \times 10^{\beta}$, so that $a = 3 + 10^{\beta+3}$, allows us to substitute a shift for a multiplication in computing aZ_{i-1}. In the binary case, $k = 2^{\beta}$ permits the same substitution. By contrast with the mixed congruential generator we also have a multiplication by 3 to perform. As before, the choice of $\beta = \alpha/2$ (β integer) ensures low correlation in (4.10).

Although comforting, the computational efficiency offered by selecting a, c, m, and Z_0 as described above is gained at a price. In the binary case there are a total of $m = 2^{\alpha}$ numbers, but only $m/4 = 2^{\alpha-2}$ of them are available for use. In other words, although the selection rule for Z_0 and a with $m = 2^{\alpha}$ guarantees a period of maximum length, it precludes a period of full length. A question arises about where in the 2^{α} numbers of the binary case the gaps due to the $2^{\alpha} - 2^{\alpha-2}$ omitted numbers appear. The choice of Z_0 relatively prime to 2^{α} guarantees that we run through the $2^{\alpha-2}$ available numbers, but it is not clear how these numbers are distributed over $(0, 2^{\alpha})$. Because of this less-than-full-period property, the full period mixed congruential generators appear relatively more attractive. However, recent work has revealed that, if the selection of m is also a design consideration, full period multiplicative congruential generators are possible. Since they are becoming more common in practice we describe them in Section 7.4.4.

7.4.3 A Bad Multiplicative Congruential Generator

As mentioned earlier, use of a pseudorandom number generator should be predicated on its first passing a battery of diverse statistical tests. One mandatory test involves possible correlation between pairs of numbers. From a theoretical point of view we have Greenberger's result (4.10), which allows us to choose a to minimize the correlation between successive variates. The following example shows how sole reliance on this result and on the efficiency consideration can seriously impair the statistical integrity of a simulation experiment. Consider the multiplicative congruential generator [97, 98]:

$$Z_{i+1} \equiv aZ_i \pmod m \tag{4.16}$$

$$a = 2^{18} + 3 \quad m = 2^{35}$$

* Other values of a satisfy the conditions also, but the values given here are the ones most commonly encountered in practice. See reference 100, p. 237.

for use on the available bits of the IBM/704 computer. It satisfies the maximum period criterion, facilitates efficient computation, and comes close to minimizing the correlation in (4.10). We may write

$$
\begin{aligned}
Z_{i+1} &\equiv (2^{18} + 3)Z_i && \pmod{2^{35}} \\
Z_{i+2} &\equiv (2^{18} + 3)Z_{i+1} && \pmod{2^{35}} \\
&\equiv (2^{18} + 3)^2 Z_i && \pmod{2^{35}} \\
&\equiv (2^{36} + 2^{19} \times 3 + 9)Z_i && \pmod{2^{35}}.
\end{aligned} \tag{4.17}
$$

Since $2 \times 2^{35} Z_i$ is an integral multiple of 2^{35} we may also write

$$
\begin{aligned}
Z_{i+2} &\equiv (6 \times 2^{18} + 9)Z_i && \pmod{2^{35}} \\
&\equiv (6 \times (2^{18} + 3) - 18 + 9)Z_i && \pmod{2^{35}} \\
&\equiv (6 \times (2^{18} + 3) - 9)Z_i && \pmod{2^{35}} \\
&\equiv 6Z_{i+1} - 9Z_i && \pmod{2^{35}}.
\end{aligned} \tag{4.18}
$$

Recalling that $U_i = Z_i/m$, we also have the congruence relationship

$$U_{i+2} \equiv 6U_{i+1} - 9U_i \pmod{1}. \tag{4.19}$$

Suppose that $0 < U_i \leq 0.1$. Then a little thought shows that

$$
[6U_{i+1} - 9U_i] = \begin{cases} [6U_i] & 0 < U_i \leq \dfrac{6U_{i+1} - [6U_{i+1}]}{9} \\[2ex] [6U_i] - 1 & \dfrac{6U_{i+1} - [6U_{i+1}]}{9} < U_i \leq 0.1. \end{cases}
$$

Since

$$U_{i+2} = 6U_{i+1} - 9U_i - [6U_{i+1} - 9U_i],$$

it is clear that for $0 < U_i \leq (6U_{i+1} - [6U_{i+1}])/9$

$$0 < U_{i+2} = 6U_{i+1} - 9U_i - [6U_{i+1}] < 6U_{i+1} - [6U_{i+1}], \tag{4.20a}$$

and for $(6U_{i+1} - [6U_{i+1}])/9 < U_i \leq 0.1$

$$1 > U_{i+2} = 6U_{i+1} - 9U_i - [6U_{i+1}] + 1 \geq 6U_{i+1} - [6U_{i+1}] + 0.1. \tag{4.20b}$$

This result shows that, for $0 \leq U_i \leq 0.1$, U_{i+2} is excluded from a subinterval on the unit interval determined by the line $6U_{i+1} \pmod 1$ and $6U_{i+1} \pmod 1 + 0.1$, as Fig. 41 shows. It is interesting to note that this theoretical investigation was prompted by an experimental study which produced a plot as in Fig. 41 of U_{i+1} and U_{i+2} for $0 \leq U_i \leq 0.1$. The dark intervals correspond to the excluded intervals in (4.20).

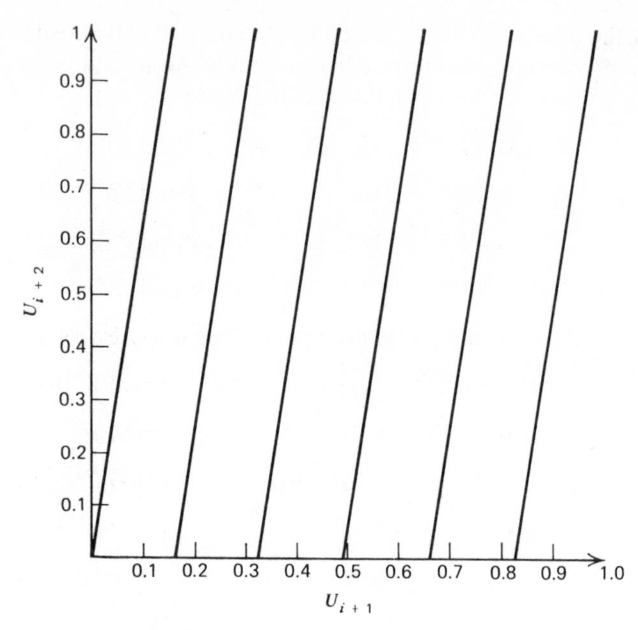

Fig. 41 Pseudorandom number plot of U_{i+1} and U_{i+2}. *Source:* [98, p. 178].

Unfortunately this bad feature of $a = 2^{18} + 3$ extends to all multiplicative generators with $m = 2^\beta$ and $a = 2^s + b$. Because of the common usage of such generators the reader is advised to check the specification on any prepackaged multiplicative generator. We wish to emphasize that such a check is instructive, regardless of the type of prepackaged generator available.

7.4.4 An Almost Full Period Multiplicative Congruential Generator

To overcome these number theoretic inadequacies, recent work has emphasized the selection of the modulus m. In particular it has been suggested that for binary computers m be the largest prime in 2^β, β being the number of available bits [101, 102]. This permits the longest possible cycle length or period. To assure the maximum period within this largest possible period, we require Z_0 to be relatively prime to m and $a^n \equiv 1 \pmod{m}$, where the smallest value of n for which this congruence can occur is* $n = m - 1$. In this case a period of $m - 1$ is attained, which is one less than full period. This eliminates the possibility of gaps such as multiplicative generators with $m = 2^\alpha$

* See reference 91 for proof.

allow. Moreover choosing the largest prime in 2^β for m eliminates the congruence difficulty that occurred in our previous example. For decimal computers m is chosen to be the largest prime in 10^β, β here being the number of digits available. The multiplier a is then chosen according to the same form of the congruence relationship as given above for binary computers.

On the IBM/360 system the number of available bits for computation is $\beta = 31$. The largest prime is conveniently $m = 2^{31} - 1$, and a choice of $a = 14^{29}$ or $a = 7^5$ satisfies the condition for maximum period. Parameters chosen in this way have performed favorably on a variety of statistical tests [101, 110, 21].*

To regain partially the computational efficiency lost by choosing $m \neq 2^\beta$ we note that

$$Z_{i+1} \equiv aZ_i \pmod{2^\alpha}$$

$$(4.21) \qquad Z_{i+1} = aZ_i - k \times 2^\alpha$$

$$k = \left[\frac{aZ_i}{2^\alpha}\right].$$

Let

$$Z'_{i+1} = Z_{i+1} + k = aZ_i - k(2^\alpha - 1).$$

If $Z_{i+1} + k < 2^\alpha - 1$ then

$$Z'_{i+1} \equiv aZ_i \pmod{2^\alpha - 1}.$$

If $Z_{i+1} + k \geq 2^\alpha - 1$ let

$$Z'_{i+1} = Z_{i+1} + k - (2^\alpha - 1) = aZ_i - (k + 1)(2^\alpha - 1)$$
$$\equiv aZ_i \pmod{2^\alpha - 1}.$$

The sequence $\{Z'_i\}$ provides us with a random number stream with $m = 2^\alpha - 1$ according to the rule

$$Z'_{i+1} = \begin{cases} Z_{i+1} + k & Z_{i+1} + k < 2^\alpha - 1 \\ Z_{i+1} + k + 1 - 2^\alpha & Z_{i+1} + k \geq 2^\alpha - 1, \end{cases}$$

where Z_{i+1} is obtained by the shift modulo reduction in (4.21). The added logical comparison consumes considerably less time than division by $2^\alpha - 1$ would require.

For the IBM/7090 computer 35 bits are available, and the largest prime in 2^{35} is $m = 2^{35} - 31$. The quantity $a = 5^5$ suffices for a maximum period

* We present some of the results for an IBM/360 system generator in Section 7.7.1.

[102]. To regain partially the computational efficiency we consider the general case of $m = 2^\alpha - \gamma$. Let

$$Z'_{i+1} = Z_{i+1} + k\gamma = aZ_i - k(2^\alpha - \gamma), \tag{4.22}$$

where Z_i, Z_{i+1}, and k are defined in (4.21). If $Z_{i+1} + k\gamma < 2^\alpha - \gamma$ then

$$Z'_{i+1} \equiv aZ_i \pmod{2^\alpha - \gamma}.$$

If $Z_{i+1} + k\gamma \geq 2^\alpha - \gamma$ let

$$\begin{aligned} Z'_{i+1} &= Z_{i+1} + k\gamma - (2^\alpha - \gamma) = aZ_i - (k+1)(2^\alpha - \gamma) \\ &\equiv aZ_i \pmod{2^\alpha - \gamma}. \end{aligned}$$

The sequence $\{Z'_i\}$ provides us with the desired pseudorandom number stream according to the rule

$$Z'_{i+1} = \begin{cases} Z_{i+1} + k\gamma & Z_{i+1} + k\gamma < 2^\alpha - \gamma \\ Z_{i+1} + (k+1)\gamma - 2^\alpha & Z_{i+1} + k\gamma \geq 2^\alpha - \gamma. \end{cases}$$

7.5 *k*-TUPLES

As mentioned earlier, we occasionally wish to generate a k-tuple of independent uniform random numbers. For example, if $k = 2$ we want two numbers U_1 and U_2, where the point U_1, U_2 is uniformly distributed over the unit square. More generally for k numbers we want the point $U_1, \ldots, U_k$ uniformly distributed over the unit k-cube.

Many situations arise in which a k-tuple plays a role. For example, each job in the simple queueing model of Chapter 3 has an interarrival time, which denotes the elapsed time between its arrival and the time of the preceding arrival, and has a service time. Together the interarrival time and service time form a 2-tuple of numerical attributes both of which are functions of random numbers. Our modeling specifies that these times be independent, and this requires that the random numbers generating each time also be independent.

Sometimes a k-tuple plays a more subtle role. Chapter 8 describes a procedure that uses two independent random numbers to generate a standardized normal variate. Hence, although the user sees only the resulting normal variate, a 2-tuple is central to its generation.

Two important statistical properties that we want k-tuples of pseudorandom numbers to have are the following:

1. the elements of a k-tuple are independent and uniformly distributed;
2. the k-tuples are independent and uniformly distributed on the k-dimensional unit hypercube.

The *serial* test, to be described in Section 7.7, is designed to reveal departures from these properties. Unfortunately the computational aspects of the test become inconvenient for $k > 2$. As a result less precise investigations such as the *gap*, *poker*, and *runs* tests, which are computationally more appealing, are occasionally employed to reveal departures. Coveyou and MacPherson [90] have developed a theory to explain the statistical properties of k-tuples from congruential generators. As with the serial test, the procedure that Coveyou and MacPherson recommend for testing presents formidable computational difficulties.* Because of the limited testing of k-tuples in practice, a prudent course is to form k-tuples of pseudorandom numbers in ways that reduce the potential for nonrandomness.

It behooves us to study initially ways in which a simulation design can contribute to reducing dependence between pseudorandom numbers in a k-tuple. The single server queueing problem offers a convenient model for demonstrating the advantages of one simulation design over another. Suppose that we generate U_i and U_{i+1} as the pseudorandom numbers from which the interarrival time and service time, respectively, of the next arrival are to be computed. Here U_i and U_{i+1} form a 2-tuple. Alternatively we may generate U_i, compute the next arrival's interarrival time from it, *wait* until the server selects this arrival, and then generate a U_j from which his service time is computed. Here $j \neq i + 1$ if pseudorandom numbers are generated for other purposes while this arrival waits. Since this method allows us to form a 2-tuple (U_i, U_j) which need not necessarily be successive pseudorandom numbers, we incline to the view that such a 2-tuple is more likely than the 2-tuple (U_i, U_{i+1}) to be a sequence of independent numbers. More generally we suggest that, when an entity has attributes that assume stochastic values, each attribute value should be determined as needed rather than determining all attribute values at once.

To encourage randomness MacLaren and Marsaglia [111] suggest a procedure which uses two multiplicative generators. The first generator serves to fill 128 locations in core storage. The second is used to choose a location randomly from among the 128 available. Whenever a number is selected from the 128 it is replaced with the next pseudorandom number generated from the first generator. For example, we may use two multiplicative generators, each with modulus m being the largest prime number that the computer can store and with respective multipliers c_1 and c_2 chosen to guarantee periods $m - 1$. Then we generate a number

$$W_{i+1} \equiv c_1 W_i \pmod{m},$$

utilize its first seven bits to generate an integer J between 1 and 128, and then use the number generated by the second generator which is stored in location

* See reference 108 for an algorithm to apply when using their test.

J. We then generate a new number from the second generator and store it in location *J*. This method doubles the number of pseudorandom numbers needed.

7.6 GPSS/360, SIMSCRIPT II, AND SIMULA PSEUDORANDOM NUMBER GENERATORS

In this section we describe the pseudorandom number generators in use in GPSS/360, SIMSCRIPT II, and SIMULA at the time of preparation of this book. For the GPSS/360 multiplicative pseudorandom number generator, we quote Felder's description [93]:

> The GPSS/360 program maintains 8×1 arrays: (1) The *base number* array contains eight words, each different. The first of these words is called the *seed.* (2) The *multiplier* array contains eight words, one for each random number generator. Each multiplier is initially one, unless an alternate initial multiplier is supplied by the analyst (by means of the RMULT card). (3) The *index* array contains eight words, one for each random number generator; each is initially zero.
>
> When a random number is called for, the following procedure is used:
>
> 1. The appropriate word in the index array (depending upon which RN# is used) points to one of the eight numbers in the base number array. Since the index array words are initially zero, the first base number used will be the seed. All eight random number generators use this common seed.
> 2. The appropriate number in the multiplier array (again depending upon which RN# is used) is multiplied by the base number chosen in step 1.
> 3. The low-order 32 bits of this product are then sometimes subjected to another transformation. If the highest-order bit of these 32 bits is 1, the two's complement of these 32 bits replace the original 32 low-order bits of the 64 bit product.
> 4. The low-order 31 bits of this possibly partially complemented product are stored in the appropriate word of the multiplier array, to be used the next time a random number is called for.
> 5. Three bits of the high-order 16 bits of the product after step 3 are stored in the appropriate word of the index array, for future use. This number (0–7) points to one of eight words of the base number to be used the next time a random number is called for.
> 6. (a) If the random number required is a fraction, the middle 32 bits of the product after step 3 are divided by 10^6, and the remainder becomes the six-digit fractional random number.
>   (b) If the random number required is an integer, the middle 32 bits of the product after step 3 are divided by 10^3 and the remainder becomes the three-digit random number.

More recent GPSS publications indicate that the default multiplier is now 37.

Several observations are worth making. Unlike the generators described in earlier sections, the GPSS generator does not use the previously generated number to produce the next number recursively. The "scrambling" described above is intended to reduce nonrandomness that may otherwise enter the sequence of pseudorandom numbers. Moreover the generator is not amenable to analytical determination of the characteristics of desirable multipliers or the period of the generator. However, it can be established that the multiplier must be odd. Statement 4 also adds skepticism about the reliability of this generator by indicating that the contents of the multiplier array are regularly changed. The writer has been unable to obtain clarification of this statement.

Felder also reports that a battery of tests revealed that the multipliers 31, 37, 743 appear satisfactory. However, the interpretation of at least one test, namely, the 2-tuple test, is open to question.

SIMSCRIPT II has a multiplicative pseudorandom number generator with $a = 14^{29}$, $m = 2^{31} - 1$, and 10 seeds:

SEED.V(1) =	524267	SEED.V(6) =	1157240309
SEED.V(2) =	683743814	SEED.V(7) =	15726055
SEED.V(3) =	964393174	SEED.V(8) =	48108509
SEED.V(4) =	1217426631	SEED.V(9) =	1797920909
SEED.V(5) =	618433579	SEED.V(10) =	477424540.

Each seed corresponds to the initial number in a sequence of 100,000 pseudorandom numbers. These sequences are consecutive. For example, the 100,000th number following 683743814 is 964393174. These seeds are automatically set before execution. The user specifies the stream to be employed for a particular purpose and, if he wishes, can read in a new seed at any time.*

The value of controlling seed selection was discussed in Section 5.5. We give another reason here. Practical experience with stream 1 has indicated that the resulting pseudorandom numbers have poor sampling properties.† Therefore it has been suggested that SEED.V(1) above be replaced by SEED.V(1) = 2116429302, which has better sampling properties. This new number is the 100,000th number following SEED.V(10) = 477424540. SIMSCRIPT II.5, the proprietary version of SIMSCRIPT II, incorporates this change. We suggest that SIMSCRIPT II users make this change.

One approach to guarding against nonrandomness is to use different streams for different purposes. For example, in Section 5.5, we used stream 1 to generate interarrival times and stream 2 to generate service times. A more conservative, but also more expensive, scheme using these streams can serve

* This contrasts with GPSS, where the seed is fixed but the multipliers can be changed.
† Personal communication from Philip Kiviat.

to reduce possible dependence. We illustrate this scheme for SIMSCRIPT II. To select a number from stream j in SIMSCRIPT II, one inserts j as the stream argument in the appropriate location in the coding. If, however, one inserts RANDI.F(2,10,1) instead of j, a pseudorandom number is generated from stream 1; this number is used to select an integer j from 2 to 10, each integer being equally likely, and then the number to be used is generated from stream j.* Since this algorithm randomizes the stream selection, one feels intuitively that the chance of dependence is reduced in k-tuples where each number is computed as described.

Unfortunately the cost of this method is a doubling of the quantity of pseudorandom numbers generated. Moreover an additional issue deserves thought. The seed in stream $j + 1$ is the same as the 100,001st pseudorandom number that would be generated by stream j. This means that if we use stream 1, as described above, we can generate at most 100,000 pseudorandom numbers before the sequence begins to repeat itself. Accordingly it is advisable to check on how many numbers are to be generated before employing the above approach.

This problem can be resolved at the expense of reducing the degree of randomization of the streams. Suppose that we require between $100{,}000 \times (k - 1)$ and $100{,}000 \times k$ pseudorandom numbers, where $k = 1, \ldots, 8$. Then insertion of RANDI.F $(k + 1, 10, 1)$ generates an integer from $k + 1$ to 10 and uses this integer to select a random number stream. Here no overlap occurs.

SIMULA has a multiplicative pseudorandom number generator with $a = 5^{2p+1}$ and $m = 2^n$, where n is the number of available bits. On the UNIVAC 1107/8 $n = 35$ and p is chosen as 6. From the discussion in Section 6.4.2 we know that this generator has period 2^{33}, provided that the seed is odd. According to reference 117, p. 14, the last 2 bits of each number remain constant while the remaining 33 take on all possible values. Hence there is no need for major concern over the uniformity requirement.

7.7 TESTS OF INDEPENDENCE AND UNIFORMITY

Many statistical tests exist that are designed to reveal departures from randomness. Knuth [108] describes ten of them with computational algorithms. Our purpose here is to familiarize the reader with the major purposes of many of these tests; it is not to provide a detailed account of the statistical foundations of the tests. A reader interested in applying them is encouraged to consult reference 108.

* See reference 107, p. 312.

7.7.1 Chi-Square Goodness-of-Fit Test

We first wish to test for uniformity. To do so we use the chi-square goodness-of-fit test. Suppose that we collect n observations $U_1, \ldots, U_n$ on a run. We divide the unit interval into r equal subintervals so that, under the hypothesis of uniformity, the probability that a number U_k, $1 \leq k \leq n$, falls in a particular interval is $1/r$. Also the expected number of observations in a particular cell is n/r. If we are testing numbers on a decimal machine then $r = 10^\alpha$ causes us to examine the α most significant digits for uniformity. On a binary machine a choice of $r = 2^\alpha$ allows us to test the α most significant bits.

Let f_j be the frequency with which numbers fall in the interval $[(j-1)/r, j/r]$. Then, if $\{U_j\}$ is a sequence of independent, uniformly distributed random variables, the statistic

$$\chi_1^2 = \frac{r}{n} \sum_{j=1}^{r} \left(f_j - \frac{n}{r} \right)^2 \tag{7.1}$$

has a distribution that converges to that of chi square with $r - 1$ degrees of freedom as n becomes large. The choice of $n > 5r$ is recommended to reduce the chance of too few observations in any one cell, a result which would conflict with the asymptotic result.

Under the hypothesis of independence and uniformity we expect

$$\text{Prob}[\chi_1^2 < Q_{1-\alpha}(r-1)] = 1 - \alpha \tag{7.2}$$

where α is a selected critical value, say 0.05 or 0.1, and $Q_{1-\alpha}(r-1)$ is the point on the cumulative chi-square distribution with $r - 1$ degrees of freedom corresponding to probability $1 - \alpha$.

Some writers prefer an alternative approach to presenting results on random number generators. They find the β for which

$$\text{Prob}(\chi^2 > \chi_1^2) = \beta. \tag{7.3}$$

The greater β is in (7.3), the greater the investigator's confidence is in the truth of the hypothesis.

In large samples χ_1^2 has mean $r - 1$ and variance $2(r-1)$. When r is sufficiently large the quantity $[\chi_1^2 - (r-1)]/\sqrt{2(r-1)}$ has approximately the normal distribution with mean zero and unit variance so that

$$\text{Prob}\left[\left|\frac{\chi_1^2 - (r-1)}{\sqrt{2(r-1)}}\right| < P_{1-\alpha/2}\right] \sim 1 - \alpha, \tag{7.4}$$

$P_{1-\alpha/2}$ now being the $1 - \alpha/2$ probability point of the normal distribution.

Alternatively one may treat $[\chi_1^2 - (r - 1)]/\sqrt{2(r - 1)}$ as the critical value and find the probability β of a normal variate falling below that value as in (7.3). Use of the normal distribution is suggested only when the degrees of freedom $r - 1$ exceed the values listed in standard tables of the chi-square distribution. Values of $r - 1$ up to 100 are usually given.

In testing a multiplicative congruential generator for the IBM/360 system with $a = 16{,}807$ and $m = 2^{31} - 1$, Lewis, Goodman, and Miller [110] set $n = 2^{16} + 5 = 65{,}541$ and $r = 2^{12} = 4096$. Consequently the first 12 bits were examined, and the expected number per cell was about 16. The 5 in n occurred because of need for 5 additional observations for a later test. Table 20 shows these authors' results for 10 independent runs.

Table 20 Test Results for a Multiplicative Congruential Generator[a]

$a = 7^5 = 16{,}807, \quad m = 2^{31} - 1, \quad n = 2^{16} + 5$

Run	Z_0	χ_1^2 $r = 2^{12}$
1	12345678	4015.25
		(−79.75)
2	855998726	4112.12
		(+17.12)
3	745681489	4125.12
		(+30.12)
4	506104362	4113.50
		(+18.50)
5	236686234	4150.75
		(+55.75)
6	1912615462	4079.87
		(−15.13)
7	481694049	4628.87
		(+172.87)
8	785044942	4114.50
		(+19.50)
9	864268549	4058.37
		(−36.63)
10	13034519	4096.87
		(+1.87)
max $\lvert Q' \rvert$		1.92

Source: [110, p.142]. [a]The quantities in parentheses are deviations from means. The last row was computed by the writer.

Each χ_1^2 has 4095 degrees of freedom. Here we may safely regard $(\chi_1^2 - 4095)/\sqrt{2 \times 4095} \sim (\chi_1^2 - 4095)/90$ as normal with mean zero and variance unity. Let

$$Q' = \frac{\chi_1^2 - 4095}{90}. \tag{7.5}$$

If $P_{1-\alpha/2}$ is the point on the cumulative normal curve corresponding to probability $1 - \alpha/2$, then under the null hypothesis we may make the probability statement

$$\text{Prob}(-P_{1-\alpha/2} < Q' < P_{1-\alpha/2}) \sim 1 - \alpha. \tag{7.6}$$

For example, a test size of $\alpha = 0.05$ gives $P_{0.975} = 1.96$. Since the largest deviation occurs on run 7 with

$$\max|Q'| = \frac{172.87}{90} \sim 1.92 \tag{7.7}$$

we accept the hypothesis for each of the 10 runs at the 0.05 level since

$$\max|Q'| \sim 1.92 < 1.96.$$

To derive a single test statistic for the results of n independent runs taken together, we make use of the fact that χ_1^2 has the chi-square distribution for large n. Let F_l be the frequency of χ_1^2 on M runs that fall into the lth quantile of the chi-square distribution, where $l = 1, \ldots, L (L \leq M/5)$. Then we use the probability statement

$$\text{Prob}(\chi_1^2 < \phi_l) = \frac{l}{L} \tag{7.8}$$

and a chi-square table for $r - 1$ degrees of freedom to find the lth quantile value ϕ_l. For M replications the expected frequency in each cell is M/L. Then the summary test statistic

$$\chi_F^2 = \frac{L}{M} \sum_{l=1}^{L} \left(F_l - \frac{M}{L}\right)^2 \tag{7.9}$$

has a distribution that converges to that of chi square with $L - 1$ degrees of freedom as the number of replications M increases.

7.7.2 Kolmogorov–Smirnov Test

This test allows us to evaluate the hypothesis that a sample of data was drawn from a specified continuous distribution F. The test is nonparametric and exact for all sample sizes in contrast to the asymptotic nature of the chi-square test. Suppose that we have n observations $X_1, \ldots, X_n$ with sample cumulative distribution function

$$(7.10) \quad F_n(X) = \begin{cases} 0 & X < X_1 \\ \dfrac{j}{n} & X_j \leq X < X_{j+1} \qquad j = 1, \ldots, n-1. \\ 1 & X \geq X_n \end{cases}$$

The test statistic is

$$(7.11) \qquad D_n(X) = \max_X |F(X) - F_n(X)|$$

which can be shown to have a distribution that does not depend on $F(X)$. This is a consequence of $F(X)$ having a uniform distribution, as discussed earlier. For the particular distribution being tested, one computes $F(X_j)$, $j = 1, \ldots, n$, the sample distribution being

$$(7.12) \qquad F_n(X_j) = \frac{j}{n} \qquad j = 1, \ldots, n.$$

Critical values D_α for significance level α and sample size n can be found, for example, in reference 113. One accepts the null hypothesis if $D_n(X) < D_\alpha$ but rejects it otherwise.

The computation of $F_n(x)$ requires us to order the observations $X_1, \ldots, X_n$. Since n may be quite large when testing pseudorandom numbers the problem of storage and ordering can become critical.* The chi-square test offers the advantage of retaining frequency counts and hence is more attractive for working with large sample sizes. The exactness of the Kolmogorov-Smirnov test is an attractive feature, however, when we wish to test distributional hypotheses with relatively small sample sizes. It is particularly useful for testing a summary test statistic based on a small number of independent replications. The 10 replications in Table 20 offer a good exercise for the reader, bearing in mind that F is the chi-square distribution here with 9 degrees of freedom.

7.7.3 Serial Test

Let $W_1 = (U_1, \ldots, U_k)$, $W_2 = (U_{k+1}, \ldots, U_{2k}), \ldots, W_n = (U_{(n-1)k+1}, \ldots, U_{nk})$ be a sequence of n k-tuples. We wish to test the hypothesis that $W_1, \ldots, W_n$ are independent and uniformly distributed points in the unit k-cube. The serial test proceeds analogously to the chi-square rest.

* This can be handled by establishing a chained list in increasing order of magnitude, as discussed in Section 4.2.3.

Let

$$W_s = (U_{(s-1)k+1}, \ldots, U_{jk}) \qquad s = 1, \ldots, n.$$

We divide the unit hypercube for which $0 \leq U_{i+l} \leq 1$, $l = 0, \ldots, k-1$, into r^k hypercubes defined by

$$\frac{j_1 - 1}{r} \leq U_i < \frac{j_1}{r}, \frac{j_2 - 1}{r} \leq U_{i+1} < \frac{j_2}{r}, \ldots,$$

(7.13)

$$\frac{j_k - 1}{r} \leq U_{i+k-1} < \frac{j_k}{r} \qquad j_l = 1, \ldots, r; \quad l = 1, \ldots, k.$$

Here each j_l can assume any integer value from 1 to r for $l = 1, \ldots, k$, so that there are r^k possible hypercubes into which a string of k observations may fall. Then $f_{j_1, \ldots, j_k}$ is the frequency with which k-tuples $\{W_j; j = 1, \ldots, n\}$ fall into the hypercube characterized by $j_1, \ldots, j_k$ in (7.13). Since there are n k-tuples W_j and there are r^k k-cubes, the expected number of k-tuples per k-cube is n/r^k under the hypothesis of independence and uniformity. Then the statistic

$$\chi_k^2 = \frac{r^k}{n} \sum_{j_1, \ldots, j_k = 1}^{r} \left(f_{j_1, \ldots, j_k} - \frac{n}{r^k} \right)^2 \tag{7.14}$$

asymptotically has the chi-square distribution with $r^k - 1$ degrees of freedom.

Two issues regarding this test deserve explicit mention. In the first place, notice that the W_j include nonoverlapping elements of the U_i. For example, $k = 2$ gives $W_1 = (U_1, U_2)$, $W_2 = (U_3, U_4)$, $W_3 = (U_5, U_6), \ldots$. In some applications of the serial test the data are $V_1 = (U_1, U_2)$, $V_2 = (U_2, U_3)$, $V_3(U_3, U_4), \ldots$. In this case χ_k^2 in (7.14) does not asymptotically have the chi-square distribution. However, Good [94] has given correct procedures for testing k-tuples in this case. Although this alternative test is described in several books on simulation, there are two reasons why its use is inappropriate. First, k-tuples with overlapping elements do not seem to occur in simulation work. Second, Marsaglia [111a] has persuasively argued from a theoretical point of view that overlapping k-tuples from congruential generators cannot be expected to exhibit random behavior. His result indicates that the degree of nonrandomness increases with k.

A second issue in regard to the serial test is that a computational problem severely limits its application. Since there are r^k k-cubes into which W_j may fall, the magnitude of r^k becomes important. If $k = 3$ and we set $r = 2^{12}$ then $r^k = 2^{36}$, a number which creates a problem in regard to both storage and search. If we reduce r to 2^3, the problem becomes more manageable but our study is of the first 3 bits of each number. It is clear that for $k > 3$ the problem is even more acute.

Because of the computational burden that the serial test imposes, it is seldom used for $k > 2$. Instead a variety of less exact tests are employed in the hope that taken together they will reveal departures from randomness. In the next four sections we briefly describe several of these tests, which exist in a variety of forms. We offer one example of each. Algorithms for applying them may be found in Knuth [108].

7.7.4 Gap Test

In one form of the gap test the user specifies a number of *gaps* n and then examines a sequence $U_1, U_2, \ldots$ until n gaps are found. A gap of length k is recorded when $\alpha < U_j < \beta$, $\alpha < U_{j+k} < \beta$, and U_{j+i} for $i = 1, \ldots, k-1$ do not fall in (α, β) for user-specified α and β. A statistic based on the sample gap lengths is then tested for consistency with the theoretical result expected when randomness prevails.

7.7.5 Poker Test

In the serial test for k-tuples, we noted that a cell was defined for each $j_1, \ldots, j_k$, where $j_i = 1, \ldots, r$, thereby producing r^k *permutations*. The *poker* test, which also examines n nonoverlapping k-tuples, is concerned with the $\binom{r}{k}$ *combinations* on $j_1, \ldots, j_k$. For example, for $k = 2$ the frequencies f_{ij} and f_{ji} would be accumulated in the same cell, thereby reducing the required number of cells from r^2 to $(r^2 - r)/2$. These counts are then used to form a statistic that is tested for consistency with theory. Alternative forms of the test exist.

7.7.6 Runs Test

Let

$$S_j = \begin{Bmatrix} 0 & U_j < U_{j+1} \\ 1 & U_j > U_{j+1} \end{Bmatrix} \qquad j = 1, \ldots, n.$$

Let V_k be the number of runs of all zeros or all ones of length k in the sequence S_j. The V_k are used to form a statistic which is tested for consistency with theory. This form of the test includes runs "up" and "down."

7.7.7 Tests of Correlation

The tests described so far are designed to use frequency counts based on the sample data to test for independence and uniformity. Since these tests do not use the actual sample values one has the feeling that they do not utilize all of the available information. The tests described in this section do use the actual sample values and emphasize the search for correlation between events in a sequence.

In Chapter 6 we described the concept of an autocovariance function R which described the extent of correlation between events in a series. Under the hypothesis of independence and uniformity the sequence $U_1, \ldots, U_n$ has mean

$$E(U_j) = \tfrac{1}{2} \tag{7.15}$$

and autocovariance function

$$R_{j-k} = E[(U_j - \tfrac{1}{2})(U_k - \tfrac{1}{2})] = \begin{cases} \frac{1}{12} & j = k \\ 0 & j \neq k. \end{cases} \tag{7.16}$$

For a lag k between observation, a common estimator of the autocovariance is

$$\hat{R}_k = \frac{1}{n-k} \sum_{j=1}^{n-k} [(U_j - \tfrac{1}{2})(U_{j+k} - \tfrac{1}{2})]. \tag{7.17}$$

For $k > 0$, $\hat{R}_k$ has mean 0 and variance $1/[144(n-k)]$. The mean is derived by noting that

$$\begin{aligned} E(\hat{R}_k) &= \frac{1}{n-k} \sum_{j=1}^{n-k} E(U'_j U'_{j+k}) = \frac{1}{n-k} \sum_{j=1}^{n-k} E(U'_j)E(U'_{j+k}) = 0 \qquad k > 0 \\ U'_j &= U_j - \tfrac{1}{2}. \end{aligned} \tag{7.18}$$

The variance follows from

$$\left.\begin{aligned} E(\hat{R}_k^2) &= \frac{1}{(n-k)^2} \sum_{i,j=1}^{n-k} E(U'_i U'_{i+k} U'_j U'_{j+k}) \\ &= \frac{1}{(n-k)^2} \sum_{i,j=1}^{n-k} E(U'_i U'_j) E(U'_{i+k} U'_{j+k}) \\ &\quad + \frac{1}{(n-k)^2} \sum_{i,j=1}^{n-k} E(U'_i U'_{j+k}) E(U'_j U'_{i+k}) \\ &\quad + \frac{1}{(n-k)^2} \sum_{i,j=1}^{n-k} E(U'_i U'_{i+k}) E(U'_j U'_{j+k}) \\ &= \frac{1}{144(n-k)} \end{aligned}\right\} \quad k > 0. \tag{7.19}$$

As n increases, the distribution of $\sqrt{n-k}\,\hat{R}_k$ converges to the normal with mean 0 and variance 1/144. Hence one simple test is to compute $\hat{R}_k$ for several values of k and test the hypothesis

$$R_k = 0 \qquad k > 0. \tag{7.20}$$

The user may easily test for zero correlation with $k = 1, 2, \ldots, K$, provided that $n \gg K$.

Although this test is simple and easy to apply it can lead to a problem in resolving favorable and unfavorable results corresponding to various k. For example, what does one do if hypothesis (7.20) is accepted for $k = 1, \ldots, 5$; rejected for $k = 6, \ldots, 8$; and accepted for $k = 11, \ldots, 20$? Since $\hat{R}_1, \ldots, \hat{R}_K$ jointly have the multivariate normal distribution with zero mean vector and easily derivable covariance matrix, one approach might be to test the sample autocovariances jointly for zero correlation. Then the question remains as to how large to make K, while retaining $n \gg K$.

An alternative approach is to test for a flat spectrum [92]. If events in the sequence are independent, then, in addition to $R_k = 0$, $k > 0$, it should be clear that $R_0 = \frac{1}{12}$. Then the spectrum of $\{U_j\}$ is

$$g(\lambda) = (2\pi)^{-1} \sum_{k=-\infty}^{\infty} R_k e^{-i\lambda k} = (24\pi)^{-1}. \tag{7.21}$$

The test of this hypothesis is based on the *periodogram* estimate of* $2\pi g(\lambda)$:

$$\begin{aligned} P_j &= \hat{R}_0 + 2 \sum_{k=1}^{n-1} \hat{R}_k \cos \lambda_j k \\ \lambda_j &= \frac{\pi j}{K} \qquad j = 1, \ldots, K = \frac{n}{2}. \end{aligned} \tag{7.22}$$

Although the periodogram is seldom used to estimate the spectrum because of its poor sampling properties, it is a convenient statistic to employ in testing for independence. Let

$$S_k = \frac{\sum_{j=1}^{k} P_j}{\sum_{j=1}^{K} P_j} \qquad k = 1, \ldots, K, \tag{7.23}$$

which we call the *normalized cumulative periodogram*. The statistic S_k is an estimate of the cumulative spectral distribution function:

$$F(\lambda_k) - F(-\lambda_k) = \frac{\int_{-\lambda_k}^{\lambda_k} g(\lambda)\, d\lambda}{\int_{-\pi}^{\pi} g(\lambda)\, d\lambda} = \frac{\int_{-\lambda_k}^{\lambda_k} g(\lambda)\, d\lambda}{R_0}. \tag{7.24}$$

In the case of g as given in (7.21), this is

$$F(\lambda_k) - F(-\lambda_k) = \frac{12}{24\pi}(2\lambda_k) = \frac{k}{K}. \tag{7.25}$$

* We assume n even for convenience.

Under fairly general conditions a sequence $\{U_j\}$ of independent events leads to the P_ks being independent, exponentially distributed variates when n is sufficiently large [112]. Moreover the order statistics S_k, $k = 1, \ldots, K$, are from the uniform distribution. But this is essentially the situation described in the Kolmogorov-Smirnov test for a uniform distribution with S_k replacing the sample cumulative probability distribution. Our cumulative test statistic is then

$$c = \max_k \left(\left| S_k - \frac{k}{K} \right| \right) \tag{7.26}$$

the significance of which may be tested by using a table of Kolmogorov-Smirnov critical values as in reference 113. It is important to note that, whereas the Kolmogorov-Smirnov test is exact when testing for a specified *probability* distribution function, it is applicable to testing for a specified *spectral* distribution function *only* when n is large.

One subtle refinement which periodogram testing offers is the ability to test for positive and negative correlation. When a sequence contains positive correlation, successive events are likely to assume more closely clustered values than if the sequence contained zero or negative correlation. This propensity to bunch produces long, irregular cycles in the sequence so that the corresponding spectrum shows considerably more variance attributable to low frequency (long period) contributions than to high frequency (short period) contributions. Since the true spectral distribution function for λ_k would exceed the uniform function of k/K, a natural statistic to consider for a test of excessive low frequency and, consequently, positive correlation is

$$c^+ = \max_k \left(S_k - \frac{k}{K} \right). \tag{7.27}$$

Critical values for this one-sided Kolmogorov-Smirnov test statistic are given in reference 113.

If a sequence contained negative correlation between events one would expect considerably more oscillation around its mean than would occur for zero or positive correlation. In this case the variance concentration would not occur at the low frequency end of the spectrum, and consequently the theoretical spectral distribution function for frequency λ_k would be below k/K. A natural test statistic is then

$$c^- = \min_k \left(S_k - \frac{k}{K} \right) = \max_k \left(\frac{k}{K} - S_k \right)$$

the critical values for which are also given in reference 113.

EXERCISES

1. You are given a generator that produces the numbers 0.1(0.1)1, each with equal probability. You wish to generate exponential deviates with $\beta = 1$.
   a. Plot the resulting "exponential" probability *mass* function and the theoretical exponential probability density function and compare them.
   b. Plot the resulting "exponential" cumulative probability distribution and the theoretical exponential cumulative probability distribution and compare them.

2. You are told that instead of the numbers 0.1(0.1)1 in Exercise 1 you can design a generator that produces on each trial one of any ten specified numbers in (0, 1), each with equal probability.
   a. Discuss how you would go about specifying the numbers from which you could sample so as to produce "better" exponential deviates. The graphs in Exercises 1a and 1b should be helpful.
   b. Would you prefer the generator in Exercise 1a to a generator that produces the numbers 0.05(0.05)1, each with equal probability? Explain your answer.

3. A generator produces deviates with the probability density function

$$f_X(x) = \begin{cases} 1/(a - b + 1) & 0 \le x \le a \\ 1/(a - b + 1) & b \le x \le 1 \\ 0 & \text{elsewhere} \end{cases} \qquad b > a > 0.$$

   a. Describe a method for producing uniform deviates on (0, 1).
   b. Would the method work for the following?

$$f_X(x) = \begin{cases} c_1/(a - b + 1) & 0 \le x \le a \\ c_2/(a - b + 1) & b \le x \le 1 \\ 0 & \text{elsewhere} \end{cases} \qquad b > a > 0,\ c_1 \ne c_2$$

   If not, describe one that would.

4. For a full period mixed congruential pseudorandom number generator

$$Z_1 \equiv aZ_{i-1} + c \pmod{m} \qquad a, c < m$$

   requires that c be relatively prime to m, $a \equiv 1 \pmod{q}$ for each prime factor q of m, and $a \equiv 1 \pmod{4}$ if 4 is a factor of m.
   a. Determine a, c, and m, intended to produce the digits 0 through 9.
   b. Starting with $Z_0 = 0$, generate Z_1 through Z_{10}.

c. The quantity

$$\rho = \frac{10 \sum_{i=1}^{10} Z_{i-1}Z_i - (\sum_{i=1}^{10} Z_i)^2}{10 \sum_{i=1}^{10} Z_i^2 - (\sum_{i=1}^{10} Z_i)^2}$$

provides one sample measure of the correlation between successive variates. Compute this quantity for the stream generated in 4b and interpret the result.

5. Consider the multiplicative congruential generator

$$Z_i \equiv aZ_{i-1} + c \pmod{m}$$

with $m = 10^3$, $a = 101$, $c = 1$, and $Z_0 = 98$.

a. Compute Z_1 and Z_2, using shifts and additions only.

b. Show that this generator has full period.

6. Consider the following sequence of 48 digits:

5	9	5	8	0	0	6	7	8	7	5	5	6	9	7	1
8	8	0	0	8	8	8	3	5	5	4	4	8	6	2	7
6	0	7	9	0	1	8	1	5	7	6	2	1	1	1	6

where the order goes from left to right.

a. Compute the Kolmogorov-Smirnov statistic and test the hypothesis that the sequence is drawn from a uniform distribution using a test size of $\alpha = 0.05$.

b. Using the serial test with $r = 2$, test the hypothesis at $\alpha = 0.05$ that triplets are composed of independent events.

c. Using the serial test with $r = 2$, test the hypothesis at $\alpha = 0.05$ that pairs are composed of independent events.

d. Under the hypothesis in Exercise 6c what is the probability of exceeding the number given by the test statistic?

7. *Monte Carlo Sampling:* Let U be a uniform variate on (0, 1) and let $X = g(U)$. Then

$$E(X) = \int_0^1 g(u)\, du.$$

For a sequence $U_1, \ldots, U_n$ of uniformly distributed variates we have

$$X_i = g(U_i), \qquad \bar{X}_n = \frac{1}{n} \sum_{i=1}^{n} g(U_i)$$

which has expectation

$$E(\bar{X}_n) = \int_0^1 g(u)\, du$$

and variance inversely proportional to n. We wish to evaluate the expression

$$A = \int_0^\infty ye^{-y}\, dy.$$

a. Rewrite A in terms of an integral over (0, 1).
b. Draw five uniform deviates and estimate A and the variance of the estimate.
c. Draw an additional five uniform deviates and combine them with the first five to estimate A and the variance of the estimate. Generate deviates with at least five digits. Students should use independent streams of random numbers.
d. Draw an additional ten uniform deviates and combine them with the first ten to estimate A and the variance of the estimate.
e. List your results for Exercise 7a, b, and c in a table with the corresponding theoretical answers.

8. *Estimation of* π: Consider the unit square $0 \leq x \leq 1$, $0 \leq y \leq 1$ and the circle

$$(x - \tfrac{1}{2})^2 + (y - \tfrac{1}{2})^2 = 1$$

superimposed on it. The square has unit area and the circle has area $\pi/4$. Suppose that we generate 2-tuples of independent uniform deviates. Then the probability that a 2-tuple falls in the circle is $\pi/4$. Using this fact, estimate π with each of the following:
a. Ten 2-tuples.
b. Twenty 2-tuples.
c. Fifty 2-tuples.
Generate deviates with at least five digits. Students should use independent streams of random numbers and should remember the value of information when presenting results.

CHAPTER 8

STOCHASTIC VARIATE GENERATION

8.1 INTRODUCTION

Having established in Chapter 7 that stochastic behavior in most simulation models can be generated by transformations on independent, uniformly distributed random numbers on (0, 1), we now present the algorithms for effecting these transformations. All the methods described here are theoretically exact. Occasionally the penalty for exactness is loss of computational efficiency. This potential loss, however, is seldom a critical issue with the current generation of computers.

In practice a user often finds that the programming language he chooses provides subroutines for generating random variables with specified probability laws. This is true of SIMULA and SIMSCRIPT II. In such cases the discussion in this chapter is academic except where the programming language offers an approximating method, for example, for the gamma distribution with nonintegral shape parameter.* When a subroutine employs an approximating method, a user may wish to substitute a subroutine based on a correspondingly exact method offered here.

As discussed in Chapter 5 GPSS does not permit subroutines for generating random variables as in SIMSCRIPT II and SIMULA. However, we emphasize that many algorithms described in this chapter can be written into GPSS programs, provided that the parameters of the distribution are numerically specified and that appropriate function tables are included. Exercises 10, 11, and 12 provide opportunities for the reader to write such programs.

* See reference 132, pp. 88–89, for a description of a commonly employed approximate method for gamma variate generation.

In addition to the inverse transformation and independent Bernoulli trials methods mentioned in Section 7.1, the *rejection* method also occurs in practice. Here repeated independent sampling without replacement is conducted until a specified condition is met. The distribution of the desired variate determines the condition and also the transformation on the generated uniform deviates that produce the actual variate. Since the distribution of the number of trials for success is easily seen to follow the geometric distribution, all one has to know is the probability that the condition is met on a single trial to compute the mean number of trials.

A variety of situations exists in simulation experiments where an investigator may wish to introduce stochastic events. In a queueing problem interarrival times and service times are often stochastic. In an inventory model the times between demands and the pipeline time to fill each demand may be stochastic. In reliability models the times to failure may be stochastic. In these examples the investigator is interested in generating nonnegative times, and he would like the prerogative of choosing from among alternative distributions the one or ones which meet his requirements.

To generate times one often uses the exponential, gamma, and Weibull distributions. The exponential is a special case of both the gamma and Weibull distributions. The principal differences among these distributions involve the locations of the modes of the p.d.f.s and the shapes of their tails for large times. In particular, the exponential has its mode at the origin, whereas the gamma and Weibull have theirs at positive values which are functions of their parameter values. The tail of the gamma distribution is exponential, whereas the tail of the Weibull declines more rapidly than that of an exponential. The Weibull has been used extensively in studying failure distributions.

We emphasize that the methods generally available for generating these times rest on the assumption that the times are independent. For example, the time to failure of an item in a reliability maintenance model is assumed to be independent of the service time necessary to repair the item. In reality, this assumption may often be tenuous. In Section 8.3 we describe a method for generating events in pairs from particular bivariate exponential and bivariate gamma distributions.

In an inventory model a demand may represent a request for several units of a particular item. The number of units demanded, X, may also be drawn from one of several probability distributions. The geometric, Poisson, and negative binomial distributions provide a range of distributional shapes that satisfy a variety of demand patterns. The geometric distribution, which is a special case of the negative binomial, has its mode at $X = 1$, given that at least one demand must occur. The negative binomial provides a long-tail distribution which often characterizes demand data. The Poisson, whose tail is generally shorter than that of the negative binomial, is a limiting form of the latter distribution.

Figure 42 shows a demand pattern which combines stochastic demand times with stochastic order sizes. When demand cannot exceed an upper limit n the binomial distribution is sometimes used. Truncated forms of the above distributions can be used alternatively to effect the appropriate distributional shape.

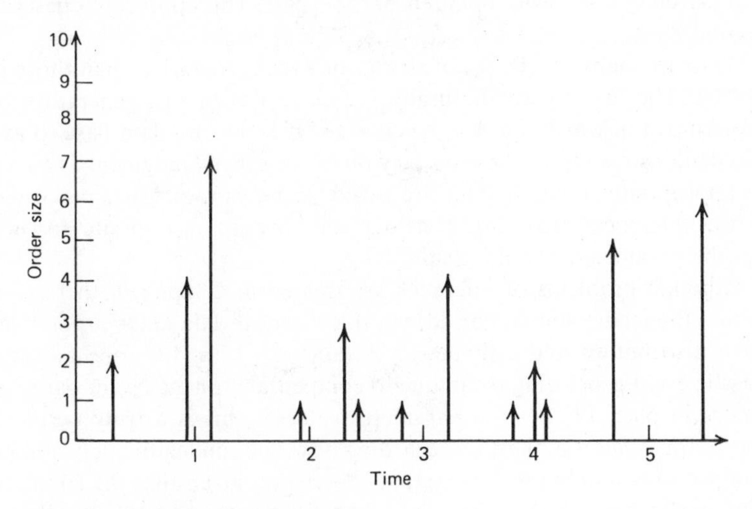

Fig. 42 Item demand through·time.

Several other distributions are useful in discrete stochastic event simulations. It is often desirable to sample a random variable from an interval (a, b). The beta distribution is helpful here by providing a variety of distributional forms on the unit interval which, with appropriate modification, can be shifted to (a, b). The uniform distribution is a special case of the beta distribution. The beta distribution is also helpful in deriving an exact method of generating gamma variates with nonintegral shape parameters.*

The hypergeometric distribution often plays a role in sampling problems. In particular, in drawing a sample of size X without replacement from a population of Y items, Z of which have a particular attribute, the distribution of the number of items in the sample with the attribute is hypergeometric.

In some experiments the investigator may choose to advance the state of the simulation at periodic intervals Δt. The geometric, Poisson, and

* See Section 8.2.5.

negative binomial distributions can then be used, for example, to generate the number of items demanded in an inventory model in a time interval Δt. More often the choice of periodic intervals characterizes econometric simulations. Occasionally, the normal and log normal distributions suffice to provide the stimuli to produce activity in an econometric model. Sometimes the stimulus consists of a sequence of independent shocks. At other times the investigator prefers the shocks to be autocorrelated. One approach to generating autocorrelated behavior is to use an autoregressive scheme which is itself driven by a sequence of independent events. This approach is described in Section 8.6.

There are many more uses of stochastic event generation than those just presented. The investigator naturally is free to design this generation into his experiment in whichever way he chooses. If he has no data base to assist him in defining his inputs, he must rely on his or others' judgment as to what is an appropriate input. If data are available he can, at least, make some statistical inferences regarding what distributions are appropriate and what values the parameters should assume.

Although problems of inference are treated in Chapter 9, there is one issue that the reader should consider at this point. If data make it possible to select a distribution and estimate its parameters, how does one adjust his stochastic event generating technique to compensate for the use of parameter estimates in place of the true parameter values? This is a truly perplexing problem, for it raises a question regarding how one can legitimately compare the output of a simulation with actual experience, given that the simulation output is influenced by the substitution of estimates for true values. Unfortunately a solution to this problem remains to be found.

8.2 CONTINUOUS DISTRIBUTIONS

In this section we present methods of generating stochastic events from a variety of continuous univariate distributions. Table 21 lists the information that characterizes each distribution, and it is these parametric representations that are used throughout the rest of the study.

8.2.1 Uniform Distribution

Suppose that X has the uniform distribution

$$f_X(x) = \begin{cases} \dfrac{1}{b-a} & a \leq x \leq b \\ 0 & \text{elsewhere,} \end{cases} \tag{2.1}$$

Table 21 Continuous Distributions

Name	Symbol	Domain	Restriction	Mean	Variance	Mode
Beta	$\mathscr{Be}(a, b)$	$0 \le x \le 1$	$a,\ b > 0$	$\frac{a}{a+b}$	$\frac{ab}{(a+b+1)(a+b)^2}$	$\frac{a-1}{a+b},\quad a \ge 1$
Cauchy	$\mathscr{C}(\alpha, \beta)$	$-\infty \le x \le \infty$	$\beta > 0$	Undefined	∞	α
Chi-square	$\chi^2(n)$	$0 \le x \le \infty$	$n = 1, 2, \ldots$	n	$2n$	$n - 2$
Exponential	$\mathscr{E}(\beta)$	$0 \le x \le \infty$	$\beta > 0$	β	β^2	0
F	$\mathscr{F}(m, n)$	$0 \le x \le \infty$	$m, n = 1, 2, \ldots$	$\frac{n}{n-2},\quad n > 2$	$\frac{2m(m+n-2)}{m(n-4)(n-2)^2},\quad n > 4$	$\frac{n(m-2)}{m(n+2)},\quad m > 2$
Gamma	$\mathscr{G}(\alpha, \beta)$	$0 \le x \le \infty$	$\alpha,\ \beta > 0$	$\alpha\beta$	$\alpha\beta^2$	$(\alpha - 1)\beta,\quad \alpha > 1$
Logistic	$\mathscr{L}(\alpha, \beta)$	$-\infty \le x \le \infty$	$\beta > 0$	α	$\frac{(\beta\pi)^2}{3}$	α
Lognormal	$\mathscr{L}\mathscr{N}(\mu, \sigma^2)$	$0 \le x \le \infty$	$\sigma^2 > 0$	μ	σ^2	$e^{\mu - \sigma^2}$
Normal	$\mathscr{N}(\mu, \sigma^2)$	$-\infty \le x \le \infty$	$\sigma^2 > 0$	μ	σ^2	μ
Student t	$t(n)$	$-\infty \le x \le \infty$	$n = 1, 2, \ldots$	0	$\frac{n}{n-2},\quad n > 2$	0
Triangular	$\mathscr{T}(a, b, c)$	$a \le x \le c$	$c > b > a$	$\frac{a+b+c}{3}$	$\frac{a(a-b) + c(c-a) + b(b-c)}{18}$	b
Uniform	$\mathscr{U}(a, b)$	$a \le x \le b$	$b > a$	$\frac{a+b}{2}$	$\frac{b-a}{12}$	(a, b)
Weibull	$\mathscr{W}(\alpha, \beta)$	$0 \le x \le \infty$	$\alpha,\ \beta > 0$	$\beta\Gamma\left(\frac{1}{\alpha} + 1\right)$	$\beta^2\left[\Gamma\left(\frac{1}{\alpha} + 2\right) - \Gamma^2\left(\frac{1}{\alpha} + 1\right)\right]$	$\left(\frac{\alpha - 1}{\alpha}\right)^{\alpha}\beta,\quad \alpha \ge 1$

so that

$$F_X(x) = \int_a^x \frac{du}{a - b} = \frac{x - a}{b - a}. \tag{2.2}$$

We denote this distribution by $\mathcal{U}(a, b)$. The random variable U or U_j is always taken to have distribution $\mathcal{U}(0, 1)$. From (7.1.2) we have that

$$U = F_X(X) = \frac{X - a}{b - a} \tag{2.3}$$

so that

$$X = a + (b - a)U. \tag{2.4}$$

To compute X, one generates U and uses (2.4).

8.2.2 Triangular Distribution

A triangular variate X has p.d.f.

$$f_X(x) = \begin{cases} \dfrac{2(x - a)}{(b - a)(c - a)} & a \le x \le b \\[2ex] \dfrac{2(c - x)}{(c - b)(c - a)} & b \le x \le c \\[2ex] 0 & \text{elsewhere} \end{cases} \tag{2.5}$$

so that

$$F_X(x) = \begin{cases} \dfrac{(x - a)^2}{(b - a)(c - a)} & a \le x \le b \\[2ex] \dfrac{b - a}{(c - a)} + \dfrac{(x - b)^2}{(c - b)(c - a)} & b \le x \le c. \end{cases} \tag{2.6}$$

We denote the triangular distribution by $\mathcal{T}(a, b, c)$. Then we generate a triangular variate X from a uniform variate U by

$$X = \begin{cases} a + \sqrt{(b - a)(c - a)U} & 0 \le U \le \dfrac{b - a}{c - a} \\[2ex] b + \sqrt{[(c - a)U - (b - a)](c - b)} & \dfrac{b - a}{c - a} < U \le 1. \end{cases} \tag{2.7}$$

8.2.3 Exponential Distribution

We review the generation of an exponential variate given in Section 7.1. Let X have p.d.f.

$$(2.8)\qquad f_X(x) = \begin{cases} \frac{1}{\beta} e^{x/\beta} & 0 \le x \le \infty \\ 0 & x < 0, \end{cases}$$

and distribution function

$$(29.)\qquad F_X(x) = \frac{1}{\beta} \int_0^x e^{-u/\beta}\, du = 1 - e^{-x/\beta}.$$

We then say X is from $\mathscr{E}(\beta)$. We generate a random number U and set

$$(2.10)\qquad U = F_X(X) = 1 - e^{-X/\beta}$$

so that

$$(2.11)\qquad X = -\beta \log(1 - U).$$

Since U is a uniform variate, it is easily seen that $1 - U$ is also a uniform variate. Hence we may save a step by choosing X as

$$(2.12)\qquad X = -\beta \log(U).$$

8.2.4 Gamma Distribution with Integral Shape Parameter

Let X have p.d.f.

$$(2.13)\qquad f_X(x) = \begin{cases} \frac{1}{\Gamma(\alpha)\beta^\alpha} e^{-x/\beta} x^{\alpha - 1} & 0 \le x \le \infty \\ 0 & x < 0. \end{cases}$$

Then X is a gamma variate from $\mathscr{G}(\alpha, \beta)$ with corresponding characteristic function

$$(2.14)\qquad \psi(s) = (1 - i\beta s)^{-\alpha}.$$

Let k be the largest integer in α, and let

$$(2.15)\qquad \gamma = \alpha - k$$

so that

$$(2.16)\qquad \psi(s) = (1 - i\beta s)^{-\gamma} \prod_{j=1}^{k} (1 - i\beta s)^{-1}.$$

From elementary probability theory we know that the sum of $k + 1$ independent random variables has a characteristic function which is the product of the $k + 1$ characteristic functions of the independent random variables. Hence we may regard X as the sum of $k + 1$ variables $X_1, \ldots, X_{k+1}$, where $X_1, \ldots, X_k$ are from $\mathscr{G}(1, \beta)$ and X_{k+1} is from $\mathscr{G}(\gamma, \beta)$. If α is an integer, so that $\gamma = 0$, then X is the sum of $X_1, \ldots, X_k$ gamma variates, each with parameters 1 and β. For now we assume α to be an integer. We return to the noninteger case in Section 8.2.6.

Since a gamma variate $\mathscr{G}(1, \beta)$ is easily seen to be $\mathscr{E}(\beta)$ we can derive a gamma variate with parameters α (integer) and β by generating α exponential variates as in (2.11) and summing them. A more efficient computation yielding an equivalent result can be accomplished by using

$$X = -\beta \log\left(\prod_{j=1}^{\alpha} U_j\right). \tag{2.17}$$

8.2.5 Beta Distribution

Let X have p.d.f.

$$f_X(x) = \begin{cases} \dfrac{\Gamma(a+b)}{\Gamma(a)\Gamma(b)} x^{a-1}(1-x)^{b-1} & 0 \leq x \leq 1 \\ 0 & \text{elsewhere.} \end{cases} \tag{2.18}$$

Then X has the beta distribution, which we denote as $\mathscr{Be}(a, b)$. There are several ways of generating beta variates exactly. The choice of the appropriate method depends on the values of the parameters of the distribution, "appropriate" denoting computational efficiency.

We first note the relationship between the gamma and beta distributions. Consider the independent gamma variates X_1 and X_2 with p.d.f.s

$$\begin{aligned} f_{X_1}(x) &= \begin{cases} \dfrac{1}{\Gamma(a)} e^{-x} x^{a-1} & 0 \leq x \leq \infty \\ 0 & x < 0 \end{cases} \\ f_{X_2}(y) &= \begin{cases} \dfrac{1}{\Gamma(b)} e^{-y} y^{b-1} & 0 \leq y \leq \infty \\ 0 & y < 0, \end{cases} \end{aligned} \tag{2.19}$$

respectively. It is known that $X = X_1/(X_1 + X_2)$ is a beta variate with p.d.f. as in (2.18). To see this we observe that the joint p.d.f. of X_1 and X_2 is

$$f_{X_1,X_2}(x, y) = \begin{cases} \dfrac{1}{\Gamma(a)\Gamma(b)} e^{-(x+y)} x^{a-1} y^{b-1} & 0 \leq x, y < \infty \\ 0 & x, y < 0. \end{cases} \tag{2.20}$$

Let $S = X_1$ and $T = X_1 + X_2$. Then the joint p.d.f. of S and T is

$$f_{S,T}(s, t) = Jf_{X_1,X_2}(x, y)$$

$$(2.21) \qquad = \begin{cases} \dfrac{1}{\Gamma(a)\Gamma(b)} e^{-t} s^{a-1}(t-s)^{b-1} & 0 \le s, t \le \infty \\ 0 & \text{elsewhere} \end{cases}$$

where the Jacobian J is defined as

$$(2.22) \qquad J = \left\| \begin{matrix} \dfrac{\partial x}{\partial s} & \dfrac{\partial x}{\partial t} \\ \dfrac{\partial y}{\partial s} & \dfrac{\partial y}{\partial t} \end{matrix} \right\|.$$

Let $W = S$ and $X = S/T = X_1/(X_1 + X_2)$. Then W and X have joint p.d.f.

$$(2.23) \quad f_{W,X}(w, x) = \begin{cases} \dfrac{1}{\Gamma(a)\Gamma(b)} e^{-w/x} w^{a+b-1} x^{-(b+1)}(1-x)^{b-1} \\ \qquad\qquad 0 \le w \le \infty, 0 \le x \le 1 \\ 0 \qquad\qquad \text{elsewhere.} \end{cases}$$

Integrating (2.23) with respect to w gives us $f_X(x)$, which is of the form (2.18).

If a and b are integers we may generate a beta variate in the following way: First generate $U_1, \ldots, U_{a+b}$ and then form

$$(2.24) \qquad \begin{aligned} X_1 &= -\log\left(\prod_{j=1}^{a} U_j\right) \\ X_2 &= -\log\left(\prod_{j=a+1}^{a+b} U_j\right) \\ X &= \frac{X_1}{X_1 + X_2}, \end{aligned}$$

X being the desired variate.

An alternative approach when a and b are integers is based on the theory of order statistics [126]. Suppose that we generate $U_1, \ldots, U_{a+b-1}$ and order them so that

$$(2.25) \qquad \begin{aligned} U_{(1)} &= \min_j (U_j) \\ &\vdots \\ U_{(a+b-1)} &= \max_j (U_j). \end{aligned}$$

Then the ath-order statistic $U_{(a)}$ is from $\mathscr{B}e(a, b)$. To see this we first note that there are $\binom{a+b-1}{a-1}\binom{b}{b-1}$ ways of grouping $a - 1$ of the U_js in one subgroup and $b - 1$ of the U_js in another subgroup. Then the probability that for one such arrangement each of the $a - 1$ U_js in the first subgroup is less than v and each of the $b - 1$ U_js in the second subgroup is greater than $v + dv$ is

$$(2.26) \qquad \left(\int_0^v du\right)^{a-1}\left(\int_v^{v+dv} du\right)\left(\int_{v+dv}^1 du\right)^{b-1} \sim v^{a-1}(1-v)^{b-1}\, dv.$$

But this is the probability that the U_j in neither group or, equivalently, the ath smallest U_j lies in the interval $(v, v + dv)$. Since there are

$$(2.27) \qquad \binom{a+b-1}{a-1}\binom{b}{b-1} = \frac{\Gamma(a+b)}{\Gamma(a)\Gamma(b)}$$

such arrangements, the p.d.f. of the ath-order statistic $U_{(a)}$ is

$$(2.28) \qquad f_{U_{(a)}}(v) = \begin{cases} \dfrac{\Gamma(a+b)}{\Gamma(a)\Gamma(b)} v^{a-1}(1-v)^{b-1} & 0 \le v \le 1 \\ 0 & \text{elsewhere,} \end{cases}$$

which is identical in form to (2.18).

To use this result for generating a beta variate with a and b integers, we generate $U_1, \ldots, U_{a+b-1}$ numbers and then find $U_{(a)}$. This requires us to find $U_{(1)}, \ldots, U_{(a-1)}$. For $U_{(j)}$ we make $a + b - 1 - j$ comparisons. Hence the total number of comparisons needed to find $U_{(a)}$ is

$$(2.29) \qquad \sum_{j=1}^{a} (a + b - 1 - j) = \frac{a}{2}(a + 2b - 3).$$

This contrasts with the two logarithmic transformations required in the more commonly employed gamma variate approach. No statistics on comparative efficiencies are known to the writer.

Occasionally either a or b is nonintegral, so that the above method for generating a beta variate do not apply. We now describe a *rejection* method that applies for all a and b and incidentally facilitates the generation of a gamma variate with nonintegral shape parameter as discussed in Section 8.2.6.* Consider the transformations

$$(2.30) \qquad Y = U_1^{1/a}, \qquad Z = U_2^{1/b}$$

* I am grateful to Bennet Fox of the University of Montreal, who called this approach in reference 130 to my attention. It is also described in reference 127.

on the independent uniform variates U_1 and U_2, respectively. If $Y + Z \leq 1$, then $X = Y/(Y + Z)$ has the beta distribution with p.d.f. as in (2.18). To see this we note that

$$f_Y(y) = \begin{cases} ay^{a-1} & 0 \leq y \leq 1 \\ 0 & \text{elsewhere} \end{cases}$$

$$f_Z(z) = \begin{cases} bz^{b-1} & 0 \leq z \leq 1 \\ 0 & \text{elsewhere} \end{cases} \tag{2.31}$$

$$f_{Y,Z}(y, z) = \begin{cases} aby^{a-1}z^{b-1} & 0 \leq y, z \leq 1 \\ 0 & \text{elsewhere.} \end{cases}$$

Let $X = Y/(Y + Z)$ and $W = Y + Z$. Then

$$f_{X,W}(x, w) = \begin{cases} abx^{a-1}(1 - x)^{b-1}w^{a+b-1} & 0 \leq x \leq 1, 0 \leq w \leq 2 \\ 0 & \text{elsewhere.} \end{cases} \tag{2.32}$$

It is easily seen that

$$f_{X,W}(x, \; 0 \leq W \leq 1) = \int_0^1 f_{X,W}(x, w)\, dw$$

$$= \begin{cases} \dfrac{ab}{a + b} x^{a-1}(1 - x)^{b-1} & 0 \leq x \leq 1 \\ 0 & \text{elsewhere} \end{cases} \tag{2.33}$$

$$\text{Prob}(0 \leq W \leq 1) = \int_0^1 f_{X,W}(x, \; 0 \leq W \leq 1)\, dx = \frac{ab}{a + b} \frac{\Gamma(a)\Gamma(b)}{\Gamma(a + b)}$$

so that

$$f_X(x \mid 0 \leq W \leq 1) = \frac{f_{X,W}(x, \; 0 \leq W \leq 1)}{\text{Prob}(0 \leq W \leq 1)}$$

$$= \begin{cases} \dfrac{\Gamma(a + b)}{\Gamma(a)\Gamma(b)} x^{a-1}(1 - x)^{b-1} & 0 \leq x \leq 1 \\ 0 & \text{elsewhere.} \end{cases} \tag{2.34}$$

The generation scheme in Fig. 43 is based on reference 127. Since the number of trials for success follows the geometric distribution, the expected number of U_j required to generate a $\mathcal{B}e(a, b)$ variate is $2(a + b)\Gamma(a + b)/ab\Gamma(a)\Gamma(b)$.* For example, for $a = b = 3$ the expected number is 40. For

* See reference 130.

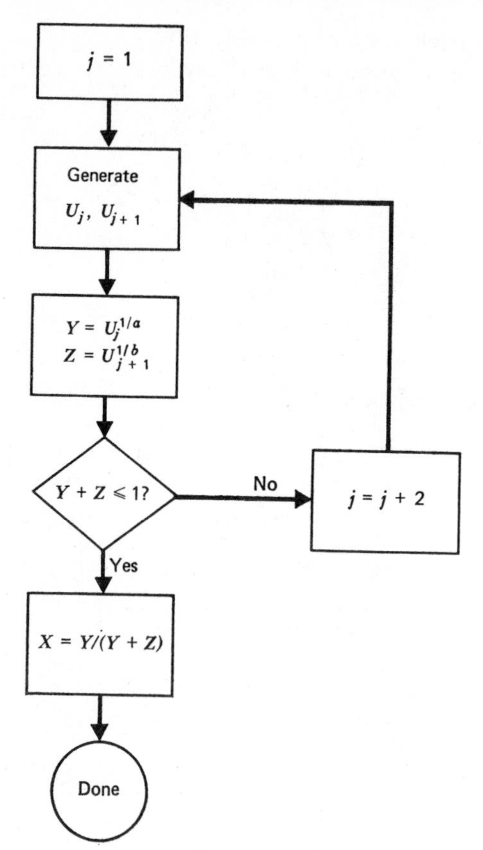

Fig. 43 Beta variate generation from $\mathscr{B}e(a, b)$.

$a = b = 4$ it is 140. Clearly when a and b are large integers the method based on the ratio of gamma variates is to be preferred.

An improvement in efficiency for nonintegral a and b can be realized by using the algorithm for generating a gamma variate with nonintegral shape parameter as described next.

8.2.6 Gamma Distribution with Nonintegral Shape Parameter

Suppose that X is from $\mathscr{G}(\alpha, \beta)$ with α nonintegral. We may consider X to be the sum of $k + 1$ independent gamma variates, all with scale parameter β, but the first k of which have unit shape parameter and the $k + 1$st has shape parameter $\gamma = \alpha - [\alpha]$. Let Y and Z be independent variates from $\mathscr{B}e(\gamma,$

$1 - \gamma)$ and $\mathscr{G}(1, 1)$, respectively. Then $W = \beta YZ$ is a variate with $\mathscr{G}(\gamma, \beta)$. To see this we note that

$$f_{Y,Z}(y, z) = \begin{cases} \dfrac{1}{\Gamma(\gamma)\Gamma(1-\gamma)} y^{\gamma-1}(1-y)^{-\gamma}e^{-z} & 0 \le y \le 1, 0 \le z \le \infty \\ 0 & \text{elsewhere.} \end{cases} \tag{2.35}$$

Then

$$f_{W,Z}(w, z) = \begin{cases} \dfrac{\beta^{-\gamma}}{\Gamma(\gamma)\Gamma(1-\gamma)} w^{\gamma-1}\left(z - \dfrac{w}{\beta}\right)^{-\gamma} e^{-z} & 0 \le w, z \le \infty \\ 0 & \text{elsewhere.} \end{cases} \tag{2.36}$$

$$f_W(w) = \int_0^\infty f_{W,Z}(w, z)\, dz = \begin{cases} \dfrac{\beta^{-\gamma}}{\Gamma(\gamma)} w^{\gamma-1}e^{-w/\beta} & 0 \le w \le \infty \\ 0 & w < 0, \end{cases} \tag{2.37}$$

which is $\mathscr{G}(\gamma, \beta)$.

We generate a variate from $\mathscr{G}(\alpha, \beta)$ as follows:

1. compute k as the largest integer in α,
2. compute $\gamma = \alpha - k$,
3. generate k variates from $\mathscr{E}(1)$ and compute V as their sum,
4. generate variate Y and Z from $\mathscr{Be}(\gamma, 1 - \gamma)$ and $\mathscr{E}(1)$, respectively,

and

5. compute the gamma variate $X = \beta(V + YZ)$, which is from $\mathscr{G}(\alpha, \beta)$.

The expected number of U_j needed to generate a variate from $\mathscr{G}(\alpha, \beta)$ in this way is approximately $k + 3$.* These steps are shown in Fig. 44.

8.2.7 Beta Distribution with Nonintegral Parameters

In Section 8.2.5 we described a method for generating $\mathscr{Be}(a, b)$ for all nonnegative a and b. The method generates $2[(a + b)\Gamma(a + b)/ab\Gamma(a)\Gamma(b)]$ U_js on average and clearly can be time consuming for large a and b. Here we describe an alternative method applicable when either a or b is nonintegral or both are. We proceed as follows:

1. compute k_1 and k_2 to be the largest integers in a and b, respectively,
2. compute $\gamma_1 = a - k_1$ and $\gamma_2 = b - k_2$,

* See reference 130.

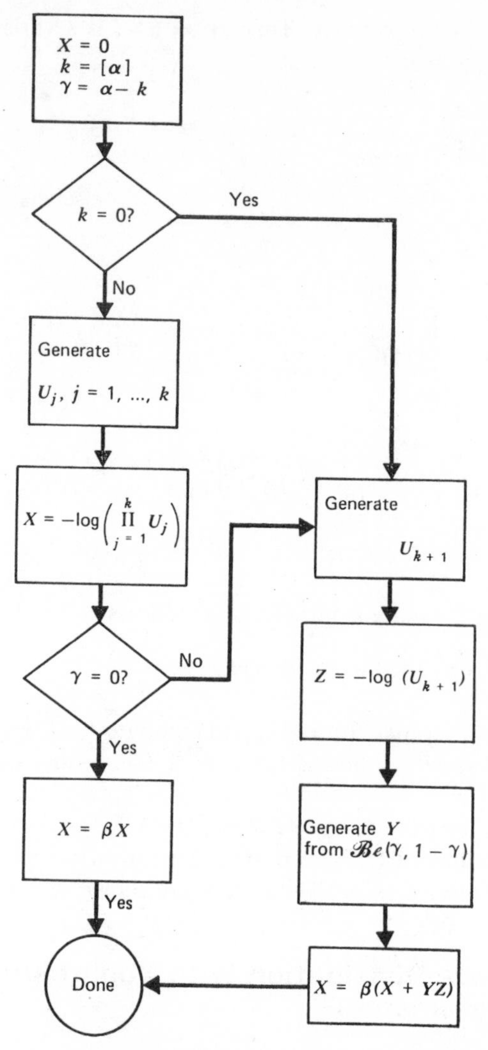

Fig. 44 Gamma variate generation from $\mathscr{G}(a, b)$. The quantity $[\alpha]$ denotes the largest integer in α.

3. generate k_1 variates from $\mathscr{E}(1)$ and compute X_1' as their sum,
4. generate k_2 variates from $\mathscr{E}(1)$ and compute X_2' as their sum,
5. generate four variates Y_1, Y_2, Z_1, and Z_2 from $\mathscr{Be}(\gamma_1, 1 - \gamma_1)$, $\mathscr{Be}(\gamma_2, 1 - \gamma_2)$, $\mathscr{E}(1)$, and $\mathscr{E}(1)$, respectively, and
6. compute the beta variate $X = (X_1' + Y_1Z_1)/(X_1' + X_2' + Y_1Z_1 + Y_2Z_2)$, which is from $\mathscr{Be}(a, b)$.

The expected number of U_js is approximately $k_1 + k_2 + 6$, which compares favorably with the $2[(a + b)\Gamma(a + b)/ab\Gamma(a)\Gamma(b)]$ expected number of steps in Section 8.2.5. The steps are shown in Figs. 44 and 45.

The beta distribution can also be used to generate a random variate Y over any interval (a_1, a_2). Suppose that X is from $\mathscr{B}e(a, b)$. Then to translate and dilate the interval from (0, 1) to (a_1, a_2) we form

$$Y = a_1 + (a_2 - a_1)X. \tag{2.38}$$

The uniform distribution described in Section 8.2.1 is a special case.

8.2.8 Weibull Distribution

Let X be a Weibull variate from $\mathscr{W}(\alpha, \beta)$ so that

$$f_X(x) = \begin{cases} \dfrac{\alpha}{\beta^\alpha} x^{\alpha-1} e^{-(x/\beta)^\alpha} & 0 \le x \le \infty \\ 0 & \text{elsewhere} \end{cases} \tag{2.39}$$

$$F_X(x) = 1 - e^{-(x/\beta)\alpha}. \tag{2.40}$$

To generate X we note that

$$\begin{aligned} U &= 1 - e^{-(X/\beta)^\alpha} \\ X &= \beta[-\log(1 - U)]^{1/\alpha}. \end{aligned} \tag{2.41}$$

Since $1 - U$ is also from $\mathscr{U}(0, 1)$ we use

$$X = \beta(-\log U)^{1/\alpha} \tag{2.42}$$

to compute X. Notice that $X^{1/\alpha}$ is an exponential variate from $\mathscr{E}(\beta^{-1/\alpha})$.

8.2.9 Normal Distribution

The generation of normal variates is based on the transformation of uniform deviates directly and not on the inverse transformation method described at the beginning of Chapter 7. Let U_1 and U_2 be independent deviates from $\mathscr{U}(0, 1)$. Then the variates [123]

$$\begin{aligned} X_1 &= (-2 \log U_1)^{1/2} \cos 2\pi U_2 \\ X_2 &= (-2 \log U_1)^{1/2} \sin 2\pi U_2 \end{aligned} \tag{2.43}$$

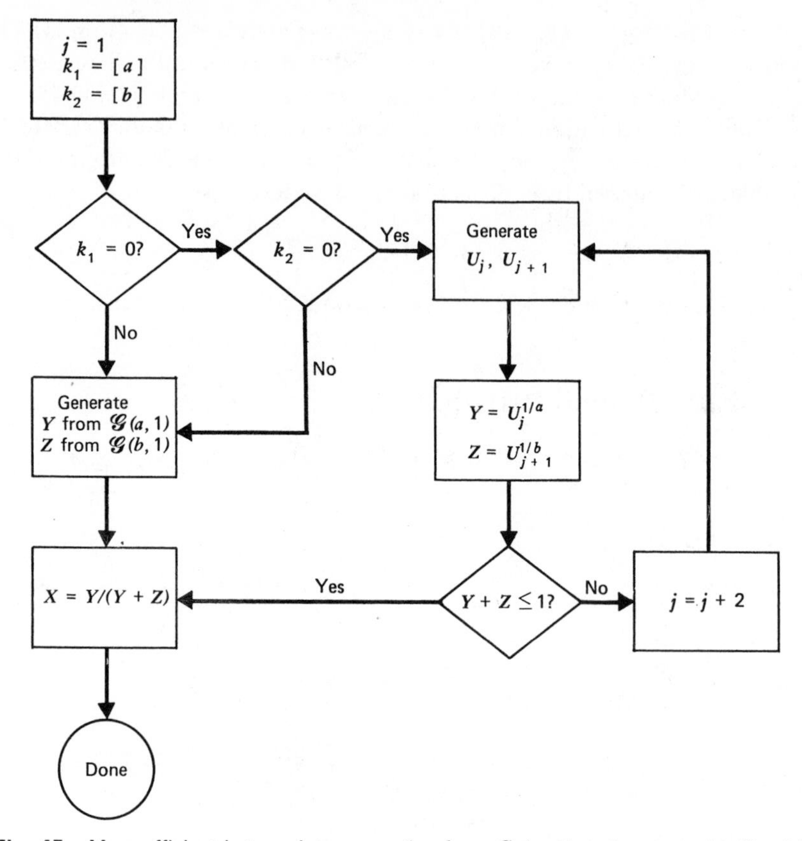

Fig. 45 More efficient beta variate generation from $\mathscr{B}e(a, b)$, to be used with Fig. 44.

are independent, and each is from $\mathcal{N}(0, 1)$. To see this we note that

$$f_{X_1,X_2}(x_1, x_2) = Jf_{U_1,U_2}(u_1, u_2) = J = \frac{1}{2\pi} e^{-(x_1^2+x_2^2)/2} \tag{2.44}$$

since

$$J = \left\| \begin{matrix} -x_1 e^{-(x_1^2+x_2^2)/2} & -x_2 e^{-(x_1^2+x_2^2)/2} \\ \dfrac{-x_2}{2\pi(x_1^2 + x_2^2)} & \dfrac{x_1}{2\pi(x_1^2 + x_2^2)} \end{matrix} \right\|. \tag{2.45}$$

The joint p.d.f. in (2.44) is that of two independent normal deviates, each with zero mean and unit variance.

To generate variates from $\mathcal{N}(\mu, \sigma^2)$ we generate U_j and U_{j+1} and compute

$$\left.\begin{aligned} X_j &= \mu + (-2\sigma^2 \log U_j)^{1/2} \cos 2\pi U_{j+1} \\ X_{j+1} &= \mu + (-2\sigma^2 \log U_j)^{1/2} \sin 2\pi U_{j+1} \end{aligned}\right\} \quad j = 1, 3, 5, \ldots. \tag{2.46}$$

8.2.10 The Chi-Square, *t*, and *F* Distributions

Let $Y_1, \ldots, Y_n$ be from $\mathcal{N}(0, 1)$. Then

$$X = \sum_{j=1}^{n} Y_j^2 \tag{2.47}$$

has the chi-square distribution with n degrees of freedom and is denoted by $\chi^2(n)$. One approach is to generate n normal variates as described in Section 8.2.9 and compute X as in (2.47). An alternative approach makes use of the fact that $\chi^2(n)$ is equivalent to $\mathcal{G}(n/2, 2)$. If n is even then X can be computed from (2.17) as

$$X = -2 \log\left(\prod_{j=1}^{n/2} U_j\right). \tag{2.48}$$

This approach requires $n/2$ U_js, compared to n for the normal approach. It also requires one logarithmic transformation, compared to n logarithmic and n cosine or sine transformations. If n is odd the method of Section 8.2.6 applies and should compare favorably with the normal approach.

Suppose that

$$T = \frac{X}{\sqrt{Z/n}} \tag{2.49}$$

where X and Z are independent random variables from $\mathcal{N}(0, 1)$ and $\chi^2(n)$, respectively. Then T is a t variate with n degrees of freedom, denoted by $t(n)$. It has mean zero and variance $n/(n - 2)$, $n > 2$. To generate T, one simply generates X as described in Section 8.2.9 and Z as described above and uses (2.49). The t distribution is symmetric about zero and has broader tails than the normal. The broader tails are a principal attraction in sampling problems in which an investigator feels that normal variates offer too limited a range of values. As in the normal case we can create a shifted variable

$$T' = \sigma T\sqrt{(n - 2)/n} + \mu \qquad n > 2 \tag{2.50}$$

so that T' has mean μ, σ^2, and distributional appearance similar to those of T.

Suppose that

$$X = \frac{Y/m}{Z/n} \tag{2.51}$$

where Y and Z are independent chi-square variates with m and n degrees of freedom, respectively. Then X has the F distribution with m and n degrees of freedom. To generate an F variate one first produces two chi-square variates and then uses (2.51).

8.2.11 Lognormal Distribution

Suppose that X is drawn from $\mathcal{N}(\mu_X, \sigma_X^2)$. Then $Y = e^X$ has the lognormal distribution with mean and variance

$$\begin{aligned} \mu_Y &= e^{\mu_X + \sigma_X^2/2} \\ \sigma_Y^2 &= e^{2\mu_X + \sigma_X^2}(e^{\sigma_X^2} - 1), \end{aligned} \tag{2.52a}$$

respectively, and is denoted by $\mathcal{LN}(\mu_Y, \sigma_Y^2)$. It is clear that $Y \geq 0$. To generate lognormal variates Y_1 and Y_2 we apply the exponential transformation to (2.43) so that

$$\begin{aligned} Y_1 &= \exp\{\mu_X + [-2\sigma_X^2 \log(U_1)]^{1/2} \cos 2\pi U_2\} \\ Y_2 &= \exp\{\mu_X + [-2\sigma_X^2 \log(U_1)]^{1/2} \sin 2\pi U_2\}. \end{aligned} \tag{2.52b}$$

Occasionally μ_Y and σ_Y^2 are specified instead of μ_X and σ_X^2. Then (2.52a) yields

$$\begin{aligned} \mu_X &= \tfrac{1}{2} \log\left(\frac{\mu_Y^4}{\sigma_Y^2 + \mu_Y^2}\right) \\ \sigma_X^2 &= \log\left(\frac{\sigma_Y^2}{\mu_Y^2} + 1\right) \end{aligned} \tag{2.53}$$

so that one may easily compute μ_X and σ_X^2 and use (2.52b) to form Y_1 and Y_2.

The lognormal distribution is often used in econometrics to characterize skewed economic data. It is also occasionally employed to represent time distributions in queueing problems. In addition to its skewness and non-negative features, the log normal distribution offers the convenience of having the maximum likelihood estimators of μ_Y and σ_Y^2 closely related to those of μ_X and σ_X^2.

8.3 BIVARIATE AND MULTIVARIATE DISTRIBUTIONS

8.3.1 Bivariate Exponential Distribution

Let X_1 and X_2 be independent from $\mathscr{G}(1 - \alpha, \beta)$ for $0 < \alpha < 1$, and X_3 be from $\mathscr{G}(\alpha, \beta)$. Then each of $Y_1 = X_1 + X_3$ and $Y_2 = X_2 + X_3$ is from $\mathscr{E}(\beta)$ and together have a correlation coefficient α. Proof of exponentiality follows from the reproductive property of gamma variates, as discussed in section 8.2.4. The correlation result follows from direct computation.

Here we specify only that the marginal p.d.f.s be exponential and that the correlation be α. However, in some problems the form of the joint p.d.f. is also specified. For example, when we consider the time to failure of two related components in a system, we would expect the failure time of the second component to be related to the time at which the first component fails [131]. Alternatively we can think of situations in which the repair time of a component is related to its time to failure [128]. In these references the marginal p.d.f.s are exponential, but the context of the problem specifies the form that the p.d.f. will assume. In such cases the generation scheme must be tailored to the specified joint p.d.f. as well as the marginal p.d.f.

8.3.2 Bivariate Gamma Distribution

Let X_1 and X_2 be independent and each from $\mathscr{G}(\alpha - \alpha_1, \beta)$, where $\alpha > \alpha_1$. Let X_3 be independent of X_1 and X_2 and from $\mathscr{G}(\alpha_1, \beta)$. Then each of $Y_1 = X_1 + X_2$ and $Y_2 = X_2 + X_3$ is from $\mathscr{G}(\alpha, \beta)$ and has correlation coefficient α_1/α. This result is also direct. Again the reader should note that this scheme produces variates with specified marginal gamma p.d.f.s and specified correlation. In contexts in which a joint p.d.f. implicitly or explicitly assumes a given form, the corresponding generation scheme must be consistent with the form.

8.3.3 Multivariate Normal Distribution

Let $\mathbf{X}' = (X_1, \ldots, X_m)$ be a multivariate normal random vector with mean $\boldsymbol{\mu}' = (\mu_1, \ldots, \mu_m)$ and covariance matrix

$$\boldsymbol{\Sigma} = \begin{bmatrix} \sigma_{11} & \sigma_{12} & \cdots & \sigma_{1m} \\ \sigma_{21} & \sigma_{22} & & \\ \vdots & & \ddots & \vdots \\ \sigma_{m1} & \cdots & & \sigma_{mm} \end{bmatrix}. \tag{3.1}$$

To generate $\mathbf{X}$ we make use of a fundamental theorem of multivariate statistical analysis which states that, if $\mathbf{Z}' = (Z_1, \ldots, Z_m)$ is multivariate normal with zero mean vector and covariance matrix

$$\mathbf{I} = \begin{bmatrix} 1 & 0 & \cdots & 0 \\ 0 & 1 & & \\ \vdots & & \ddots & \vdots \\ 0 & & \cdots & 1 \end{bmatrix}, \tag{3.2}$$

then $\mathbf{X}$ with mean vector $\boldsymbol{\mu}$ and covariance matrix $\boldsymbol{\Sigma}$ can be represented as

$$\mathbf{X} = \mathbf{CZ} + \boldsymbol{\mu} \tag{3.3}$$

where C is a unique lower triangular matrix satisfying [120, p. 19]

$$\boldsymbol{\Sigma} = \mathbf{CC}'. \tag{3.4}$$

The elements of $\mathbf{C}$ can be computed by the square root method given in reference 134 and presented in Fig. 46.

The generation of $\mathbf{X}$ now takes three steps: (1) the computation of $\mathbf{C}$ using Fig. 43, (2) the generation of m independent normal variates using (2.46), and (3) the application of (3.3).

8.4 DISCRETE DISTRIBUTIONS

Discrete random variables can be used to describe a variety of random phenomena in which counting on integers takes place. The inventory demand pattern is one example; the number of defective items in a lot is another. The distributions listed in Table 22 can be made to fit widely varying behavior by an appropriate choice of parameters.

At this point we wish to emphasize that we can usually apply the inverse transformation method to generate discrete variates without requiring the inverse transformation explicitly. Let

$$\text{Prob}(X = k) = p_{.}$$

$$p_{k+1} = A_{k+1} p_k$$

$$q_{k+1} = q_k + p_{k+1}.$$

The quantity A_{k+1} is distribution dependent. To generate a variate we proceed as follows:

1. generate U,
2. set $k = 0$,
3. compute p_0 and $q_0 = p_0$,
4. if $U \leq q_k$, $X = k$,

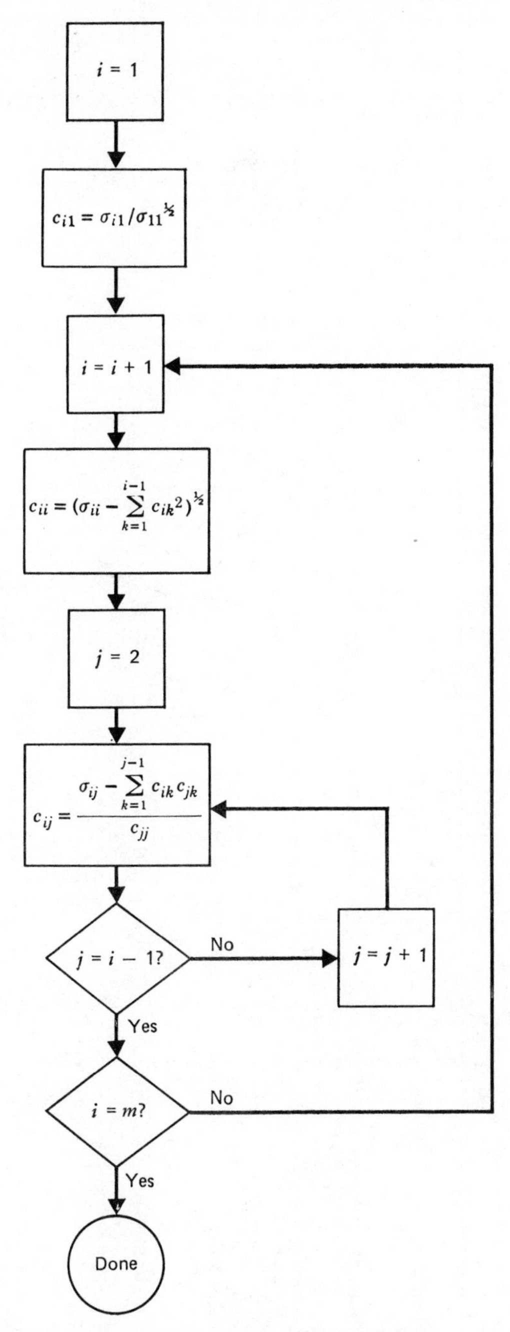

Fig. 46 Computation of lower triangular matrix C for generation of multivariate normal variates.

Table 22 Discrete Distributions

Distribution	Symbol	Domain	Restriction	Mean	Variance
Beta-binomial	$\mathscr{BB}(a, b, n)$	$a, b > 0; \quad n = 1, 2, \ldots$	$x = 0, ., \ldots, n$	$\frac{na}{a+b}$	$\frac{nab(n+a+b)}{(a+b)^2(a+b+1)}$
Binomial	$\mathscr{B}(n, p)$	$0 < p < 1; \quad n = 1, 2, \ldots$	$x = 0, 1, \ldots, n$	np	$np(1-p)$
Geometric	$\mathscr{G}e(p)$	$0 < p < 1$	$x = 0, ., \ldots, \infty$	$\frac{p}{1-p}$	$\frac{p}{(1-p)^2}$
Hypergeometric	(n_1, n_2, x)	$x = 1, \ldots, n_1 + n_2$	n_1, n_2 integer	$\frac{xn_1}{n_1+n_2}$	$\frac{xn_1n_2}{(n_1+n_2)^2}\left(\frac{n_1+n_2-x}{n_1+n_2-1}\right)$
Negative binomial	$\mathscr{NB}(r, p)$	$0 < p < 1, \quad r > 0$	$x = 0, 1, \ldots, \infty$	$\frac{rp}{1-p}$	$\frac{rp}{(1-p)^2}$
Poisson	$\mathscr{P}(\beta)$	$\beta > 0$	$x = 0, ., \ldots, \infty$	β	β
Uniform	$\mathscr{U}_d(a, b)$	$b > a$	$x = a, \ldots, b$ a and b integer	$\frac{b+a}{2}$	$\frac{(b-a)(b-a+2)}{12}$

5. otherwise, compute A_{k+1},
6. compute $p_{k+1} = A_{k+1} p_k$,
7. compute $q_{k+1} = q_k + p_{k+1}$,
8. compute $k = k + 1$,
9. go to step 4.

For a binomial variate from $\mathscr{B}(n, p)$ we must add a check for $k = n$ after step 4.

As we shall see when we come to binomial, Poisson, geometric, and negative binomial generation, the inverse transformation method essentially substitutes the computation of each set of A_k, p_k, and q_k for the generation of a uniform deviate required by the Bernoulli trial approach. A statement of the relative efficiencies of the two alternative methods seems to be lacking in general.

8.4.1 Discrete Uniform Distribution

Suppose that the probability of selecting any integer between r_1 and r_2 is equally likely, so that

$$f_X(x = r) = p \qquad r = r_1, \ldots, r_2$$

$$(r_2 - r_1 + 1)p = 1 \tag{4.1}$$

$$p = \frac{1}{r_2 - r_1 + 1}.$$

To generate X we first generate U and set

$$X = r_1 + \left[\frac{U}{p}\right] = r_1 + [U(r_2 - r_1 + 1)] \tag{4.2}$$

where the brackets indicate the largest integer. This step corresponds to

$$X = \begin{cases} r_1 & 0 \leq U < p \\ r_1 + 1 & p \leq U < 2p \\ \vdots & \vdots \\ r_1 + i & pi \leq U < p(i + 1) \\ \vdots & \vdots \\ r_2 & p(r_2 - r_1) \leq U < p(r_2 - r_1 + 1) = 1. \end{cases} \tag{4.3}$$

It is evident that the probability of U falling into a particular interval pi, $p(i + 1)$, is simply p, so that X is uniformly distributed on the integers.

8.4.2 Binomial Distribution

Let X have the binomial distribution with probability mass function

$$f(x) = \binom{n}{x} p^x(1-p)^{n-x} \qquad x = 0, \ldots, n. \tag{4.4}$$

We denote the distribution of X by $\mathscr{B}(n, p)$. To generate X we first note that we can represent X as

$$X = \sum_{i=1}^{n} Y_i \tag{4.5}$$

where each Y_i is sampled on a Bernoulli trial with

$$Y_i = \begin{cases} 1 & p \\ 0 & 1-p \end{cases} \qquad i = 1, \ldots, n. \tag{4.6}$$

To see this we note that Y_i has moment generating function (m.g.f.)

$$M_{Y_i}(z) = 1 - p + pz. \tag{4.7}$$

Since the Y_is are independent the m.g.f. of X is

$$M_X(z) = (1 - p + pz)^n \tag{4.8}$$

which is the m.g.f. of a random variable from $\mathscr{B}(n, p)$.

To reiterate our discussion in Chapter 7 regarding the generation of Bernoulli variates we generate U_i and assign $Y_i = 1$ if $U_i > p$ and $Y_i = 0$ if $U_i \leq p$, so that the long run frequency of $Y_i = 1$ is p. Figure 47 presents a flowchart of steps for generating X from $\mathscr{B}(n, p)$.

To apply the inverse transformation approach as described at the beginning of Section 8.4, we use $p_0 = 1 - p^n$ and $A_{k+1} = (n-k)p/[(k+1)(1-p)]$. We must also check after step 4 for $k = n - 1$ which, if true, terminates the procedure with $k = n$.

Since a binomial variate can assume but a finite set of values we can create a table of the cumulative distribution function for use with the inverse transformation method. We discuss this approach in Section 8.5.

When n is large these methods of generating binomial variates can be time consuming. Here an alternative approach, which relies on an asymptotic result, can significantly reduce the generating time. As n increases, the distribution of $Z = (X - np)/\sqrt{np(1-p)}$ approaches $\mathscr{N}(0, 1)$. To obtain a binomial variate we generate Z from $\mathscr{N}(0, 1)$, form $X = \sqrt{np(1-p)}\, Z + np$, and use X, rounded to an integer. A negligible error occurs [129, p. 109] using this asymptotic result when $np > 5$ for $p \leq \frac{1}{2}$ and $n(1-p) > 5$ for $p > \frac{1}{2}$.

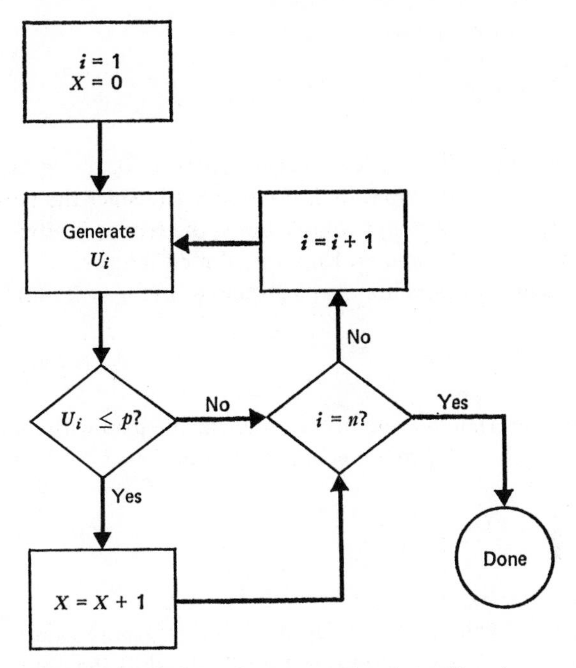

Fig. 47 Binomial variate generation from $\mathscr{B}(n, p)$.

8.4.3 Beta-Binomial Distribution

Occasionally an investigator finds it desirable to allow the probability p to be a random variable. This may occur when the sampling mechanism on individual trials is subject to random variation. In particular, if p is a beta variate from $\mathscr{B}e(a, b)$, where $a, b > 0$, then

$$f(X = x \mid p) = \binom{n}{x} p^x(1 - p)^{n-x} \tag{4.9}$$

$$\begin{aligned} f(X = x) &= \int_0^1 f(X = x \mid p) \frac{\Gamma(a + b)}{\Gamma(a)\Gamma(b)} p^{a-1}(1 - p)^{b-1}\, dp \\ &= \binom{n}{x} \frac{\Gamma(a + b)\Gamma(x + a)\Gamma(n - x + b)}{\Gamma(a)\Gamma(b)\Gamma(n + a + b)} \end{aligned} \tag{4.10}$$

$$x = 0, \ldots, n.$$

We then say that X has the beta-binomial distribution, denoted by $\mathscr{B}\mathscr{B}(a, b, n)$.

Notice that for $a = b = 1$

$$f(X = x) = \frac{1}{n + 1} \qquad x = 0, \ldots, n. \tag{4.11}$$

which is the uniform distribution over the integers $0, \ldots, n$, a special case of the discrete uniform distribution in Section 8.4.1. Since the present method requires n uniform variates and comparisons, the reader is advised to use the method in Section 8.4.1 for uniformly distributed integers.

Several other examples are also instructive. Let $a = 2$ and $b = 1$. Then

$$f(X = x) = \frac{2(x + 1)}{(n + 1)(n + 2)} \qquad x = 0, \ldots, n. \tag{4.12}$$

which is a discrete triangular distribution of the integers with mode at $x = n$. For $a = 1$ and $b = 2$ the mode is at $x = 0$. Let $a = b = 2$, so that

$$f(X = x) = \frac{6(x + 1)(n - x + 1)}{(n + 1)(n + 2)(n + 3)} \qquad x = 0, \ldots, n \tag{4.13}$$

with mode at $n/2$ when n is even.

To generate a beta-binomial variate one first generates a beta variate p (Fig. 43) and then follows the steps in Fig. 47. To shift the domain of X, say, to $X = q, q + 1, \ldots, n + q$, one need only generate X as in Fig. 47 and set $X = X + q$. The inverse transform method can alternatively be used.

8.4.4 Geometric Distribution

Let X have a geometric distribution with

$$f(X = x) = p^x(1 - p) \qquad x = 0, \ldots, \infty, \tag{4.14}$$

which we denote by $\mathscr{G}e(p)$. This distribution has its mode at $x = 0$ and is the discrete analog of the exponential distribution. We can think of X as the number of successful Bernoulli trials before a failure on the $X + 1$ trial. To generate X we follow the steps in Fig. 48. For the inverse transformation approach at the beginning of Section 8.4, we use $p_0 = 1 - p$ and $A_{k+1} = p$.

An alternative technique based on the relationship between the exponential and geometric distributions can also be used. Let Z be from $\mathscr{E}(\beta)$. We note that

$$\begin{aligned} \text{Prob}\,(r \leq Z < r + 1) &= \frac{1}{\beta} \int_r^{r+1} e^{-z/\beta}\, dz \\ &= e^{-r/\beta}(1 - e^{-1/\beta}) \qquad r = 0, 1, \ldots, \infty \end{aligned} \tag{4.15}$$

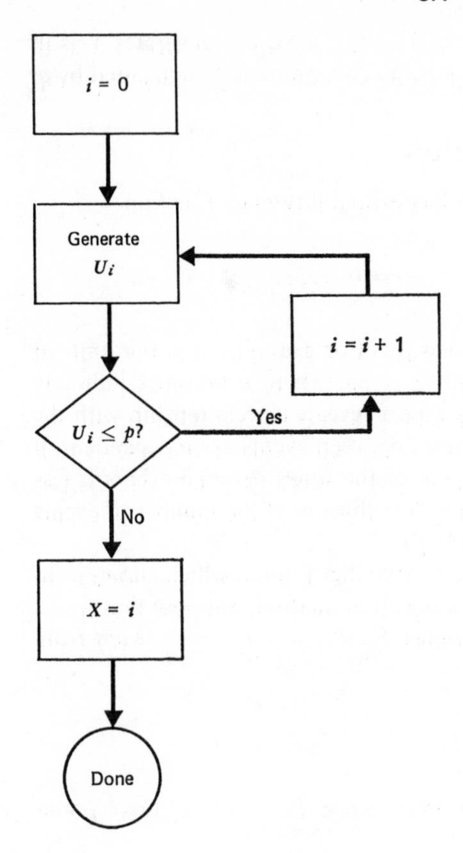

Fig. 48 Geometric variate generation from $\mathcal{G}e(p)$.

which is the probability that a random variable X from $\mathcal{G}e(e^{-1/\beta})$ assumes the value r. To generate X we use (4.15) with $\beta = -1/\log p$ in

$$X = [Z] = [-\beta \log U] = [\log U/\log p], \tag{4.16}$$

the brackets indicating the largest integer.

Although this technique requires only one uniform deviate, it also requires two logarithmic transformations and a truncation. Consequently the question of which of the three techniques suggested in this section gives the greatest computational efficiency remains to be answered. However, one observation is clear. Computation time for this technique is constant whereas the inverse transformation and Bernoulli trials methods have a component of computation time proportional to $p/(1 - p)$.

To shift the domain of X to $x = q, q + 1, \ldots, \infty$ one generates X as in Fig. 48 and sets $X = X + q$, the mean being concomitantly augmented by q.

8.4.5 Poisson Distribution

The Poisson distribution $\mathscr{P}(\lambda)$, which has probability mass function

$$f(X = x) = \frac{e^{-\lambda}\lambda^x}{x!} \qquad x = 0, 1, \ldots, \infty, \tag{4.17}$$

theoretically arises in several different ways. For example, it is the limit of the binomial distribution $\mathscr{B}(n, p)$ with $\lambda = np$, where n becomes infinitely large and p infinitely small.* Another aspect reveals its relationship with the exponential distribution $\mathscr{E}(\beta)$. If the times between events are independent, if each is drawn from $\mathscr{E}(\beta)$, and if the sum of the times between events is less than or equal to Δt, then the probability distribution of the number of events occurring in the interval $(0, \Delta t)$ is $\mathscr{P}(\Delta t/\beta)$.

It is this relationship with the exponential distribution which allows us to generate X as a Poisson variate using a rejection method. Suppose that we let $\Delta t = 1$ and generate exponential variates $T_0, T_1, \ldots, T_X, T_{X+1}$, each from $\mathscr{E}(1/\lambda)$, until

$$\sum_{j=0}^{X} T_j \leq 1 < \sum_{j=0}^{X+1} T_j \qquad X = 0, 1, \ldots. \tag{4.18}$$

Then X has the Poisson distribution $\mathscr{P}(\lambda)$. Since $T_j = -(1/\lambda) \log U_j$ condition (4.18) can be written as

$$\begin{gathered} -\sum_{j=0}^{X} \log U_j \leq \lambda < -\sum_{j=0}^{X+1} \log U_j \\ \prod_{j=0}^{X+1} U_j < e^{-\lambda} \leq \prod_{j=0}^{X} U_j \qquad X = 0, 1, \ldots. \end{gathered} \tag{4.19}$$

As an alternative approach, the inverse transformation method at the beginning of Section 8.4 uses $p_0 = e^{-\lambda}$ and $A_{k+1} = \lambda/(k + 1)$.

In some simulation experiments we may choose to generate X at periodic intervals of length Δt. In this case we replace $e^{-\lambda}$ in (4.19) by $e^{-\lambda \Delta t}$, noting that, over any interval Δt, X has mean and variance $\lambda \, \Delta t$. The choice of this approach, rather than that of generating each event from $\mathscr{E}(1/\lambda)$, means that the investigator foregoes the knowledge of precisely when each event occurs.

* See reference 121, pp. 67–69.

This lack may reduce the accuracy of measurement if the experiment is designed to determine the time response to events, unless the time response is considerably greater than Δt. Figure 49 shows the necessary steps for generating X from $\mathscr{P}(\lambda\, \Delta t)$.

When λ is large, $e^{-\lambda}$ is small so that the number of U_j needed to satisfy (4.19) may be large and the procedure therefore time consuming. The inverse transform approach would also be time consuming. To effect a time saving here we make use of a limiting characteristic of X from $\mathscr{P}(\lambda)$. As λ increases, the distribution of $Z = (X - \lambda)/\sqrt{\lambda}$ converges to $\mathscr{N}(0, 1)$. To produce a Poisson variate we generate Z from $\mathscr{N}(0, 1)$, compute $X = \sqrt{\lambda}\, Z + \lambda$, and use X, rounded to an integer. It is not clear, however, how large λ should be to induce negligible error.

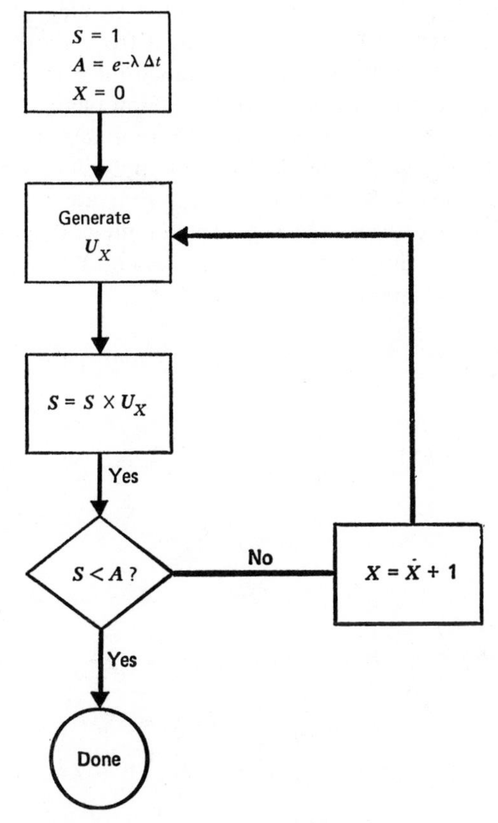

Fig. 49 Poisson variate generation from $\mathscr{P}(\lambda\, \Delta t)$.

8.4.6 Negative Binomial Distribution

The negative binomial distribution has probability mass function

$$(4.20) \qquad f(X = x) = \frac{\Gamma(r + x)}{\Gamma(x + 1)\Gamma(r)} p^x(1 - p)^r \qquad x = 0, \ldots, \infty$$

and is denoted by $\mathcal{NB}(r, p)$. When $r = 1$, we have the geometric distribution $\mathcal{G}e(p)$. The negative binomial distribution has a longer tail and hence is more skewed than the Poisson distribution. It has been found to be representative of a variety of demand patterns, especially the demand for aircraft spare parts [124].

When r is integer this distribution bears a relationship to the elementary concept of Bernoulli trials in that X from $\mathcal{NB}(r, p)$ is the number of successes encountered in a sequence of independent Bernoulli trials, with probability p of success, before the rth failure. Therefore we need only generate Bernoulli trials and apply the steps in Fig. 50 to generate X from $\mathcal{NB}(r, p)$, when r is an integer. In this case the distribution is also called the *Pascal* distribution. As an alternative approach the inverse transformation method at the beginning of Section 8.4 uses $p_0 = 1 - p^r$ and $A_{k+1} = (r + k)p/(k + 1)$.

When r is not an integer Fig. 50 does not apply since the concept of the rth success is meaningless. Nevertheless we can generate X exactly by making use of a relationship between the negative binomial, Poisson, and gamma distributions. Suppose that X is from $\mathcal{P}(Y)$, where Y is a random variable from $\mathcal{G}[r, p/(1 - p)]$. That is,

$$f(X = x \mid Y) = \frac{e^{-Y}Y^x}{x!} \qquad x = 0, 1, \ldots$$

$$(4.21) \qquad f_Y(y) = \frac{\lambda^r y^{r-1} e^{-\lambda y}}{\Gamma(r)} \qquad 0 \leq y \leq \infty$$

$$\lambda = \frac{1 - p}{p}.$$

Then

$$f(X = x) = \int_0^\infty f(X = x \mid Y) f_Y(y)\, dy$$

$$(4.22) \qquad = \frac{\Gamma(x + r)}{\Gamma(x + 1)\Gamma(r)} \left(\frac{\lambda}{1 + \lambda}\right)^r \left(\frac{1}{1 + \lambda}\right)^x$$

$$= \frac{\Gamma(x + r)}{\Gamma(x + 1)\Gamma(r)} p^x(1 - p)^r \qquad x = 0, 1, \ldots.$$

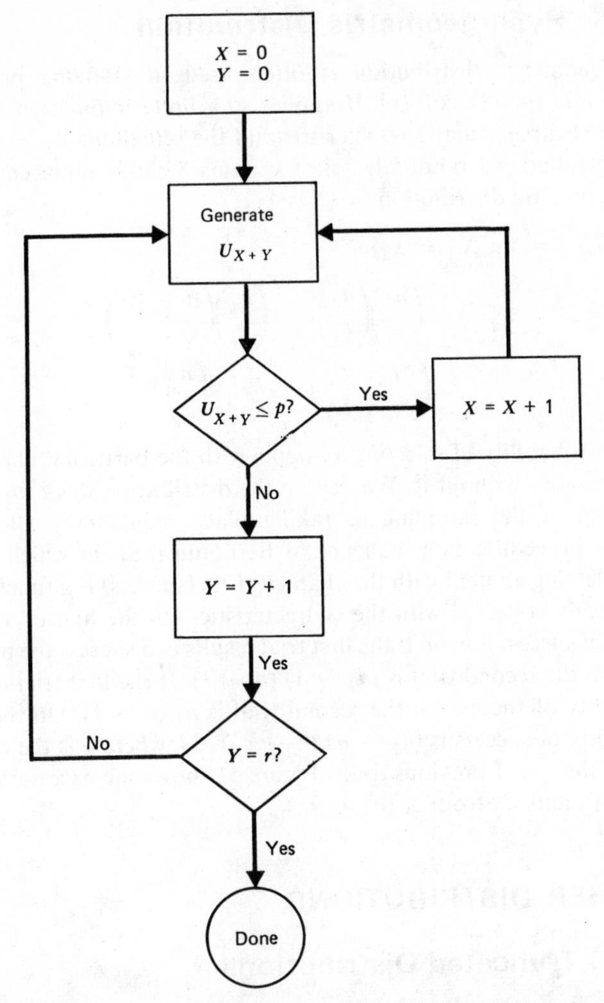

Fig. 50 Negative binomial variate generation from $\mathcal{NB}(r, p)$, $r = 1, 2, \ldots$.

A negative binomial variate X is generated as follows:

1. generate Y from $\mathcal{G}[r, p/(1-p)]$ (Fig. 44) and
2. generate X from $\mathcal{P}(Y)$ (Fig. 49) or from the inverse transformation approach.

The domain of generation can be shifted to $x = q, q+1, \ldots$, simply by adding q to X and noting that the new mean is $q + rp/(1-p)$.

8.4.7 Hypergeometric Distribution

The hypergeometric distribution is often used in studying problems in production and quality control. It applies to a finite population of n items, where n_1 have a particular characteristic and the remaining $n_2 = n - n_1$ do not. Suppose that we randomly select x items without replacement. Then the hypergeometric distribution

$$f(X_1 = x_1, X_2 = x_2)$$

$$= \frac{\binom{n_1}{x_1}\binom{n_2}{x_2}}{\binom{n_1+n_2}{x_1+x_2}} = \frac{\binom{n_1}{x_1}\binom{n-n_1}{x-x_1}}{\binom{n}{x}} \qquad \begin{array}{l} x_1 = 0, \ldots, x, \\ x_2 = x - x_1 \end{array} \tag{4.23}$$

tells us the probability of selecting x_1 items with the particular characteristic and $x_2 = x - x_1$ without it. We denote the distribution by $\mathscr{H}(n_1, n_2, x)$.

Thinking of the sampling as taking place sequentially allows us to consider the procedure as a sequence of Bernoulli trials in which the probability of selecting an item with the characteristic (success) is a function of the number already collected with the characteristic. On the first of x trials the probability of success is n_1/n. If the first trial results in a success the probability of success on the second trial is $(n_1 - 1)/(n - 1)$. If the first trial is a failure the probability of success on the second trial is $n_1/(n - 1)$. On the rth trial the probability of success is $(n_1 - q)/(n - r + 1)$, where q is the number of successes in the $r - 1$ previous trials. Figure 51 shows the necessary steps for generating X_1 and X_2 from $\mathscr{H}(n_1, n_2, x)$.

8.5 OTHER DISTRIBUTIONS

8.5.1 Truncated Distributions

Occasionally we wish to generate a random variable X from a specified interval (a, b) although F_X, the probability distribution function of X, is defined over a larger interval. In other words, we wish to *truncate* the range of X for sampling purposes. If X is continuous with p.d.f. f_X then the truncated p.d.f. is

$$f_X(x \mid a \le x \le b) = \begin{cases} \dfrac{f_X(x)}{F_X(b) - F_X(a)} & a \le x \le b \\ 0 & \text{elsewhere} \end{cases} \tag{5.1}$$

so that

$$F_X(x \mid a \le x \le b) = \frac{F_X(x) - F_X(a)}{F_X(b) - F_X(a)}. \tag{5.2}$$

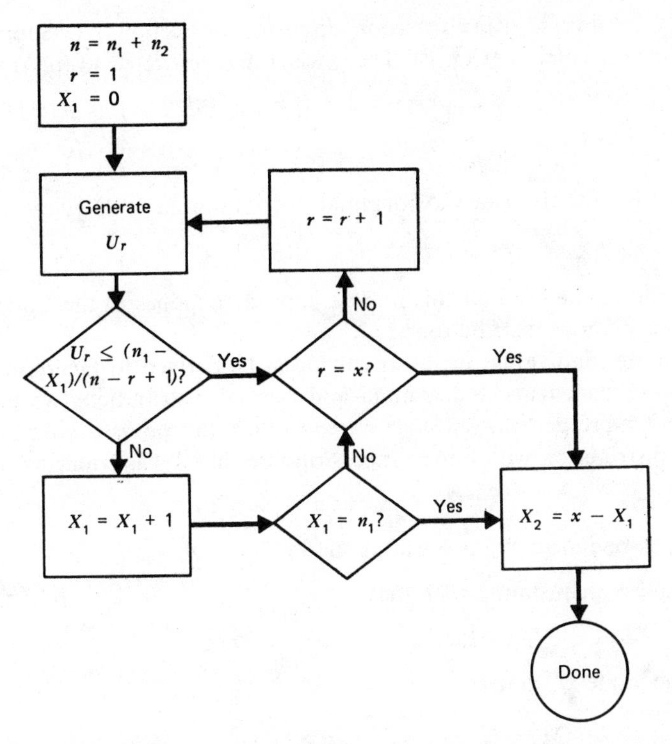

Fig. 51 Hypergeometric variate generation from $\mathscr{H}(n_1, n_2, x)$.

One approach to sampling from a truncated distribution is to generate deviates from the unrestricted distribution and to accept the variates that fall in (a, b). With this *gross* method the expected number of variates needed to get one in (a, b) is $1/[F_X(b) - F_X(a)]$, which can be large when the unrestricted probability mass in (a, b) is small. This approach is most appealing when we wish to truncate the tails of a distribution. For example, a loss of 0.1 probability in the tails raises the expected number of iterations to 1.11; a loss of 0.2 raises the number to 1.25.

A second approach is limited to distributions whose inverse transformations are easily specified. We note that

$$U = F_X(X \mid a \leq X \leq b) \tag{5.3}$$

so that

$$\begin{aligned} F_X(X) &= F_X(a) + U[F_X(b) - F_X(a)] \\ X &= \Phi[F_X(X)], \end{aligned} \tag{5.4}$$

Φ being the inverse transformation discussed in Section 7.1. Suppose the unrestricted X is from $\mathscr{W}(\alpha, \beta)$. Then, when X is restricted to (a, b) we have

$$\begin{aligned} 1 - e^{-(X/\beta)^\alpha} &= 1 - e^{-(a/\beta)^\alpha} + U[1 - e^{-(b/\beta)^\alpha} - 1 + e^{-(a/\beta)^\alpha}] \\ X &= \beta\{-\log[e^{-(a/\beta)^\alpha} - U(e^{-(a/\beta)^\alpha} - e^{-(b/\beta)^\alpha})]\}^{1/\alpha}. \end{aligned} \tag{5.5}$$

In particular, the truncated exponential distribution ($\alpha = 1$) gives

$$X = -\beta \log[e^{-a/\beta} - U(e^{-a/\beta} - e^{-b/\beta})]. \tag{5.6}$$

In addition to the Weibull this inverse approach applies to the Cauchy and logistic continuous distributions.*

Although individual methods can often be derived to facilitate variate generations from truncated gamma and normal distributions we prefer to describe a more general *rejection method* which can be used with any continuous distribution with finite range. Suppose that X has truncated p.d.f.

$$\bar{f}_X(x) = f_X(x \mid a \le x \le b) \tag{5.7}$$

where f_X is bounded. We proceed as follows:

1. select a constant c such that
$$\bar{f}(x) \le c \qquad a \le x \le b,$$
2. generate U_1 and U_2,
3. form
$$Y = a + (b - a)U_1,$$
4. accept $X = Y$ as the desired deviate if $U_2 \le \bar{f}(Y)/c$, or
5. otherwise reject Y and perform these steps again for U_3 and U_4, etc.

To see that X has the desired p.d.f. $\bar{f}_X$, we note that in step 3 unconditionally Y is from $\mathscr{U}(a, b)$ so that

$$\text{Prob}(Y = x) = dx \qquad a \le x \le b. \tag{5.8}$$

Moreover U_2, being from $\mathscr{U}(0, 1)$, has

$$\text{Prob}\left(U_2 \le \frac{\bar{f}(x)}{c} \,\middle|\, Y = x\right) = \frac{\bar{f}(x)}{c} \tag{5.9}$$

so that for any x

$$\begin{aligned} \text{Prob}\left(U_2 \le \frac{\bar{f}(x)}{c}\right) &= \int_a^b \text{Prob}\left(U_2 \le \frac{\bar{f}(x)}{c} \,\middle|\, Y = x\right) dx \\ &= \frac{1}{c}\int_a^b \bar{f}_X(x)\, dx = \frac{1}{c}. \end{aligned} \tag{5.10}$$

* See Exercises 6 and 7.

But

$$\text{Prob}\left(Y = x \mid U_2 \le \frac{\bar{f}(x)}{c}\right) = \frac{\text{Prob}(U_2 \le \bar{f}(x)/c \mid Y = x)\, dx}{\text{Prob}(U_2 \le \bar{f}(x)/c)} \tag{5.11}$$

$$= \bar{f}(x)\, dx.$$

It is easily seen that the probability of an acceptance on the Nth trial is $(1/c)(1 - 1/c)^{N-1}$, so that the expected number of uniform variates required is

$$2E(N) = 2 \sum_{n=1}^{\infty} \frac{n}{c}\left(1 - \frac{1}{c}\right)^{n-1} = 2c. \tag{5.12}$$

To guarantee that $1/c$ is a probability we set

$$c = \max[1, f(x)] \qquad a \le x \le b. \tag{5.13}$$

Suppose that we wish to restrict a variate Z from $\mathcal{N}(\mu, \sigma^2)$ to (a, b). This is equivalent to restricting a standardized variate X from $\mathcal{N}(0, 1)$ to (a', b'):

$$a' = \frac{a - \mu}{\sigma}, \qquad b' = \frac{b - \mu}{\sigma}. \tag{5.14}$$

Then

$$\max f_X(x) \le \frac{1}{\sqrt{2\pi}} \qquad a' \le x \le b' \tag{5.15}$$

so that

$$2c \le \frac{0.8}{F_X(b') - F_X(a')} \tag{5.16}$$

which compares favorably with $1/[F_X(b') - F_X(a')]$ in the first gross approach in this section.

For a gamma variate from $\mathcal{G}(\alpha, \beta)$ we note, using Table 21, that

$$\max f_X(x) \le \frac{[(\alpha - 1)/e]^{\alpha-1}}{\Gamma(\alpha)\beta} \qquad a \ge 1, a \le x \le b. \tag{5.17}$$

This upper limit decreases as α, β, or both increase. We wish to emphasize that the rejection method applies when $0 < \alpha < 1$, provided that $a > 0$. This lower bound on a keeps $\bar{f}_X$ and hence c finite.

In contrast to all the algorithms mentioned earlier in this chapter, the rejection method as described here requires us to be able to compute $\bar{f}_X$ only. This property makes the technique attractive for more general use, provided that we are willing to truncate our unrestricted distributions at appropriately chosen a and b, so that $F_X(b) - F_X(a) \sim 1$. However, efficiency considerations generally dictate against use of the method for such purposes, especially when $\max f_X(x)$ is large. The generation of beta variates using this rejection method is left to the reader as an exercise.

The inverse transform approach also works for discrete random variables where the range limits a and b are now taken as integers. Using the notation at the beginning of Section 8.4, we have

(5.18)
$$q_x = \sum_{j=0}^{x} p_j$$

so that

(5.19)
$$Q(X \leq x \mid a \leq x \leq b) = \frac{q_x - q_{a-1}}{q_b - q_{a-1}}.$$

The procedure to generate X is as follows:

1. compute p_j for $j = 0, \ldots, b$,
2. compute q_x using (5.18) for $x = a - 1, \ldots, b$,
3. compute $Q(X \leq x \mid a \leq x \leq b)$ from (5.8) for $x = a - 1, \ldots, b$,
4. generate U, and
5. set $X = r$, where the quantity r satisfies $Q(X \leq r - 1 \mid a \leq x \leq b) < U \leq Q(X \leq r \mid a \leq x \leq b)$.

In Section 8.5.3 we describe an alternative method for generating variates from a truncated distribution that eliminates the need for the comparisons in step 5 of the procedure presented here.

8.5.2 Empirical Distributions

Suppose that we are given the cumulative distribution function

(5.20)
$$G(x) = \frac{1}{n} \sum_{j=1}^{n} I_{v(j-1),\infty}(x)$$

where

(5.21)
$$I_{v(j-1),\infty}(x) = \begin{cases} 1 & v(j-1) \leq x \\ 0 & v(j-1) > x \end{cases}$$

$v(j); j = 0, \ldots, n - 1$ being the boundaries of the intervals. Assuming X is continuous and $v(0)$ and $v(n)$ finite, we take

(5.22)
$$F(x) = G[v(j-1)] + \frac{x - v(j-1)}{[v(j) - v(j-1)]n} \qquad v(j-1) \leq x < v(j)$$

as the empirical cumulative distribution function of X. To generate X we proceed as follows:

1. compute $G[v(j-1)]$ for $j = 1, \ldots, n$,
2. generate U,
3. determine j^* so that

$$G[v(j^* - 1)] \le U < G[v(j^*)], \qquad \text{and}$$

4. determine X from

$$X = n[v(j^*) - v(j^* - 1)]\{U - G[v(j^* - 1)]\} + v(j^* - 1). \tag{5.23}$$

Interpolation methods other than linear are, of course, possible. If the random variable of interest assumes only integer values then we take the largest integer in X as our variate.

Empirical distributions play a central role when we study how well the output of a simulation model compares with the output of the real system being simulated. For example, if we have input and output data for the real system, we may run the simulation model, driving it with stimuli sampled from relevant empirical input distributions. A close statistical match between real and simulated output would give us confidence in the validity of the model as a representation of reality.*

Apart from comparisons of real and simulated systems, we wish to emphasize the desirability of using a fitted theoretical distribution whenever the empirical function is based on a sample. If we restrict ourselves to the empirical function we can, at best, reproduce statistical behavior in the sample. Use of a fitted theoretical distribution, however, enables us to generate behavior characteristic of the parent population of which the sample represents but a part. This extension from the particular to the general is a principal objective of statistical inference. The fitting of theoretical distributions is discussed in Chapter 9.

8.5.3 Tabled Distributions

Our discussion of truncated distributions made clear the fact that logical comparisons enter into the generation of each random variate. For example, if a discrete distribution has

$$\text{Prob}(X = i) = p_i \qquad i = 0, \ldots, n$$

$$\sum_{i=0}^{n} p_i = 1,$$

* See Section 11.8.

then on average the mean number of comparisons is

$$\sum_{i=0}^{n-1} (i + 1)p_i + np_n = \sum_{i=0}^{n} ip_i + \sum_{i=0}^{n-1} p_i.$$

In the binomial case $\mathscr{B}(n, p)$ this is $np - p^n + 1$. Although to determine a variate may be less time consuming than the Bernoulli trial approach of Section 8.4.2, which requires n comparisons and n pseudorandom numbers, it is clear that $np - p^n + 1$ can be large. To reduce the time required, a tabling approach can be used.

Suppose that the p_i are specified to j digits, and let X be a $10^j \times 1$ array. We load zero into $X(1), \ldots, X(10^j p_0)$, one into $X(10^j p_0 + 1), \ldots, X(10^j(p_0 + p_1)), \ldots, k$ into $X(10^j \sum_{i=0}^{k-1} p_i + 1), \ldots, X(10^j \sum_{i=0}^{k} p_i)$ for $k = 2, \ldots, n$. To produce a desired variate, we proceed as follows:

1. generate U and
2. obtain $X([10^j U] + 1)$ as the desired variate.*

This tabling approach consumes more space but less time. Although this tradeoff is almost always a cost-saving one, the fact that this method requires a separate table for each discrete random variate makes it less than completely preferable to the alternative approaches already mentioned. Marsaglia [130a] offers an example of how to reduce the space requirement at the cost of a marginal increase in time. However, his suggestion forces the user to do even more programming to take advantage of the properties of the distribution of interest. In summary, a user must consider the costs to him of computer time, space, and programming before he can adjudge which approach to generating a variate from a particular distribution serves him best.

8.6 AUTOCORRELATED SEQUENCES

In simulation experiments designed to exercise a dynamic econometric model, it is customary to use stochastic disturbances to stimulate the model at periodic intervals Δt. If one sequence $\{Y_t\}$ of disturbances is used and if the disturbances are independent, the investigator need only select the sampling distribution he prefers and periodically sample from it to generate the stream of disturbances. If the model has n simultaneous equations, there is usually a multivariate stochastic disturbance sequence $\{\mathbf{Y}_t\}$, where $\{Y_{j,t}\}$ is the stream corresponding to the stimulus for the jth equation. If the sequences are jointly independent in addition to the events in each being jointly independent,

* The quantity $[10^j U]$ denotes the largest integer in $10^j U$.

then at each update time t the investigator simply generates $Y_{j,t}$ from its corresponding sampling distribution.

Occasionally the sequences are statistically dependent even though the events in any one sequence are independent. Then, if $Y_{1,t}, \ldots, Y_{n,t}$ jointly have the multivariate normal distribution with mean vector $\boldsymbol{\mu}$ and covariance matrix $\boldsymbol{\Sigma}$, one may follow the algorithm described in Section 8.3.3 to generate $\mathbf{Y}_t$. If $\log Y_{1,t}, \ldots, \log Y_{n,t}$ jointly have the multivariate normal distribution, one may generate $Y_{1,t}, \ldots, Y_{n,t}$ from the multivariate log normal distribution simply by taking the exponential transformations of $\log Y_{1,t}, \ldots, \log Y_{n,t}$.

Sometimes an investigator would like to use *exogenous* economic processes to provide stimuli for his model. An exogenous process is one whose behavior is in no way determined by the econometric model under study. In this case an economic process usually exhibits autocorrelated behavior since a present event generally shows a statistical association with past events in the same sequence. Alternatively the model under study may be incompletely specified so that the disturbance sequences themselves exhibit autocorrelated behavior for which the explanatory processes in the model are unable to account. Hence the investigator may wish to induce at least a facsimile of this temporal association in his simulated driving processes. Here we describe two ways of generating autocorrelated behavior, one based on a *moving average* representation and the other on an *autoregressive* scheme.

Consider the moving average

$$X_t = \sum_{s=0}^{p} a_s Y_{t-s} \qquad t = 0, \pm 1, \pm 2, \ldots, \pm\infty \tag{6.1}$$

where $\{Y_t\}$ is a sequence of independent, identically distributed random variables with mean zero and variance σ^2. Then $\{X_t\}$ has the autocovariance function

$$\begin{aligned} R_\tau = E(X_t X_{t+\tau}) &= \sum_{r,s=0}^{p} a_r a_s E(Y_{t-r} Y_{t+\tau-s}) \\ &= \begin{cases} \sigma^2 \sum_{s=|\tau|}^{p} a_s a_{s-|\tau|} & |\tau| \leq p \\ 0 & \text{elsewhere.} \end{cases} \end{aligned} \tag{6.2}$$

Expression (6.2) implies that the autocorrelation in $\{X_t\}$ vanishes between events that are more than $|\tau|$ units apart.

Suppose that $a_s = c$ for $s = 0, \ldots, p$. Then the resulting autocovariance function has the form

$$R_\tau = \begin{cases} \sigma^2 c^2 (p + 1 - |\tau|) & |\tau| \leq p \\ 0 & \text{elsewhere.} \end{cases} \tag{6.3}$$

To form X_t with an autocovariance function as in (6.3) we generate $Y_t, \ldots, Y_{t-p}$ and apply (6.1). To form X_{t+1} we use

$$X_{t+1} = X_t + c(Y_{t+1} - Y_{t-p}) \tag{6.4}$$

and more generally

$$X_{t+q} = X_{t+q-1} + c(Y_{t+q} - Y_{t-p+q}). \tag{6.5}$$

As an alternative model suppose that

$$a_s = \begin{cases} \alpha^s & 0 \le s \le p \\ 0 & \text{elsewhere} \end{cases} \tag{6.6}$$

with $|\alpha| < 1$. Then X has the autocovariance function

$$R_\tau = \begin{cases} \sigma^2 \sum_{s=|\tau|}^{p} \alpha^{2s-|\tau|} = \sigma^2 \alpha^{|\tau|} \left[\dfrac{1 - \alpha^{2(p-|\tau|)}}{1 - \alpha^2} \right] & |\tau| \le p \\ 0 & |\tau| > p \end{cases} \tag{6.7}$$

where the term $\alpha^{|\tau|}$ dominates.

It often occurs that the correlation between events in a sequence does not dissipate over a finite interval as (6.6) and (6.7) suggest but becomes progressively weaker as the time between events increases. One way to accommodate this form of correlation is to let $p \to \infty$ in (6.6) so that

$$R_\tau = \frac{\sigma^2 \alpha^{|\tau|}}{1 - \alpha^2} \qquad -\infty \le \tau \le \infty. \tag{6.8}$$

This leads to an infinite moving average

$$X_t = \sum_{s=0}^{\infty} \alpha^s Y_{t-s} \tag{6.9}$$

whose use would require an infinite number of Y_ts to generate X_t. Fortunately this approach can be avoided by noting the recursive expression

$$\begin{aligned} X_t - \alpha X_{t-1} &= \sum_{s=0}^{\infty} \alpha^s Y_{t-s} - \alpha \sum_{s=0}^{\infty} \alpha^s Y_{t-1-s} \\ &= Y_t. \end{aligned} \tag{6.10}$$

To generate X_t one simply adds Y_t to αX_{t-1}. The reader should note that (6.10) is a first-order autoregressive scheme, which we discussed in Chapter 6 in connection with output representations.

There is a problem in using this recursive expression at the beginning of the experiment. Since

$$X_1 = \alpha X_0 + Y_1 \tag{6.11}$$

we need a method for generating the initial random variable X_0. Setting $X_0 = 0$ will not do, since $X_1 = Y_1$ would then have a different marginal probability distribution than each of the remaining $X_2, X_3, \ldots$ has. If, however, X is a normal process we may generate X_0 from $\mathcal{N}[0, \sigma^2/(1 - \alpha^2)]$ since (6.8) implies that

$$\operatorname{var}(X_t) = \frac{\sigma^2}{1 - \alpha^2} \qquad t = 0, \pm 1, \pm 2, \ldots, \pm\infty. \tag{6.12}$$

Because $\{Y_t\}$ is a sequence of independent events and the dependence of X_t on the past is entirely summarized in αX_{t-1}, $\{X_t\}$ is called a first-order Markov process. The autoregressive scheme in (6.10) produces an autocorrelation function as in Fig. 35. Using the spectrum representation (6.4.11), we see that the autocorrelation function produces a spectral density function

$$f(\lambda) = \frac{1}{2\pi(1 - 2\alpha \cos \lambda + \alpha^2)}.$$

We note that any semblance to a regular cycle in (6.10) is restricted to frequency π when $-1 < \alpha < 0$. Since economic phenomena often produce autocorrelated behavior with several cyclic components we would prefer to work with a less restrictive model than (6.10). The more general pth-order autoregressive scheme described in Chapter 6 can serve this purpose.

Consider the pth-order autoregressive scheme

$$\sum_{s=0}^{p} b_s X_{t-s} = Y_t \qquad b_0 \equiv 1. \tag{6.13}$$

As mentioned in Chapter 6 this expression can produce a wide variety of autocorrelated behavior; in particular, if p is even, we may show that up to $p/2$ frequency concentrations can occur in the spectrum in the interval* $0 < \lambda < \pi$.

To generate X_t we first generate Y_t and add it to $-\sum_{s=1}^{p} b_s X_{t-s}$. A problem arises again, however, when generating X_1, namely, how can we produce $X_{1-p}, \ldots, X_0$ so that each of $X_1, X_2, \ldots$ has the same marginal distribution? In other words, how can we eliminate the initial startup *bias* in $X_1, X_2, \ldots$? Suppose that $\{X_t\}$ is a normal process so that $\{X_t, \ldots, X_{t+p-1}\}$ jointly have

* See Section 6.6.

the multivariate normal distribution with zero mean vector and covariance matrix

$$\mathbf{R}_p = \begin{bmatrix} R_0 & R_1 & & R_{p-1} \\ R_1 & R_0 & & \vdots \\ \vdots & & \ddots & \vdots \\ R_{p-1} & & \cdots & R_0 \end{bmatrix}, \tag{6.14}$$

the elements of which are given by (6.6.9a). Since we can show that the autocovariance function of a covariance stationary sequence with an autoregressive representation as in (6.13) is positive definite, $\mathbf{R}_p$ is nonsingular. To generate $X_{1-p}, \ldots, X_0$ we use the method of Section 8.3.3, first computing the lower triangular matrix $\mathbf{C}$, for which $\mathbf{R}_p = \mathbf{CC}'$.

An autoregressive scheme can also be used to generate nonstationary behavior. For example, suppose that $\{X_t\}$ has the mean

$$E(X_t) = c_1 t + c_2 \tag{6.15}$$

but satisfies the second-order scheme

$$X_t + b_1 X_{t-1} + b_2 X_{t-2} = Y_t. \tag{6.16}$$

Then we require

$$\begin{aligned} E(X_t &+ b_1 X_{t-1} + b_2 X_{t-2}) \\ &= c_1 t + c_2 + b_1[c_1(t-1) + c_2] + b_2[c_1(t-2) + c_2] \\ &= 0, \end{aligned} \tag{6.17}$$

which holds when $1 + b_1 + b_2 = 0$ and $b_1 = -2b_2$, so that $b_1 = -2$, $b_2 = 1$. Notice that this is precisely the case in which the roots of the characteristic equation (6.14) lie on the unit circle. Also notice that $\{X_t - (c_1 t + c_2);\ t = 0, \pm 1, \pm 2, \ldots, \pm\infty\}$ is a covariance stationary sequence. In generating the starting values X_0 and X_{-1} we use the multivariate normal distribution, when $\{X_t\}$ is a normal process, with means c_2 and $c_2 - c_1$, respectively, and covariance matrix $\mathbf{R}_p$, $p = 2$.

It turns out that a pth-order autoregressive scheme can generate a $(p - 1)$-order polynomial trend in the mean provided that all the roots of (6.14) lie on the unit circle. The solution to (6.14) is not always direct, especially for $p > 2$. Consequently it is often more convenient (but less computationally efficient) to add the desired trend to X_t after it has been generated as a zero mean variate from (6.13).

The autoregressive representation also offers convenience with regard to estimating the autocorrelation function, spectrum, and variance of a sample mean. In Chapter 9 we present a method for estimating the autoregressive parameters and describe some of their sampling properties.

EXERCISES

1. *Gamma Approximation:* One approximate method of generating a gamma variate from $\mathcal{G}(\alpha, \beta)$ when α is nonintegral is as follows:

   1. compute $k = [\alpha]$;
   2. generate k variates from $\mathcal{E}(\beta)$ and sum them as X;
   3. generate a uniform variate U_{k+1};
   4. if $U_{k+1} < \alpha - k$, (i) generate a variate from $\mathcal{E}(\beta)$, (ii) add it to X, and (iii) use X as the desired gamma variate;
   5. if $U_{k+1} > \alpha - k$, use X as formed in step 2 as the desired gamma variate. A gamma variate has mean $\alpha\beta$ and variance $\alpha\beta^2$.

   a. Show that the above algorithm gives the correct mean.
   b. Show that the above algorithm overstates the variance (i.e., find the variance first).
   c. Show that the approximation improves as α increases.

2. *Normal Approximation:* For a variate drawn from $\mathcal{N}(0, 1)$ the skewness and kurtosis are 0 and 3, respectively. One suggested method for generating a variate approximately distributed as $\mathcal{N}(0, 1)$ is as follows:

   1. generate $U_1, \ldots, U_j$ from $\mathcal{U}(0, 1)$ and
   2. use $X_n = \sqrt{12/n}\sum_{j=1}^{n}(U_j - \frac{1}{2})$ as the desired variate.

   One approach to assessing how good this method is is to compare the resulting central moments with those of a standardized normal variate.
   a. Show that X_n has mean zero and variance unity.
   b. Find the skewness $E(X_n^3)/[E(X_n^2)]^{3/2}$ of X_n.
   c. Find the kurtosis $E(X_n^4)/[E(X_n^2)]^2$ of X_n.
   d. Find n so that the relative error in the kurtosis is less than 0.05, that is,

   $$\frac{E(X_n^4)}{3[E(X_n^2)]^2} \geq 0.95.$$

   Hint:

   $$E(U_j - \tfrac{1}{2})^k = (\tfrac{1}{2})^{k+1}\left[\frac{1 + (-)^k}{k + 1}\right].$$

3. *Maxima and Minima:* Consider n independent, identically distributed random variables $X_1, \ldots, X_n$ with probability density function f_X.

From the theory of order statistics one may show that

$$W = \max(X_1, \ldots, X_n)$$

has the p.d.f.

$$f_W(w) = n[F_X(w)]^{n-1} f_x(w)$$

$$F_X(w) = \int_{-\infty}^{w} f_X(x)\, dx$$

and

$$Z = \min(X_1, \ldots, X_n)$$

has the p.d.f.

$$f_Z(z) = n[1 - F_X(z)]^{n-1} f_X(z)$$

$$F_X(z) = \int_{-\infty}^{z} f_X(x)\, dx.$$

Use a single variate to find the following.

a. W when $X \sim \mathscr{U}(0, 1)$.
b. Z when $X \sim \mathscr{U}(0, 1)$.
c. Z when $X \sim \mathscr{E}(\beta)$.

4. Let X have p.d.f.

$$f(x) = \begin{cases} (x - 3)^2/18 & 0 \leq x \leq 6 \\ 0 & \text{elsewhere} \end{cases}$$

a. Describe in detail two ways of generating X.
b. How many uniform variates does each method require?

5. Let X have p.d.f.

$$f_X(x) = \begin{cases} p\beta_1 e^{-\beta_1 x} + (1 - p)\beta_2 e^{-\beta_2 x} & 0 \leq x \leq \infty \\ 0 & x < 0 \end{cases} \qquad 0 \leq p \leq 1.$$

Devise an algorithm for generating X, using at most two uniformly distributed variates.

6. The Cauchy distribution has p.d.f.

$$f_X(x) = \frac{\beta}{\pi} \frac{1}{\beta^2 + (x - \alpha)^2} \qquad -\infty \leq x \leq \infty,\ \beta \geq 0.$$

Describe a method for generating Cauchy variates.

7. The logistic distribution has p.d.f.

$$f_X(x) = \frac{e^{-(x-\alpha)/\beta}}{\beta[1 + e^{-(x-\alpha)/\beta}]^2} \qquad -\infty \leq x \leq \infty, \beta > 0.$$

Describe a method for generating logistic variates.

8. Suppose that we have two independent variates W_1 and W_2, one from $\mathscr{U}(a_1, b_1)$ and the other from $\mathscr{U}(a_2, b_2)$. We form the variate $W = c_1 W_1 + c_2 W_2$ and want W to be from $\mathscr{T}(a, b, c)$. Determine a_1, a_2, b_1, b_2, c_1, and c_2 in terms of a, b, and c.

9. a. Use the rejection method to derive a beta variate $\mathscr{B}e(a, b)$.
   b. Compare the efficiency of the rejection approach with the method described in Section 8.2.7. Efficiency is a function of the number of uniform variates generated and the number of logarithmic transformations. Assume that an exponentiation consumes as much computer time as does a logarithmic transformation.

10. Write a program in GPSS/360 to generate variates from $\mathscr{G}(1.3, 1)$ and using the technique of Section 8.2.6. Run the program to produce 1000 such variates and use the TABULATE option, with intervals 0.1, to create the sample cumulative probability distribution. Check your results with the quartiles of the theoretical distribution.

11. Write a program in GPSS to generate variates from $\mathscr{N}(0, 1)$. As a reasonable approximation assume a range $-3 < x < 3$. Run the program to produce 1000 such variates, and use the TABULATE option, with intervals 0.1, to create the sample cumulative probability distribution. Check your results with the quartiles of the theoretical distribution.

12. Write a program in GPSS to generate variates from $\mathscr{N}\mathscr{B}(2, 0.37)$. Run the program to produce 1000 such variates, and use the TABULATE option with unit intervals to create the sample cumulative probability distribution. Check your results with the quartiles of the theoretical distribution. Section 8.5.1 provides a recursive formula for computing the quartiles, if necessary.

CHAPTER 9

INPUT PARAMETERS

9.1 INTRODUCTION

Although the generating algorithms in Chapter 8 allow us to specify a wide variety of stochastic input behavior, there remain at least two input problems whose solutions strongly influence the scientific quality of a simulation experiment. These problems concern the selection of probability laws that appropriately characterize the input to a particular simulation and the assignment of numerical values to the parameters of the distributions corresponding to these laws. Their solutions affect strongly the validity of generalizations made about the real world from the particular results of the experiment.

In some simulations an investigator specifies the probability laws and a range of parameter values based solely on theoretical considerations about the environment being modeled. Suppose, for example, that one wishes to determine the sampling distribution of a particular estimator of the parameters of a specified distribution as a function of sample size n. Then the input distribution is the specified one, and the investigator selects ranges of parameter values and sample sizes n for which he is interested in studying the sampling distribution as output. The governing tenet here is simply that parameter values and sample sizes be chosen sufficiently dense so that reasonably accurate interpolations can be made for other values in the range. In this example there exists no conflict with the specifications of input based on theoretical considerations.

Many other simulation models are designed to resemble large systems presently in existence or anticipated for the future. These systems often have multiple inputs whose characteristics vary greatly. When the simulation represents a future system, we can only conjecture as to the characteristics of its input. One useful approach is to infer these characteristics from the inputs of similar systems currently in existence. If the simulation represents a present system then inferences from the inputs to the corresponding real system are appropriate building blocks.

In order to make the characterization problem tractable, these inferences for large systems are inevitably based to a great extent on theoretical considerations. Whenever data are available or can be collected, the investigator can effect a considerable improvement in the quality of his input and hence the meaningfulness of his experimental results by using statistical methods to augment this theoretical considerations in the selection of input distributions and the specification of parameter values. The balance between theoretical considerations and empirical evidence naturally depends on the comprehensiveness of the theory regarding the input, on the volume and quality of available data, and on the body of statistical theory available to make useful statistical inferences from the data.

Data costs and time constraints also play roles. The more expensive data collection and processing are, the more reluctant an investigator is to devote effort to developing empirical evidence. However, the magnitude of the sum of money involved in the policies to be recommended on the basis of the output from the simulation should determine what is a reasonable data cost. If the simulation is to reveal information regarding a capital expenditure of several hundred million dollars, then a $100,000 data analysis program does not seem unreasonable. On the other hand, if the simulation is to recommend a policy for saving $10,000 over 5 years, a present data analysis cost of $2000 is questionable.

Data collection and analysis take time, a rare commodity when answers are needed in a hurry. When the time constraint precludes all but the most cursory inferences from the data, it is incumbent on the investigator to state the conclusions of his simulation study with qualifications regarding the limited analysis of the input. Otherwise his conclusions may mislead his readers about the extent to which the simulation output is a consequence of the characteristics of the real system.

In this chapter we describe methods for estimating parameters from a number of the distributions given in Chapter 8, and we discuss goodness-of-fit tests in which estimated parameters are substituted for true but unknown parameters. The chapter also includes a discussion of statistical inference for autoregressive processes. As mentioned in Chapter 8, autoregressive representations offer a convenient way of generating autocorrelated input data in fixed time increment simulations. It turns out that such representations are quite helpful also in describing the autocorrelated characteristics of the output of a simulation. Because of the potentially broad usage and infrequent appearance of autoregressive representations in general statistics courses, we present a detailed account of their estimation and use.

9.2 ESTIMATION*

There are many ways of estimating the parameters of a distribution, each method having its own virtues and shortcomings. For example, the method of moments is easy to apply for most distributions but often has low *efficiency*. By efficiency we mean the ratio of the variance of the minimum variance estimator and the variance of a chosen estimator. The minimum chi-square method offers the convenience of asymptotically giving an exact chi-square test of fit, but the computation of the parameter estimates is often formidable. The computation of the maximum likelihood estimates (MLE) can also be troublesome, but because of their highly desirable large sample properties methods of computation have been devised for a wide range of distributions and considerable study has been devoted to their small sample properties. The Weibull distribution examined in reference 156 is a good example.

Suppose that we have n independent observations $X_1, \ldots, X_n$, each with p.d.f. $f(x; \theta_1, \ldots, \theta_q)$ where $\boldsymbol{\theta} = (\theta_1, \ldots, \theta_q)$ are the parameters to be estimated. As useful tools in deriving estimates we define the likelihood function

$$L = \prod_{j=1}^{n} f(X_j; \theta_1, \ldots, \theta_q) \tag{2.1}$$

and the log-likelihood function

$$\log L = \sum_{j=1}^{n} \log f(X_j; \theta_1, \ldots, \theta_q). \tag{2.2}$$

If L is twice differentiable, it is known that any estimator of $\boldsymbol{\theta}$ has a covariance matrix whose determinant (generalized variance) is greater than the determinant (generalized variance) of†

$$\boldsymbol{\Sigma} = -\begin{bmatrix} E\dfrac{\partial^2 \log L}{\partial \theta_1^{\,2}} & E\dfrac{\partial^2 \log L}{\partial \theta_1\, \partial \theta_2} & \cdots & E\dfrac{\partial^2 \log L}{\partial \theta_1\, \partial \theta_q} \\ E\dfrac{\partial^2 \log L}{\partial \theta_2\, \partial \theta_1} & E\dfrac{\partial^2 \log L}{\partial \theta_2^{\,2}} & & \vdots \\ \vdots & & \ddots & \\ E\dfrac{\partial^2 \log L}{\partial \theta_q\, \partial \theta_1} & \cdots & & E\dfrac{\partial^2 \log L}{\partial \theta_q^{\,2}} \end{bmatrix}^{-1} \tag{2.3}$$

* See references 144 and 145 for alternative estimation techniques applied to a variety of continuous and discrete distributions.

† See, for example, reference 146, p. 56.

We say that any estimator with the covariance matrix in (2.3) is a minimum variance estimator of $\boldsymbol{\theta}$, a clearly desirable property.

Differentiating log L with respect to $\boldsymbol{\theta}$, we have a set of q equations

$$\frac{\partial \log L}{\partial \theta_k} = \sum_{j=1}^{n} \frac{\partial f / \partial \theta_k}{f(X_j; \theta_1, \ldots, \theta_k)} \qquad k = 1, \ldots, q. \tag{2.4}$$

If $\hat{\boldsymbol{\theta}} = (\hat{\theta}_1, \ldots, \hat{\theta}_q)$ sets these partial derivatives to zero (maximizes L) we call $\hat{\boldsymbol{\theta}}$ the *maximum likelihood* estimator (MLE) of $\boldsymbol{\theta}$. The MLE is unique and has the property that, as n increases, the joint distribution of $\sqrt{n}\,(\hat{\theta}_1 - \theta_1), \ldots, \sqrt{n}\,(\hat{\theta}_q - \theta_q)$ converges to the multivariate normal distribution with zero mean and covariance matrix $n\Sigma$. The result implies that the MLE is asymptotically unbiased and is a minimum variance estimator. Although the unbiased and minimum variance properties do not generally hold in small and moderate-size samples, the MLE often continues to have relatively better sampling properties in limited samples than do many other commonly encountered estimators. In addition, the MLE of any function $h(0)$ is $h(\hat{\boldsymbol{\theta}})$. An occasional drawback in using the MLE is the difficulty in solving (2.4) for $\hat{\theta}_1, \ldots, \hat{\theta}_q$. Iterative techniques are often necessary, but high-speed computers make this step considerably less onerous than it was a decade ago.

Table 23 presents the MLEs for several distributions considered in Chapter 8, along with their asymptotic covariance matrices. Known small sample properties are also described and referenced. Knowledge of the sampling properties can greatly assist us in the description of input data. Since the MLEs are point estimates, there exist possible differences between the estimates and true parameters due to bias and sampling variation. If we use only point estimates we run the risk that these differences may significantly affect the output of the simulation. If, however, we specify intervals in which we expect the true parameters to lie with high probability we can run the simulation for input parameters over the ranges of the intervals and determine the sensitivity of the output to possible variation in input parameters. If we can establish insensitivity our confidence is significantly enhanced about overlooking the substitution of estimates for the true parameters.

Suppose that we estimate $\hat{\mu} = 1.315$ and $\hat{\sigma}^2 = 1.803$ for a sample of size 10 from a normal distribution. Then a 95 per cent confidence interval for μ, using the t distribution with 9 degrees of freedom, yields

$$\mu_l = \hat{\mu} - t_{9,0.975}\sqrt{\hat{\sigma}^2/n} = 0.355, \qquad \mu_u = \hat{\mu} + t_{9,0.975}\sqrt{\hat{\sigma}^2/n} = 2.27.$$

A 95 per cent confidence interval for σ^2, using the chi-square distribution with 9 degrees of freedom, yields

$$\sigma_l^2 = \frac{n\hat{\sigma}^2}{\chi^2_{9,0.975}} = 0.853, \qquad \sigma_u^2 = \frac{n\hat{\sigma}^2}{\chi^2_{9,0.025}} = 6.010.$$

Table 23 Maximum Likelihood Estimators for Selected Distributions

$$a = \frac{1}{n}\sum_{j=1}^{n} X_j, \quad b = \sum_{j=1}^{n} X_j^2, \quad c = \frac{1}{n}\sum_{j=1}^{n} \log X_j, \quad d = \frac{1}{n}\sum_{j=1}^{n} \log(1 - X_j)$$

Distribution	ML Equations	ML Estimators	Sampling Properties	Asymptotic Covariance Matrix
Exponential	$\hat{\beta} - a = 0$	$\hat{\beta} = a$	$\frac{n\hat{\beta}}{\beta} \sim \chi^2(2n)$	$\frac{\beta^2}{n}$
Gamma $\mathscr{G}(\alpha, \beta)$	$\Psi(\hat{\alpha}) + \log \hat{\beta} = c$ $\hat{\alpha}\hat{\beta} = a$	See Table A1 for $\hat{\alpha}$ $\hat{\beta} = \frac{a}{\hat{\alpha}}$	See ref. 138 for small sample bias and variance.	$\frac{1}{n[\alpha\Psi'(\alpha) - 1]} \times \begin{bmatrix} \alpha & -\beta \\ -\beta & \beta^2\Psi'(\alpha) \end{bmatrix}$
Weibull $\mathscr{W}(\alpha, \beta)$	$f(\hat{\alpha}) = \frac{n}{\hat{\alpha}} + nc - \frac{n\sum_{j=1}^{n} X_j^{\hat{\alpha}} \log X_j}{\sum_{j=1}^{n} X_j^{\hat{\alpha}}}$ $f'(\hat{\alpha}) = -\frac{n}{\hat{\alpha}^2} - \frac{n\sum_{j=1}^{n} X_j^{\hat{\alpha}}(\log X_j)^2}{\sum_{j=1}^{n} X_j^{\hat{\alpha}}} - \frac{n(\sum_{j=1}^{n} X_j^{\hat{\alpha}} \log X_j)^2}{(\sum_{j=1}^{n} X_j^{\hat{\alpha}})^2}$	$\hat{\alpha}_k = \hat{\alpha}_{k-1} - \frac{f(\hat{\alpha}_{k-1})}{f'(\hat{\alpha}_{k-1})}$ $\hat{\alpha}_0 = \frac{\sqrt{(b - na^2)/(n-1)}}{a}$ Iterate until convergence. $\hat{\beta} = \left(\frac{1}{n}\sum_{j=1}^{n} X_j^{\hat{\alpha}}\right)^{1/\hat{\alpha}}$	See Tables A2 and A3 for distributions of $\hat{\alpha}/\alpha$ and $\hat{\alpha} \log(\hat{\beta}/\beta)$. See Table A4 for unbiasing $\hat{\alpha}$.	$\frac{1}{n}\begin{bmatrix} 1.109\beta^2/\alpha^2 & 0.257\beta \\ 0.257\beta & 0.608\alpha^2 \end{bmatrix}$
Beta[a] $\mathscr{B}e(\alpha, \beta)$	$-\Psi(\hat{\alpha} + \hat{\beta}) + \Psi(\hat{\alpha}) + c = 0$ $-\Psi(\hat{\alpha} + \hat{\beta}) + \Psi(\hat{\beta}) + d = 0$	See Table A5 for "rough" estimates. More detailed table is given in ref. 142.		$\frac{1}{n\{\Psi'(\alpha)\Psi'(\beta) - \Psi'(\alpha+\beta) \times [\Psi'(\alpha) + \Psi'(\beta)]\}} \times \begin{bmatrix} \Psi'(\alpha+\beta) - \Psi'(\beta) & -\Psi'(\alpha+\beta) \\ -\Psi'(\alpha+\beta) & \Psi'(\alpha+\beta) - \Psi'(\alpha) \end{bmatrix}$

Distribution	ML Equations	ML Estimators	Sampling Properties	Asymptotic Covariance Matrix
Normal $\mathcal{N}(\mu, \sigma^2)$		$\hat{\mu} = a$ $\hat{\sigma}^2 = \frac{b - na^2}{n}$ $\frac{n\hat{\sigma}^2}{n-1}$ is unbiased estimate of σ^2.	$\frac{\hat{\mu} - \mu}{\sqrt{\hat{\sigma}^2/n}} \sim t(n-1)$ $\frac{n\hat{\sigma}^2}{\sigma^2} \sim \chi^2(n-1)$	Exact $\begin{bmatrix} \sigma^2/n & 0 \\ 0 & 2\sigma^4/n \end{bmatrix}$
Lognormal $\mathcal{L}\mathcal{N}(\mu_Y, \sigma_Y{}^2)$	$Y_j \sim \mathcal{L}\mathcal{N}(\mu_Y, \sigma_Y{}^2)$ $X_j = \log Y_j \sim \mathcal{N}(\mu, \sigma^2)$	$\hat{\mu}_Y = e^{\hat{\mu} + \hat{\sigma}^2/2}$ $\hat{\sigma}_Y{}^2 = \hat{\mu}_Y{}^2(e^{\hat{\sigma}^2} - 1)$	Unbiased estimates see refs. 141, 146, 154. $\tilde{\mu}_Y = e^{\hat{\mu}} f\left(\frac{\hat{\sigma}^2}{2}\right)$ $\hat{\sigma}_Y{}^2 = e^{2\hat{\mu}}\left[f(2\hat{\sigma}^2) - f\left(\frac{n-2}{n-1}\hat{\sigma}^2\right)\right]$ $f(t) = 1 + t + \left(\frac{n-1}{n+1}\right)\frac{t^2}{2}$	$\text{var}(\hat{\mu}_Y) = \text{var}(\tilde{\mu}_Y) = \left(\sigma^2 + \frac{\sigma^4}{2}\right)\frac{e^{2\mu+\sigma^2}}{n}$ $\text{var}(\hat{\sigma}_Y{}^2) = \frac{2\sigma^2}{n} e^{4\mu+\sigma^2} \times [2(e^{\sigma^2} - 1) + \sigma^2(2e^{\sigma^2} - 1)^2]$
Binomial $\mathcal{B}(r, p)$, r given		$\hat{p} = \frac{a}{r}$	$E(\hat{p}) = p$ $nr\hat{p} \sim \mathcal{B}(nr, p)$	Exact $\text{var}(\hat{p}) = \frac{p(1-p)}{nr}$
Poisson $\mathcal{P}(\lambda)$		$\hat{\lambda} = a$	$E(\hat{\lambda}) = \lambda$ $n\hat{\lambda} \sim \mathcal{P}(n\lambda)$	Exact $\text{var}(\hat{\lambda}) = \frac{\lambda}{n}$

Table 23 (Continued)

Distribution	ML Equations	ML Estimators	Sampling Properties	Asymptotic Covariance Matrix
Geometric $\mathscr{G}(p)$		$\hat{p} = \dfrac{a}{1+a}$	$E(a) = \dfrac{p}{1-p}$ $na \sim \mathscr{NB}(n, p)$	$\text{var}(\hat{p}) = \dfrac{p(1-p)^2}{n}$
Negative binomial[b] $\mathscr{NB}(r, p)$	$\hat{m} = \dfrac{\hat{r}\hat{p}}{1-\hat{p}}$ $f(\hat{r}) = -n\psi(\hat{r}) + n\log\left(\dfrac{\hat{r}}{\hat{r}+\hat{m}}\right) + \sum_{j=1}^{n} \Psi(\hat{r} + X_j)$ $f'(r) = -n\Psi'(\hat{r}) + \dfrac{n}{\hat{r}} - \dfrac{n}{\hat{r}+\hat{m}} + \sum_{j=1}^{n} \Psi'(\hat{r} + X_j)$	$\hat{m} = a$ $\hat{r}_k = \hat{r}_{k-1} - \dfrac{f(\hat{r}_{k-1})}{f'(\hat{r}_{k-1})}$ $\hat{r}_0 = \dfrac{\hat{m}^2}{(b - na^2)/(n-1) - a}$ Iterate until convergence. N.B. Applicable only if $\hat{r}_0 > 0$. $\hat{p} = \dfrac{1}{1 + \hat{r}/\hat{m}}$	See refs. 135–137 for small sample bias and variance.	$\text{var}(\hat{p}) = \dfrac{m\,\text{var}(\hat{r}) + r^2\,\text{var}(\hat{m})}{(m+r)^4 n}$ $\text{var}(\hat{m}) = \dfrac{m(1 + m/r)}{n}$ $\text{var}(\hat{r}) = \dfrac{1}{n[\Psi'(m+r) - \Psi'(r) + m/(mr + r^2)]}$ $\text{cov}(\hat{m}, \hat{r}) = 0$

[a] Ψ and Ψ' denote the digamma and trigamma functions, respectively.
[b] See reference 139a, p. 685, for computational formulae for Ψ and Ψ'.

The confidence region of interest is then $0.355 < \mu < 2.27$, $0.853 < \sigma^2 < 6.010$. A user may now employ the point estimates $\hat{\mu}$ and $\hat{\sigma}^2$ in his simulation and then, for example, rerun the simulation with μ_l, σ_l^2 and μ_u, σ_u^2 with the same random number stream to determine the effects of variation in input parameters.

The MLE sampling distribution can also assist us in another way. Suppose that we are interested in how the real system responds to new activity levels, and we wish to gain insight by first simulating these activity levels in our simulation model and then comparing the results. Let β_1, β_2, and β_3 denote input parameters corresponding to the activity levels. Suppose that data available on the current activity level give an MLE $\hat{\beta}$ for the exponential distribution. The question then arises whether or not the hypothetical input activity levels differ significantly from the current activity level. One quick check would be to see whether the confidence interval for β includes any of the hypothetical β_is. If it includes β_1 but not β_2 and β_3, for example, we are skeptical of treating β_1 as an activity level that differs significantly from the present one.

Using the relationship between the chi-square and gamma distributions, we note that in the exponential case $n\hat{\beta}/\beta$ has the chi-square distribution with $2n$ degrees of freedom, a convenience for hypothesis testing. For two-parameter distributions we use the asymptotic normality property and the asymptotic covariance matrix. Let $\hat{\theta}_1$ and $\hat{\theta}_2$ be the MLE for θ_1 and θ_2, respectively, with $\text{var}(\hat{\theta}_1) = \sigma_1^2/n$, $\text{var}(\hat{\theta}_2) = \sigma_2^2/n$, and $\text{cov}(\hat{\theta}_1, \hat{\theta}_2) = \sigma_{12}/n$. When n is large, the statistic

$$\chi_2^2 = \frac{n}{\sigma_1^2\sigma_2^2 - \sigma_{12}^2} \times [\sigma_2^2(\hat{\theta}_1 - \theta_1)^2 - 2\sigma_{12}(\hat{\theta}_1 - \theta_1)(\hat{\theta}_2 - \theta_2) + \sigma_1^2(\hat{\theta}_2 - \theta_2)^2]$$

has the chi-square distribution with 2 degrees of freedom.

If we wish to test the hypothesis that $\hat{\theta}_1$ and $\hat{\theta}_2$ are not significantly different from θ_1^* and θ_2^*, respectively, we test χ_2^2, with θ_1^* and θ_2^* replacing θ_1 and θ_2, respectively. We accept the hypothesis if $\chi_2^2 < \chi_{2,1-\alpha}^2$ and reject it otherwise. Here $\chi_{2,1-\alpha}^2$ is the critical value of chi square for 2 degrees of freedom for significance level α.

9.3 FITTING DISTRIBUTIONS

The chi-square test in Section 7.7 gave us a basis for determining the appropriateness of assuming that all random numbers from a particular generator have the uniform distribution $\mathscr{U}(0, 1)$. The suitability of the test stems from the general result that, if we have $X_1, \ldots, X_n$ independent data points, each with probability distribution function F, and if f_j are the number of sample points in the interval (θ_{j-1}, θ_j) for $j = 1, \ldots, k$, the statistic*

(3.1) $$\chi^2_{k-1} = \sum_{j=1}^{k} \frac{(f_j - np_j)^2}{np_j}$$

(3.2) $$p_j \equiv \int_{\theta_{j-1}}^{\theta_j} dF(x)$$

asymptotically has the chi-square distribution with $k - 1$ degrees of freedom. This remarkable fact facilitates testing the goodness-of-fit of a variety of distributions *when* the parameters of the distribution under consideration are known. When the parameters are unknown the result does not hold and requires modification. It turns out that, if the data are grouped into the intervals (θ_{j-1}, θ_j), $j = 1, \ldots, k$, and the MLE of the, say, s parameters of F are derived from the grouped data and substituted into the p_j in (3.2), then (3.1) asymptotically has the chi-square distribution with $k - s - 1$ degrees of freedom. Grouping the data, however, leads to a loss of information, and consequently we prefer, whenever possible, to work with MLE based on the ungrouped data as described in Section 9.2. If these estimators are substituted into the p_j, then (3.1) asymptotically has a distribution function that lies between that for chi square with $k - s - 1$ and that for chi square with $k - 1$ degrees of freedom [146].

For example, if $k = 11$ and $s = 2$, the critical value for a test of size $\alpha = 0.05$ lies between 15.51 and 18.31, as Table 24 shows. Notice also in Table 24 that if $k = 8$ and $s = 2$ the critical value lies between 9.24 and 12.02 for $\alpha = 0.05$; this is a larger interval relative to the end points than is the case for $k = 11$.

If the null hypothesis is rejected when (3.1) exceeds $\chi^2_{k-s-1,1-\alpha}$, then the probability of rejecting the null hypothesis when it is true is greater than size

Table 24 Selected Critical Values of $\chi^2_{j,1-\alpha}$

	j			
α	5	7	8	10
0.01	15.09	18.48	20.09	23.21
0.05	11.07	14.07	15.51	18.31
0.10	9.24	12.02	13.36	15.99
0.25	6.63	9.04	10.22	12.55

* See reference 146 for a detailed description of the test.

α. Alternatively rejection when (3.1) exceeds $\chi^2_{k-1,1-\alpha}$ means that the probability of rejecting the null hypothesis when it is true is less than α. The net conclusion of these observations is that one accepts the null hypothesis when (3.1) is less than $\chi^2_{k-s-1,1-\alpha}$, rejects it when (3.1) exceeds $\chi^2_{k-1,1-\alpha}$, and, if one is conservative, accepts it if (3.1) falls between the two critical values, thereby reducing α, the size of the test. Of course, this gray area becomes smaller as k increases. It should be noted that, if the method of moments is used to estimate the s parameters of a distribution, the resulting statistic in (3.1) asymptotically has a distribution bounded on the left by that of $\chi^2(k-s-1)$ but not necessarily bounded on the right by that of $\chi^2(k-1)$ [149].

It is recommended that $f_j > 5$ for acceptable results with the chi-square test [15]. This suggests that we choose $np_j > 5$, thus implying that the θ_j should be chosen so that

$$\frac{5}{n} < \int_{\theta_{j-1}}^{\theta_j} dF(x) = F(\theta_j) - F(\theta_{j-1}). \tag{3.3}$$

One approach is to set

$$p_1 = p_2 \cdots = p_k \tag{3.4}$$

so that the intervals are equally probable. Then we have

$$\frac{5j}{n} < \frac{j}{k} = F(\theta_j) \qquad j = 1, \ldots, k \tag{3.5}$$

so that $n > 5k$.

For the Weibull distribution, for example, we have

$$F(x) = 1 - e^{-(x/\beta)^\alpha} \tag{3.6}$$

so that

$$\begin{gathered} \frac{j}{k} = 1 - e^{-(\theta_j/\beta)^\alpha} \\ \theta_j = \beta^\alpha \sqrt{\log\,[k/(k-j)]}. \end{gathered} \tag{3.7}$$

In the case of $\alpha = 1$, we have the exponential $\mathscr{E}(\beta)$, so that

$$\theta_j = \beta \log\left(\frac{k}{k-j}\right) \tag{3.8}$$

for example, $k = 10$ leads to the θ_j in Table 25.

For $\mathscr{G}(\alpha, \beta)$ it is easily seen that the θ_j/β are determined by

$$\gamma\left(\alpha, \frac{\theta_j}{\beta}\right) = \frac{\Gamma(\alpha)j}{k} \tag{3.9}$$

Table 25 Intervals for Chi-Square Test of the Exponential Distribution ($k = 10$)

j	$\frac{\theta_j}{\beta}$	j	$\frac{\theta_j}{\beta}$
0	0	6	0.916
1	0.104	7	1.203
2	0.223	8	1.609
3	0.351	9	2.302
4	0.513	10	∞
5	0.693		

where $\gamma(.,.)$ is the incomplete gamma function, tables of which may be found in reference 153. For the normal distribution

$$\text{Prob}\left[-\infty < y < \frac{\theta_j - \mu}{\sigma}\right] = \frac{j}{k} \tag{3.10}$$

where y has $\mathcal{N}(0, 1)$. One needs only a table of the unit normal distribution to determine $(\theta_j - \mu)/\sigma$ and then θ_j. For $k = 10$ the intervals are given in Table 26.

As an example, we wish to test the hypothesis that a sample of $n = 100$ observations are drawn from the normal distribution $\mathcal{N}(\mu, \sigma^2)$. We do not know μ and σ^2, but their MLEs are $\hat{\mu} = 9.22616$ and $\hat{\sigma}^2 = 10.358159$, respectively. We select $k = 10$, and, using the results in Table 26, compute the 10 intervals in column 2 of Table 27. The frequency counts are given in column 3, and since the first and ninth intervals do not have at least 5 observations we combine intervals 1 and 2 and intervals 9 and 10. We now have

Table 26 Intervals for Chi-Square Test of the Normal Distribution ($k = 10$)

j	$\frac{\theta_j - \mu}{\sigma}$	θ_j	j	$\frac{\theta_j - \mu}{\sigma}$	θ_j
0	$-\infty$	$-\infty$	6	0.25	$\mu + 0.25\sigma$
1	-1.28	$\mu - 1.28\sigma$	7	0.52	$\mu + 0.52\sigma$
2	-0.84	$\mu - 0.84\sigma$	8	0.84	$\mu + 0.84\sigma$
3	-0.52	$\mu - 0.52\sigma$	9	1.28	$\mu + 1.28\sigma$
4	-0.25	$\mu - 0.25\sigma$	10	∞	∞
5	0	μ			

Table 27 Sample Tableau for Chi-Square Test of the Normal Distribution

$\mu = 9.22616, \quad \hat{\sigma}^2 = 10.358159, \quad n = 100, \quad k = 10$

j	Interval	f_j	np_j	$f_j - np_j$	$(f_j - np_j)^2$	$\frac{1}{np_j}$	$\frac{(f_j - np_j)^2}{np_j}$
1	$-\infty < X < 5.107$	0	20	-14	196	0.05	9.8
2	$5.107 < X < 6.523$	6					
3	$6.523 < X < 7.553$	25	10	15	225	0.1	22.5
4	$7.553 < X < 8.423$	13	10	3	9	0.1	0.9
5	$8.423 < X < 9.226$	18	10	8	64	0.1	6.4
6	$9.226 < X < 10.031$	8	10	-2	4	0.1	0.4
7	$10.031 < X < 10.900$	11	10	1	1	0.1	0.1
8	$10.900 < X < 11.926$	8	10	-2	4	0.1	0.4
9	$11.926 < X < 13.346$	3	20	-9	81	0.05	4.05
10	$13.346 < X < \infty$	8					
		100	100	0			$\chi_7^2 = 44.55$

only 8 intervals, for which we list the quantities of interest in columns 4 through 8. The test statistic is $\chi_7^2 = 44.55$. For $\alpha = 0.05$ the critical values that concern us are $\chi^2_{5,0.95} = 11.070$ and $\chi^2_{7,0.95} = 14.067$. Clearly the hypothesis is rejected.

For discrete distributions it is seldom possible to define exactly equally probable intervals. One may, however, use

$$\text{(3.11)} \qquad \text{Prob}(X \leq \theta_j) \leq \frac{j}{k} < \text{Prob}(X \leq \theta_j + 1)$$

to determine θ_j. To generate the cumulative probability distributions we may sum the probabilities obtained using the recursive formulae in Section 8.5.

Occasionally the use of (3.11) to compute the θ_j will run into difficulty. If a distribution has a sharp peak it may occur that

$$\text{(3.12)} \qquad \text{Prob}(X \leq \theta_j) \leq \frac{j}{k} \leq \frac{j+1}{k} < \text{Prob}(X \leq \theta_j + 1).$$

When this happens one solution is to work with unequal probability intervals. For example, one may have $0, 1/k, \ldots, (j-1)/k, (j+1)/k, (j+2)/k, \ldots, 1$ as the cumulative probability function over $k - 1$ intervals. Alternatively one may choose k small enough to preclude (3.12) from materializing. However, reducing k increases concern about the appropriate critical value, as mentioned earlier.

The Kolmogorov-Smirnov (K–S) test, described in Section 6.5, is also used to study how well a specified continuous distribution fits a set of data. As mentioned, this test is exact compared to the asymptotic character of the chi-square test, but, like the latter, the K–S test does not hold precisely when sample values are substituted for the parameters of the distribution being fitted. Nor have bounds on the distribution of the test statistic been established as in the chi-square case. However, modified critical values for the test statistic have been given by Lilliefors for testing for the exponential [148] and normal [147] distributions with MLEs substituted for the unknown parameters. The critical values are presented in Tables A6 and A7 in Appendix A.

The K–S test was also applied to the set of data tested earlier for normality using the chi-square test. The test statistic* was

$$D_{100} = 0.14.$$

From Table A6 the critical value at $\alpha = 0.05$ is 0.0886, so that the hypothesis of normality is again rejected.

The convenience of the modified K–S test is that it can be used to test for exponentiality and normality for all n, whereas the modified chi-square test applies only for large n. However, the fact that it is limited to these two distributions at present restricts its range of application, and in practice the modified chi-square test continues to receive wider use. Incidentially, use of the K–S test without adjustment for parameter substitution can seriously undermine the validity of any conclusion drawn from the test [147]. In the normal case, for example, the critical value for $\alpha = 0.06$ in Table A6 is slightly smaller than the corresponding critical value for $\alpha = 0.20$ in regular K–S tables.

9.4 ESTIMATION OF AUTOREGRESSIVE SCHEMES

In Section 8.6 we noted that an autoregressive scheme can be used to generate autocorrelated behavior as input to a simulation. As later sections show, the autoregressive representation is also a convenient form for describing a variety of interesting statistical facts about the output of a simulation. This relevance in both input and output suggests that we devote ample space to a description of the autoregressive representation and its estimation.

From Section 6.6 we have the Yule-Walker equations

$$\sum_{s=0}^{p} b_s R_{s-r} = \begin{cases} \sigma^2 & r = 0 \\ 0 & r > 0. \end{cases} \tag{4.1}$$

* See Section 7.7. Critical values for the K–S test are given in reference 151.

For $r > 0$ we can rewrite (4.1) as

$$\sum_{s=1}^{p} b_s R_{s-r} = -R_r \qquad r = 1, \ldots, p. \tag{4.2}$$

Substituting the sample autocovariances

$$\hat{R}_s = \frac{1}{n} \sum_{t=1}^{n-s} (X_t - \bar{X}_n)(X_{t+s} - \bar{X}_n) \qquad s > 0$$

$$\bar{X}_n = \frac{1}{n} \sum_{t=1}^{n} X_t \tag{4.3}$$

$$\hat{R}_{-s} = \hat{R}_s$$

into (4.2), we can solve this set of equations for the Yule-Walker estimates $\{\hat{b}_s\}$. Since for large n these estimates are close to the linear least-squares estimates (LLSE) which would be obtained by minimizing

$$\sum_{t=1}^{n} \left[\sum_{s=0}^{p} b_s (X_{t-s} - \mu)^2 \right],$$

we often refer to the $\{\hat{b}_s\}$ as the LLSE. Noting (4.1), we may estimate σ^2 by

$$\hat{\sigma}^2 = \sum_{s=0}^{p} \hat{b}_s \hat{R}_s. \tag{4.4}$$

We now have a way to estimate the quantities of interest. Unfortunately it requires a matrix inversion, an operation we hope to avoid whenever possible. Moreover we do not know the autoregressive order p in advance. We deal with both these problems shortly, but first we describe several sampling properties of $\{\hat{b}_s\}$.

Because of the autocorrelation between $X_t, X_{t-1}, \ldots, X_{t-p}$ we cannot generally treat the resulting least-squares estimates in the same way as we do the estimates in a classical regression problem. However, Mann and Wald [150] have shown that, when n is large and $\{\varepsilon_t\}$ is a sequence of independent, identically distributed random variables, the LLSE assume the sampling properties of classical regression estimates. Table 28 lists several of these large sample properties. The joint asymptotic normality result is due to reference 150, and the covariance result in line 2 is the extension of a result in references 152, 155, and 157, p. 73.

Suppose that we set $p = r > 0$, compute the estimates $\hat{b}_1, \ldots, \hat{b}_r$ and $\hat{\sigma}_r^2$, and want to test the hypothesis

H_0: the autoregressive order is r

against the alternative hypothesis;

H_1: the autoregressive order is greater than r but no greater than q.

Table 28 Sampling Properties of Autoregressive Estimates ($r, s = 1, \ldots, p$)

Line	Property
1	$\lim_{n\to\infty} \sqrt{n}\, E(\hat{b}_s - b_s) = 0$
2	$\lim_{n\to\infty} n \operatorname{cov}(\hat{b}_r, \hat{b}_s) = \sum_{j=0}^{\infty} (b_{r-j}b_{s-j} - b_{p+j-r}b_{p+j-s})$
3	$\lim_{n\to\infty} [\sqrt{n}\,(\hat{b}_1 - b_1), \ldots, \sqrt{n}\,(\hat{b}_p - b_p)]$ are jointly normal with covariances as in line 2
4	$\lim_{n\to\infty} \sqrt{n}\, E(\hat{\sigma}^2 - \sigma^2) = 0$
5	$\lim_{n\to\infty} n \operatorname{var}(\hat{\sigma}^2) = (2 + \kappa)\sigma^4; \quad \kappa + 3 = E(\varepsilon_t^4) < \infty$
6	$\lim_{n\to\infty} n \operatorname{cov}(\hat{b}_s, \hat{\sigma}^2) = 0$
7	$\lim_{n\to\infty} n \operatorname{cov}(\hat{b}_s, \bar{X}_n) = 0$
8	$\lim_{n\to\infty} n \operatorname{cov}(\hat{\sigma}^2, \bar{X}_n) = \frac{\gamma}{b}; \quad \gamma = E(\varepsilon_t^3) < \infty, \quad b \equiv \sum_{s=0}^{p} b_s$

Table 29 Determination of Autoregressive Order

r (1)	$\hat{b}_{r,0}$ (2)	$\hat{b}_{r,1}$ (3)	$\hat{b}_{r,2}$ (4)	$\hat{b}_{r,3}$ (5)	$\hat{b}_{r,4}$ (6)	$\hat{b}_{r,5}$ (7)	$\hat{b}_{r,6}$ (8)	$\hat{b}_{r,7}$ (9)
0	1							
1	1	−0.9440						
2	1	−0.8586	−0.0905					
3	1	−0.8569	−0.0746	−0.0185				
4	1	−0.8566	−0.0736	−0.0065	−0.0140			
5	1	−0.8559	−0.0732	−0.0027	0.0313	−0.0528		
6	1	−0.8558	−0.0732	−0.0027	0.0313	−0.0520	−0.0009	
7	1	−0.8558	−0.0718	−0.0035	0.0314	−0.0499	0.0230	−0.0281
8	1	−0.8556	−0.0720	−0.0011	0.0311	−0.0499	0.0237	−0.0207
9	1	−0.8555	0.0719	−0.0033	0.0315	−0.0501	0.0237	−0.0203

When H_0 is true we may show that the distribution of the statistic

$$F_{q-r,n-q} = \left(\frac{\hat{\sigma}_r^2}{\hat{\sigma}_q^2} - 1\right)\left(\frac{n-q}{q-r}\right) \tag{4.5}$$

converges to that of the F distribution with $q - r$ and $n - q$ degrees of freedom as n increases [143, p. 336]. This means that if r is, in fact, the correct autoregressive order and n is large we can make the approximate probability statement

$$\text{Prob}(F_{q-r,n-q} < F_{1-\alpha,q-r,n-q}) \sim 1 - \alpha. \tag{4.6}$$

If the inequality holds we accept H_0 and take $p = r$ as the autoregressive order. If it does not hold we reject H_0 and go on to estimate coefficients for a scheme of higher order than r. This test is described in detail in reference 143.

To illustrate how the test works we use the data described in Section 6.6. To repeat, observations were collected at unit intervals on the queue length process during a simulation run of the single server queueing problem described in Section 6.5 with $\omega_1 = 4.5$ arrivals and $\omega_2 = 5$ jobs serviced per hour. We choose $q = 50$ and computed the LLSE of $\{b_s\}$ and $F_{q-r,n-q}$ for $r = 0, \ldots, q$. Table 29 lists the sample statistics of interest for $r = 0, \ldots, 9$. Notice the sharp drops in $F_{q-r,n-q}$ in column 13 as r increases initially. For $\alpha = 0.025$ we see that a scheme of order $r = 2$ fits the data adequately. This

r	$\hat{b}_{r,8}$ (10)	$\hat{b}_{r,9}$ (11)	$\hat{\sigma}_r^2$ (12)	$q - r$	$F_{q-r,3950}$ (13)	Approximate[a] $F_{0.975,q-r,\infty}$ (14)	m_4 (15)
0			72.2910	50	659.91	1.44	72.29
1			7.8699	49	1.47	1.44	2509.52
2			7.8053	48	0.82[b]	1.44	3009.62
3			7.8027	47	0.83	1.45	3122.52
4			7.8011	46	0.80	1.45	3211.70
5			7.7793	45	0.57	1.46	3570.13
6			7.7793	44	0.59	1.46	3577.21
7			7.7733	43	0.53	1.47	3782.90
8	−0.0086		7.7727	42	0.53	1.47	3848.63
9	−0.0031	−0.0064	7.7724	41	0.52	1.48	3898.39

[a] The critical values were computed by linear interpolation between 40 and 60 degrees of freedom.
[b] Testing at $\alpha = 0.025$ determines $p = 2$.

scheme was used to produce the correlogram and sample spectrum in Section 6.6. The quantity m_4 in column 15 assumes meaning in Chapter 10.

It turns out that we can compute autoregressive estimates of successively higher order from recursive formulae. If $\hat{b}_{r,s}$ is the linear least-square estimate of the sth coefficient in a scheme of order $r \geq 1$, then we may show that the relevant quantities can be computed as in Table 30.* The ANALYSIS and ANALYS computer programs in Appendix B follow the steps in Table 30.

Since $n - q$ is usually large in simulation applications it is worth while to note that one may show that the distributions of $n(1 - \hat{\sigma}_q{}^2/\hat{\sigma}_r{}^2)$ converges to that of chi square with $q - r$ degrees of freedom as n increases.

Table 30 Formulae for Determining Sample Autoregressive Scheme

Mean	$\bar{X}_n = \frac{1}{n}\sum_{s=1}^{n} X_s$	
Autocovariances	$\hat{R}_s = \frac{1}{n-s}\sum_{t=1}^{n-s} [(X_t - \bar{X}_n)(X_{t+s} - \bar{X}_n)]$	$s = 0, \ldots, q$
Autoregressive coefficients	$\hat{b}_{0,0} = 1$	
	$v_{r-1} = \sum_{s=0}^{r-1} \hat{b}_{r-1,s}\hat{R}_s$	
	$w_{r-1} = \sum_{s=0}^{r-1} \hat{b}_{r-1,s}\hat{R}_{r-s}$	
	$\hat{b}_{r,r} = -\frac{w_{r-1}}{v_{r-1}}$	
	$\hat{b}_{r,s} = \begin{cases} 1 \\ \hat{b}_{r-1,s} + b_{r,r}\hat{b}_{r-1,r-s} \end{cases}$	$s = 0$; $r > 1;\ s = 1, \ldots, r-1$
Residual variance	$\hat{\sigma}_r^2 = \sum_{s=0}^{r} \hat{b}_{r,s}\hat{R}_s$	
Test statistic	$F_{q-r,n-q} = \left(\frac{\hat{\sigma}_r{}^2}{\hat{\sigma}_q{}^2} - 1\right)\left(\frac{n-q}{q-r}\right)$	$r = 0, \ldots, q-1$
Critical value	$F_{1-\alpha,q-r,n-q}$ is determined from F distribution table for significance level α.	
Alternative test statistic	$n\left(1 - \frac{\hat{\sigma}_q{}^2}{\hat{\sigma}_r{}^2}\right)$	$r = 0, \ldots, q-1$
Critical value	$\chi^2_{1-\alpha}(q - r)$	

* See reference 140.

Situations may occur in which successive testing for $r = 0, \ldots, q - 1$ does not produce an acceptance. Of the several circumstances that may be responsible for this we discuss four. First, there is always the statistical chance of erring. Second, the choice of q may not be sufficiently large to account for all the variation. With the computational speeds presently available, however, a q of 50 is not unreasonable when n exceeds 500.

Third, the presence of periodicities in the time series, regardless of how irregular, may lead to difficulties unless q is sufficiently large. Every peak in the spectrum requires at least two roots, one being the complex conjugate of the other. For example, ten peaks require at least a p of 20 in order to be revealed. Since we seldom are aware of the extent of periodicity in advance, a large q increases the probability of detecting an acceptable p.

The fourth reason also relates to model specification. If in reality $\{X_t\}$ has a finite rather than an infinite moving average representation, then its autoregressive order is infinite with autoregressive coefficients that decline geometrically. To capture an adequate amount of the structure of mean-square variation, p must be large. Again a large q enables us to determine an adequate p.

These reasons encourage us to make q large, even though we know that a small q meets the needs of many time series, Table 29 being an example. Since a more tailored treatment of processes having periodicities or a finite moving average scheme would require the investigator to be familiar with a considerably broader range of time series literature, we consider a larger-than-necessary q a small price to pay for quick answers.

We may use the results in Table 28 and in Whittle [157] to derive a rough measure of variance for the spectrum estimator $\hat{g}(\lambda)$, based on substitution into (6.6.15). Let

$$B(\lambda) = \sum b_s e^{-i\lambda s} \tag{4.7}$$

$$\hat{B}(\lambda) = \sum_{s=0} \hat{b}_s e^{-i\lambda s}$$

and let the asterisk superscript denote a complex conjugate. Assuming n large and using the Taylor series expansion, we have

$$2\pi\hat{g}(\lambda) = \frac{\hat{\sigma}^2}{|\hat{B}(\lambda)|^2} \sim \frac{\sigma^2}{|B(\lambda)|^2} + \frac{\hat{\sigma}^2 - \sigma^2}{|B(\lambda)|^2} - \frac{\sigma^2[\hat{B}(\lambda) - B(\lambda)]}{B(\lambda)|B(\lambda)|^2} - \frac{\sigma^2[B^*(\lambda) - B^*(\lambda)]}{B^*(\lambda)|B(\lambda)|^2} + \cdots \tag{4.8}$$

$$\frac{\hat{g}(\lambda)}{g(\lambda)} \sim 1 + \frac{\hat{\sigma}^2 - \sigma^2}{\sigma^2} - B^*(\lambda)[\hat{B}(\lambda) - B(\lambda)] + \frac{B(\lambda)[\hat{B}^*(\lambda) - B^*(\lambda)]}{|B(\lambda)|^2} + \cdots.$$

Then we have

$$\operatorname{var}\left[\frac{\hat{g}(\lambda)}{g(\lambda)}\right] \sim \operatorname{var}\left[\frac{\hat{\sigma}^2}{\sigma^2}\right] + \left[\frac{2\pi g(\lambda)}{\sigma^2}\right]^2 [B^*(\lambda)]^2 \operatorname{var}[\hat{B}(\lambda)]$$

$$+ [B(\lambda)]^2 \operatorname{var}[\hat{B}^*(\lambda)] + \frac{4\pi g(\lambda)}{\sigma^2} \operatorname{cov}[\hat{B}(\lambda), \hat{B}^*(\lambda)]$$

$$= \frac{2}{n}\left\{1 + \frac{\kappa}{2} + \frac{[\sum_{r,s=0}^{p} b_r b_s \cos\lambda(p-r-s)] \times \sum[_{r,s=0}^{p} b_r b_s \sin\lambda(p-r-s)]}{\sin\lambda \sum_{r,s=0}^{p} b_r b_s \cos\lambda(r-s)} + p - 2\frac{\sum_{r,s=0}^{p} r b_r b_s \cos\lambda(r-s)}{\sum_{r,s=0}^{p} b_r b_s \cos\lambda(r-s)}\right\}. \tag{4.9}$$

This expression follows from Table 28 and reference 157, p. 73. If one estimates $g(\lambda)$ by $\hat{g}(\lambda)$ and plots these sample ordinates for selected λ in $(0, \pi)$, it is instructive to compute an estimate of $\operatorname{var}[\hat{g}(\lambda)/g(\lambda)]$ by neglecting κ and substituting $\hat{b}_1, \ldots, \hat{b}_p$ and $\hat{\sigma}^2$ for the population parameters in (4.9). We return to this topic later when we discuss the desirability of using an autoregressive model to characterize a simulation output process.

EXERCISES

1. One input to a simulation experiment is a stream of independent gamma variates from $\mathcal{G}(\alpha, \beta)$ The statistical design of the experiment calls for running the model at two settings:

   1. $\alpha = 3.5$, $\beta = 1.5$, and
   2. $\alpha = 3.03$, $\beta = 0.95$.

   Sample data are available for this input process. It is decided to make a third run of the model, using the MLE of α and β derived from the sample data.

   a. Given $a = 2.9047$, $c = 0.8982$, and $n = 200$, compute the MLE of α and β using Table A1.
   b. At the 0.10 significance level, test the hypothesis that the values $\alpha = 3.5$ and $\beta = 1.5$ give a significantly different parameter setting from that which the MLE would give.
   c. Test $\alpha = 3.03$ and $\beta = 0.95$ in the same way.
   d. In view of the statistical testing evidence in Exercises 1b and c which setting or settings would be statistically redundant?

2. a. Compute the MLE of the covariance matrix of $\hat{\alpha}$ and $\hat{\beta}$ in Exercise 1.
   b. Compute a confidence region for α and β using the sample covariance matrix as the true covariance matrix.
   c. Compute the sample correlation between $\hat{\alpha}$ and $\hat{\beta}$. Would independent confidence intervals as exemplified in the normal case (Section 9.2) be appropriate?

3. The following data are available on the interarrival times between jobs in a service shop:

   17.88 28.92 33.00 41.52 42.12 45.60 48.48 51.84
   51.96 54.12 55.56 67.80 68.64 68.64 68.88 84.12
   93.12 98.64 105.12 105.84 127.92 128.04 173.40.

   a. Fit these data to an exponential distribution by finding the MLE of β.
   b. Using the Kolmogorov-Smirnov test, test the hypothesis that these data come from an exponential distribution with significance level 0.05.

4. a. In fitting the data in Exercise 3 to a Weibull distribution the MLE were computed to be $\hat{\alpha} = 2.102$ and $\hat{\beta} = 81.99$. Find a 0.05 exact confidence interval for α.
   b. Find a 0.05 confidence interval for β.
   c. Find an unbiased estimate of α.

5. Assuming that n is sufficiently large so that normality holds, derive a confidence interval for λ in the Poisson case in terms of $\hat{\lambda}$ and n.

6. Assuming that n is sufficiently large so that normality holds, derive a confidence interval for p in the binomial case in terms of $\hat{p}$, r, and n.

7. You are given $X_1, \ldots. X_n$ from $\mathscr{U}(a, b)$.
   a. Write down the corresponding likelihood function.
   b. Derive the MLE of a and b. (*Hint:* consider the order statistics $X_{(1)}, \ldots, X_{(n)}$.)
   c. Derive $E(\hat{a})$, $E(\hat{b})$, $\text{var}(\hat{a})$, and $\text{var}(\hat{b})$. (*Hint:* The ith-order statistic from n uniformly distributed variates on (0, 1) is from $\mathscr{B}e(i, n - i + 1.)$
   d. Compare the MLE $\hat{a}$ and $\hat{b}$ with the estimators

$$\tilde{a} = \frac{nX_{(1)} - X_{(n)}}{n - 1}, \qquad \tilde{b} = \frac{nX_{(n)} - X_{(1)}}{n - 1}$$

   from the viewpoints of bias, variance, and mean-square error.

CHAPTER 10

OUTPUT ANALYSIS

10.1 INTRODUCTION

The purpose of a simulation is to produce numbers whose interpretation leads to an improved understanding of the system being simulated. Unfortunately circumstances can easily arise in a simulation study that lead to a misinterpretation of the data and consequently to a misunderstanding of the system. These circumstances include the following:

1. poorly chosen pseudorandom number generator,
2. inappropriate approximate random variate generation techniques,
3. input parameter misspecification,
4. programming errors,
5. model misspecification,
6. data collection errors in simulation,
7. poor choice of descriptors (parameters) to estimate,
8. peculiarities of the estimation method,
9. numerical calculation errors,
10. influence of initial conditions on data,
11. influence of final conditions on data and on estimation method, and
12. misuse of estimates.

Formidable as this list appears, there is much that can be done to combat the threat from these sources. In particular, a conscientious investigator can take two steps. First, he can design his simulation to dilute the potential influence of several of these items. In Chapters 7, 8, and 9 design considerations aimed at such dilution for circumstances 1, 2, and 3, respectively, were described. Moreover the benefit of looking at the spectrum and autocorrelation function of a simulation output process to verify that its behavior is consistent with the investigator's intentions was discussed in Section 6.4. Identifying unwanted behavior in this way reduces the chance of model misspecification (item 5). In addition, we discuss in Chapter 11 model *validation*, which is intended to check for consistency between a real world system and the simulation model that supposedly represents the system. All these are *preventive* measures, as is the search for programming errors (item 4).

Second, an investigator can study the implications on output of establishing alternative sets of prevailing conditions in his simulation. In particular, this study applies to circumstances 10 and 11. He can also analyze the output in a way that reveals the influence of certain errors and demarcates the ranges over which numerical results can be used without substantive additional error. Here *confidence intervals* for population parameters provide an assessment of how representative corresponding sample parameters are. Studying the implications of prevailing conditions and computing confidence intervals are *qualifying* measures, although we should regard action taken to combat any implications as preventive in character.

The purpose of this chapter is to describe ways of employing additional preventive measures in experimental design and also methods for properly qualifying simulation output. Ideally we would like to treat each potential cause of error in our list separately, but this is not possible. For example, the influence of initial conditions (item 10) depends on the choice of descriptors (7) and on the way we estimate these descriptors (8). Moreover, study of these three problems is fruitless unless programming errors (4) and model misspecifications (5) have been removed.

Although the principal emphasis in this book is on dynamic simulation, we can benefit from a review in Section 10.2 of statistical considerations in static simulation, where time evolution plays no role. The value of this review becomes apparent when we study analogous problems in a dynamic setting in the rest of the chapter.

10.2 STATIC SIMULATION OUTPUT ANALYSIS

Sampling experiments designed to learn about the properties of statistical estimators, when these properties are unavailable analytically, belong to the category of static simulation. For example, in [174] Thoman, Bain, and Antle study the statistical properties of the maximum likelihood estimator of the parameters of the Weibull distribution in small samples. The statistical analysis of PERT networks such as in Fig. 18 also belongs to the static category. In most static stochastic simulations, output consists of statistically independent events, each of which follows the same probability law. Given n independent observations $X_1, \ldots, X_n$ generated in a particular run, we can use these data to infer the characteristics of the underlying probability law in three ways. First, we can estimate one or several population descriptors generic to all but a few probability laws. Table 31 lists the most common of

Table 31 Population Descriptors of Probabilistic Behavior[a]

Descriptor	Definition	Describes	Estimator
Mean	$\mu = E(X)$	Average value	$\bar{X}_n = \frac{1}{n}\sum_{t=1}^{n} X_t$
Median	$x_0: \int_{-\infty}^{x_0} dF_X(x) = \frac{1}{2}$	Value exceeded half the time	$\begin{cases} X_{[(n+1)/2]} & n \text{ odd} \\ \frac{X_{(n/2)} + X_{(n/2+1)}}{2} & n \text{ even} \end{cases}$
Mode[b]	$x_0: dF(x_0) = \max dF(x)$	Most likely value	$\frac{z_{i*} + z_{i*+1}}{2}, \quad H_{i*} = \max H_i$
Variance	$\sigma^2 = E[X - E(X)]^2$	Dispersion around mean	$s_n{}^2 = \frac{1}{n-1}\sum_{t=1}^{n} (X_t - \bar{X}_n)^2$
Skewness	$\gamma_1 = \frac{E[X - E(X)]^3}{(\sigma^2)^{3/2}}$	Asymmetry of p.d.f. (0 for symmetric p.d.f.)	$\gamma_1 = \frac{1}{n}\sum_{t=1}^{n} \frac{(X_t - X_n)^3}{(s_n{}^2)^{3/2}}$
Excessive kurtosis	$\gamma_2 = \frac{E[X - E(X)]^4}{(\sigma^2)^2} - 3$	Tails of p.d.f. (0 for normal p.d.f.)	$\gamma_2 = \frac{1}{n}\sum_{t=1}^{n} \frac{(X_t - \bar{X}_n)^4}{(s_n{}^2)^2} - 3$

Quantile	$x_q;\ q = 1, \ldots, Q \quad \int_{x_{q-1}}^{x_q} dF_X(x) = \frac{1}{Q}$; $x_0 = -\infty, \quad x_Q = \infty$	Subranges of x for equiprobable intervals	$X_{(nj/Q)}$	$j = 1, \ldots, Q$; n/Q integer
Cumulative p.d.f.[c]	$F_X(x) = \int_{-\infty}^{x} dF_X(u)$	Prob $(X \leq x)$	$\hat{F}_X(x) = \frac{1}{n} \sum_{j=1}^{n} I_{(X_{(j)}, \infty)}(x)$	
Histogram	$F_X(z_{i+1}) - F_X(z_i)$; $z_i < z_{i+1} \quad i = 0, \ldots, N \ll n$; $z_0 = -\infty \quad z_N = \infty$	Relative frequency	$H_i = \frac{1}{n} \sum_{j=1}^{n} I_{(z_i, z_{i+1})}(X_{(j)})$	

[a] The quantities $X_{(j)}, \ldots, X_{(n)}$ are the order statistics of the sample $X_1, \ldots, X_n$ so that $X_{(1)}$ is the smallest observation, $X_{(2)}$ the next smallest, ..., and $X_{(n)}$ the largest.
[b] See histogram for definition of bounds $\{z_i\}$.
[c] The indicator function I is defined as

$$I_{(a,b)}(\theta) = \begin{cases} 1 & a < \theta \leq b \\ 0 & \text{elsewhere.} \end{cases}$$

these descriptors. Occasionally one or more moments of a distribution are undefined or infinite, as in the Cauchy case, where we cannot expect the sample mean and variance to be meaningful statistics. In such a case the quantiles often play the role of principal descriptors.

The second way of inferring the true output behavior is to fit the data to a known p.d.f. The choice of the distribution to be fitted is usually made after the histogram and cumulative histogram have been plotted. An example of these sample functions is shown in Fig. 52. Then a test of fit, as described in Section 9.3, is applied to check the adequacy of the choice.

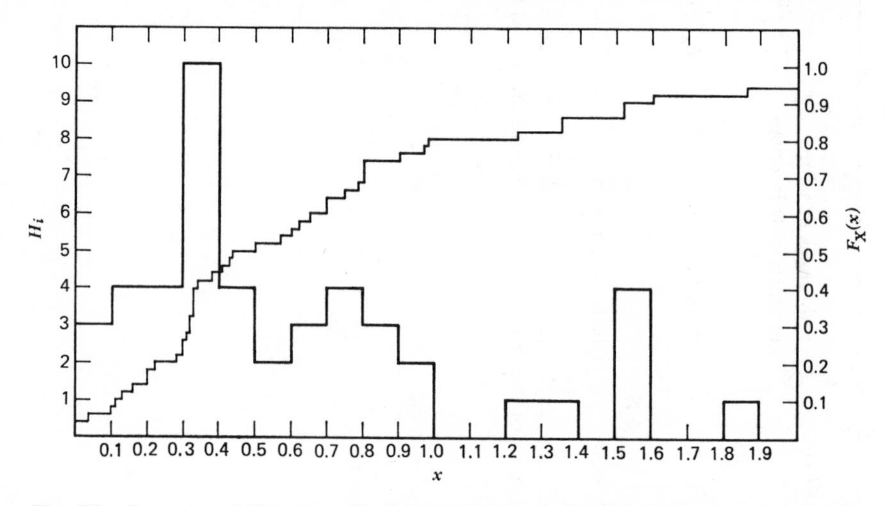

Fig. 52 Examples of histogram H_i and sample cumulative distribution function $F_X(x)$.

The third way applies, for example, to the study of sampling distributions. In reference 174 the Weibull distribution $\mathscr{W}(\alpha, \beta)$ generates the n events which are then used to obtain the maximum likelihood estimate of α and β. By replicating the experiment for fixed n we are able to generate information about the distribution of the MLE for α and β. By performing replications for different n we can, for example, generate estimates of the 0.05, 0.1, 0.90, and 0.95 points of $\hat{\alpha}$ and $\hat{\beta}$ as a function of n.

All the computed quantities discussed so far are sample statistics that converge to their corresponding population parameters as n increases. However, our simulations are necessarily of finite length so that we must content ourselves with working with the sample quantities for finite n. One question we wish to answer is how closely our numerical-valued estimate of a quantity approaches the true unknown value of that quantity. The sample mean provides a useful example.

Since the observations $X_1, \ldots, X_n$ are random variables their mean $\bar{X}_n$ is also a random variable. Then the best we can do is derive a probability statement such as

$$\text{Prob}(w_{\alpha,n} < \mu < v_{\alpha,n}) = 1 - \alpha \qquad 0 < \alpha < 1 \tag{2.1}$$

where $w_{\alpha,n}$ and $v_{\alpha,n}$ are numerical-valued constants determined by $X_1, \ldots, X_n$, $\bar{X}_n$, n, and the confidence level α. Expression (2.1) implies that the probability is $1 - \alpha$ that the interval from $w_{\alpha,n}$ to $v_{\alpha,n}$ includes the mean μ. Expression (2.1) provides us with a *confidence interval* for μ.

Since $\bar{X}_n$ is a random variable, it has a variance

$$\begin{aligned} \text{var}(\bar{X}_n) &= E(\bar{X}_n - \mu)^2 = \frac{1}{n^2} E \sum_{s,t=1}^{n} (X_s - \mu)(X_t - \mu) \\ &= \frac{1}{n^2}\left[\sum_{t=1}^{n} \text{var}(X_t) + \sum_{\substack{s,t=1 \\ s \neq t}}^{n} \text{cov}(X_t, X_s)\right], \end{aligned} \tag{2.2}$$

$\text{var}(X_t)$ denoting the variance of X_t, and $\text{cov}(X_t, X_s)$ denoting the covariance between X_t and X_s. In elementary statistics courses we study the case where $X_1, \ldots, X_n$ are independent and identically distributed, each with variance σ^2. Then

$$\text{cov}(X_t, X_s) = 0 \qquad s, t = 1, \ldots, n;\ s \neq t \tag{2.3}$$

and

$$\begin{aligned} \text{var}(\bar{X}_n) &= \frac{1}{n^2} \sum_{t=1}^{n} \text{var}(X_t) \\ &= \frac{n\sigma^2}{n^2} = \frac{\sigma^2}{n}. \end{aligned} \tag{2.4}$$

Moreover, the central limit theorem tells us that, as the sample size n becomes large, the distribution of the statistic $\sqrt{n}(\bar{X}_n - \mu)/\sigma$ converges to $\mathcal{N}(0, 1)$, provided that $\{X_t\}$ obeys some mild regularity conditions. Then $\bar{X}_n$ approximately has the normal distribution with mean μ and variance σ^2/n. If $\{X_t\}$ is a normal sequence, $\bar{X}_n$ has the exact distribution $\mathcal{N}(\mu, \sigma^2/n)$ and $(\bar{X}_n - \mu)/\sqrt{s_n^2/n}$ has the t distribution with $n - 1$ degrees of freedom, where

$$s_n^2 = \frac{1}{n-1} \sum_{t=1}^{n} (X_t - \bar{X}_n)^2. \tag{2.5}$$

Then we compute the exact confidence interval

$$
\begin{aligned}
w_{\alpha,n} &= \bar{X}_n - t_{n-1,1-\alpha/2}\sqrt{s_n^2/n} \\
v_{\alpha,n} &= \bar{X}_n + t_{n-1,1-\alpha/2}\sqrt{s_n^2/n},
\end{aligned}
\tag{2.6}
$$

$t_{n-1,1-\alpha/2}$ being the $1 - \alpha/2$ point from $t(n - 1)$.

The quantity $(\bar{X}_n - \mu)/\sqrt{s_n^2/n}$ continues to play an important role when $\{X_t\}$ is not normal. It has been shown that, if $X_1, \ldots, X_n$ are independent and have a symmetric p.d.f., approximating the distribution of $(\bar{X}_n - \mu)/\sqrt{s_n^2/n}$ by that of $t(n - 1)$ produces a relatively small error in computing a confidence interval for μ.† Moreover, as n increases it can be shown that the distribution of $(\bar{X}_n - \mu)/\sqrt{s_n^2/n}$ converges to that of $(\bar{X}_n - \mu)/\sqrt{\sigma^2/n}$, which is $\mathcal{N}(0, 1)$. This last result holds also when X_t has an asymmetric p.d.f. It is common practice therefore to treat $(\bar{X}_n - \mu)/\sqrt{s_n^2/n}$ as a t variate and to compute a confidence interval for μ using (2.6), bearing in mind that the error of approximation diminishes as n increases.

The ease of working with the sample mean derives from the fact that it is asymptotically normal and we can easily derive an estimate of its variance from the same n observations. Now asymptotic normality also holds for the sample variance s_n^2 with mean σ^2, *but* with variance

$$
\operatorname{var}(s_n^2) = \sigma^4 \left(\frac{2}{n-1} + \frac{\gamma_2}{n} \right),
\tag{2.7}
$$

where γ_2 is the excess kurtosis as defined in Table 31. If X has the normal distribution no problem occurs because $\gamma_2 = 0$ and $(n - 1)s_n^2/\sigma^2$ has the chi-square distribution with $n - 1$ degrees of freedom. But if X is not normal then it is not possible to apply known theory, for neglect of the fourth moment would underestimate the width of the confidence interval for σ^2.

In addition to the sample variance the problem of inference for the remaining statistics in Table 31 is usually made difficult by the absence of measures of dispersion. We do not have a readily available way to estimate, say, the variance of the mode, based on n observations from the same data on a single run.

To solve the problem we turn to *replication*. Suppose that we collect n' observations on each of k independent replications of the experiment. Let

† See, for example, reference 172.

X_{jt} be the tth observation on the jth replication. We estimate the variance on the jth replication to be

$$s_{j,n'}^2 = \frac{1}{n'-1} \sum_{t=1}^{n'} (X_{jt} - \bar{X}_{j,n'})^2 \tag{2.8}$$

$$\bar{X}_{j,n'} = \frac{1}{n'} \sum_{t=1}^{n} X_{jt} \qquad j = 1, \ldots, k. \tag{2.9}$$

Then our overall estimate of σ^2 is

$$\tilde{s}_n{}^2 = \frac{1}{k} \sum_{j=1}^{k} s_{j,n'}^2 \qquad n = n'k \tag{2.10}$$

which is unbiased. The variance of $\tilde{s}_n{}^2$ is

$$\operatorname{var}(\tilde{s}_n{}^2) = \frac{1}{k^2} \sum_{j=1}^{k} \operatorname{var}(s_{j,n'}^2) = \sigma^4 \left(\frac{2/(n'-1) + \gamma_2/n'}{k} \right) \tag{2.11}$$

which we estimate by V_n/k, where

$$V_n = \frac{1}{k-1} \sum_{j=1}^{k} (s_{j,n'}^2 - \tilde{s}_n{}^2)^2. \tag{2.12}$$

In the present case we see that $s_{j,n'}^2$, $\tilde{s}_n{}^2$, V_n, and k take the places of X_t, $\bar{X}_n$, $s_n{}^2$, and n, respectively, in our earlier discussion of the mean. In particular the $s_{j,n'}^2$ are now the independent observations with which we are working. As n' becomes large the distributions of $\sqrt{n'}\,(s_{j,n'}^2 - \sigma^2)$ and consequently, for fixed k, $\sqrt{kn'}\,(\tilde{s}_n{}^2 - \sigma^2)$ converge to normality. Then $\sqrt{k}\,(\tilde{s}_n{}^2 - \sigma^2)/\sqrt{V_n}$ has the t distribution with $k - 1$ degrees of freedom, and we can construct a confidence interval for σ^2 such that

$$\operatorname{Prob}(\tilde{s}_n{}^2 - t_{k-1,1-\alpha/2}\sqrt{V_n/k} < \sigma^2 < \tilde{s}_n{}^2 + t_{k-1,1-\alpha/2}\sqrt{V_n/k}) \sim 1 - \alpha. \tag{2.13}$$

Before we can extend the above approach to the remaining statistical measures in Table 31, we must clarify certain issues regarding bias and asymptotic normality. Unlike the sample mean and variance, some of the estimators in Table 31 produce biased estimates of their corresponding parameters. Since, however, all these estimators are known to converge in probability to their population values, except in special cases, we know that the bias diminishes to zero as n' increases.

It is not always clear, though, that each statistic in Table 31 has a distribution that converges to normality as n' increases. However, if we take as our overall estimate of the population parameter of interest the average of the k corresponding statistics computed on the independent replications this overall estimate is asymptotically unbiased as n' increases and has a distribution that converges to normality as k increases. We may estimate the variance of this overall estimate in a manner analogous to (2.12) for the variance and then apply the t distribution with $k - 1$ degrees of freedom to approximate the true distribution.

Using the replication approach to induce normality involves one danger that deserves attention. One might suppose that an investigator could take many replications k of relatively short length n'. This would be unfortunate, however, since the bias in the estimator decreases as n', not k, increases. Therefore we should be prepared to take large samples n' to eliminate bias and an acceptable number of replications k to provide a distributional theory for confidence intervals.

To illustrate the danger of small n' we offer the following example. Suppose that we wish to estimate the median of the distribution corresponding to independent variates generated during a sampling experiment. Unknown to the investigator is the fact that this distribution is $\mathscr{E}(1)$. Now the unit exponential has median $\log 2 \sim 0.69315$. Let n' be odd so that the median estimator from Table 31 is $X_{((n'+1)/2)}$, which for $\mathscr{E}(1)$ has expectation [171] $\sum_{i=1}^{(n'+1)/2} [1/(n' - i + 1)]$. For $n' = 3, 5, 7, 9$, and 15 these expectations are, respectively, 0.8333, 0.7833, 0.7595, 0.7456, and 0.7254. Clearly $X_{((n'+1)/2)}$ is a biased estimator of the median. If we were to take k independent replications of this experiment for a particular n' listed above we would in no way diminish the bias as k increased.

As we see in the next several chapters, the conflict between sample size n' on each independent replication and the number of replications k continues to concern us when we analyze dynamic simulation results. It therefore behooves the reader to accommodate himself to the idea that simulation data analysis is not an exercise in elementary statistics. It requires an understanding of bias, variance, and distributional considerations and, most important, the ways in which these considerations interact with one another.

10.3 DYNAMIC SIMULATION OUTPUT ANALYSIS

The population parameters listed in Table 31 continue to interest us in dynamic simulations. However, emphasis usually focuses on average system behavior and on the relationship between events in processes of interest. Again we must develop methods of inferring population parameters of interest, but this time we must, more often than not, acknowledge the dependence between events. As with the input discussed in Chapter 9 and the static output in Section 10.2, the use of statistical theory can considerably improve our ability to make useful inferences regarding the underlying nature of the output.

To solve these problems we must first determine what is to be measured. Certain processes interest us in a simulation designed for a specific purpose, and presumably we would like to acquire a quantitative characterization of each of these processes. The mean or average value of a process generally serves as the mathematical descriptor, and it is this quantity that we often hope to measure during the course of an experiment. Three different simulations illustrate the point.

1. In a job-shop environment we may be interested in studying the effect on job waiting time of using different priority rules for choosing jobs for service. The total waiting time during the simulation run divided by the number of jobs that arrive for service yields an estimated mean waiting time. The most desirable rule would minimize the mean.
2. For an inventory system our concern may be the relationship between orders filled without delay and orders placed, when studying different inventory stockage policies. Total orders filled without delay divided by total orders placed gives us an estimated mean fill rate, the most desirable stockage policy being the one that minimizes this mean.
3. For a financial environment our interest may lie in selecting operating rules that maximize profits. Total profits during a simulation run divided by the simulated time interval of the run yields average profit, and we prefer the operating rule that maximizes this quantity.

In exploratory simulations designed to help us understand a system we generally study several processes simply to determine the kind of information each provides and the value of this knowledge in regard to our objectives. For example, the mean length of the waiting line in the job shop, the mean number of stockout days in the inventory policy study, and the mean cash position in the financial environment case all provide additional information.

At the outset it is instructive to distinguish between two different kinds of means. Let $\{X(t)\}$ be a covariance stationary stochastic process that has been evolving over a long period of time, and suppose that we observe this process at an arbitrarily selected point in time s during its evolution. Then $E[X(s)]$, the expected value of the process at time s, is its *steady-state* mean. All the descriptors described in the preceding paragraphs are steady-state means, and as such each provides one global view of its respective process.

As an alternative concept, let $\{X(t)\}$ be a process that begins its evolution at time t_0 with an initially specified value $X(t_0)$. If we observe this process at an arbitrarily selected time $s > t_0$ then $E[X(s) \mid X(t_0)]$ is the *conditional* mean of the process. The term conditional implies that the expectation is a function of the initial condition $X(t_0)$.

In some simulations the objective may be to study the conditional mean of a system. For example, an outpatient clinic serves patients between 8:00 A.M. and noon. Fifty patients are given appointments for 8:00 A.M. (initial condition). The objective of the study is to find a queue discipline for patient selection that minimizes patient waiting time during the clinic session.

Here the mean waiting time per clinic session is conditional on the fifty patient initial condition. Since clinic sessions are independent one way to analyze this problem is to collect *sample* waiting times in a sequence of clinical sessions and then use the techniques described in Section 10.2.

Regardless of an investigator's interest in the steady-state mean or the conditional mean, *initial conditions* play a role in output data analysis considerations. To understand why this is so, we describe a typical sequence of events that occur when beginning a simulation run, collecting data, and estimating population descriptors.

10.4 INITIAL CONDITIONS

At the beginning of a simulation run an investigator sets the initial states of certain entities. For example, a queueing system requires specification of the following:

1. number of jobs in the system,
2. status of each job,
3. status of each server,
4. completion time of each job in service, and
5. arrival time of the next job.

We denote the initial conditions as $\mathbf{I}$ and let X_t be an observation collected on a process of interest at time t, where $t = 1, \ldots, n$. We also assume that the simulation starts at time $t = 0$.

There are many ways in which X_t can depend on $\mathbf{I}$. For expository purposes we consider the case in which the conditional mean can be represented as

$$\mu_t = E(X_t \mid \mathbf{I}) = \mu + \Delta(\mathbf{I})\alpha^t \qquad 0 \leq \alpha < 1. \tag{4.1}$$

Here $\Delta(\mathbf{I})$ is a function of the initial conditions, whose influence diminishes geometrically as time evolves. Geometric dependence is a characteristic commonly encountered in the study of queueing processes. The steady-state mean is

$$\lim_{t \to \infty} E(X_t \mid \mathbf{I}) = \mu. \tag{4.2}$$

A first-order autoregressive scheme with $\mathbf{I} = X_0$, $\Delta(\mathbf{I}) = \mathbf{I} - \mu$, and $b_1 = -\alpha$ (see Section 6.6) produces the conditional mean in (4.1) and the steady-state mean in (4.2).

In most simulations the objective is to estimate μ, not μ_t. If we form the sample mean

$$\bar{X}_n = \frac{1}{n} \sum_{t=1}^{n} X_t \tag{4.3}$$

we have

$$E(\bar{X}_n \mid \mathbf{I}) = \mu + \frac{\Delta(\mathbf{I})\alpha(1 - \alpha^n)}{n(1 - \alpha)}. \tag{4.4}$$

Our estimate μ has a bias that depends on three factors: the initial conditions $\mathbf{I}$, the damping factor α, and the sample size n. If possible, we would like to select initial conditions in such a way as to minimize their effects on the estimation of μ. For example, in the first-order autoregressive case in Section 8.6 we generate X_0 from the correct marginal distribution so that $\Delta(\mathbf{I}) = 0$ and no bias enters. In general, a judicious selection of initial conditions is beyond our ability because we know so little about the processes of interest.

The damping factor α reflects the rate at which the influence of initial conditions diminishes in X_t as time evolves. From a design point of view α is intrinsic to the process under study and cannot be altered. However, the specified sample size n can be increased to diminish the bias. Note, however, that is we ran k independent replications and collected X_{jt}, $t = 1, \ldots, n'$, on the jth replication, where $\mathbf{I}_j$ are the initial conditions, then the grand mean

$$\bar{X}'_n = \frac{1}{n} \sum_{j=1}^{k} \sum_{t=1}^{n'} X_{jt} \qquad n = kn' \tag{4.5}$$

has expectation

$$E(\bar{X}'_n \mid \mathbf{I}_1, \ldots, \mathbf{I}_k) = \mu + \frac{\alpha(1 - \alpha^{n'})}{n(1 - \alpha)} \sum_{j=1}^{k} \Delta(\mathbf{I}_j). \tag{4.6}$$

For example, with $\mathbf{I}_1 = \mathbf{I}_2 = \cdots = \mathbf{I}_k$ we have

$$E(\bar{X}'_n \mid \mathbf{I}_1, \ldots, \mathbf{I}_k) = \mu + \frac{\alpha(1 - \alpha^{n'})\,\Delta(\mathbf{I}_1)}{n'(1 - \alpha)} \tag{4.7}$$

whose dominant bias term in $1/n'$ is k times greater than that in (4.4).

There remains at least one more design consideration that can affect bias due to initial conditions. We note that as t increases X_t becomes less dependent on **I**. Suppose that we elect to delete t^* observations near the beginning of the simulation from the mean estimator (4.3) so that our new sample mean is

$$\bar{X}_{t^*,n} = \frac{1}{n - t^*} \sum_{t=t^*+1}^{n} X_t \qquad t^* < n \tag{4.8}$$

with expectation

$$E(\bar{X}_{t^*,n} \mid \mathbf{I}) = \mu + \frac{\Delta(\mathbf{I})(\alpha^{t^*} - \alpha^n)\alpha}{(n - t^*)(1 - \alpha)} \tag{4.9}$$

Comparing (4.9) with (4.4), we have

$$|E(\bar{X}_n \mid \mathbf{I})| - |E(\bar{X}_{t^*,n} \mid \mathbf{I})| = \frac{|\Delta(\mathbf{I})|\alpha[n(1 - \alpha^{t^*}) - t^*(1 - \alpha^n)]}{n(n - t^*)(1 - \alpha)} > 0 \tag{4.10}$$

since the term in brackets is positive.† This means that $\bar{X}_{t^*,n}$ has less absolute bias due to initial conditions than $\bar{X}_n$ in (4.3) does.

One approach to choosing t^* is to make several pilot runs and to collect observations at relatively short intervals [159, p. 49]. Then one plots these sample records against time to determine a reasonable deletion time. For example, one may elect to truncate a time series at a point where the first observation in the new series is neither a maximum nor a minimum in that series. For several records the truncation point might be the largest t^* for all series for which the first observation is neither a maximum nor a minimum. An alternative approach is described in reference 162.

At this point we must emphasize that deletion affects not only the bias of an estimator but also its variance. The effect on variance is to increase it. Consequently the verdict on the desirability of deletion must wait until Section 10.7, where we consider the tradeoff between a bias reduction and a variance increase.

As an alternative approach to diluting the influence of initial conditions, consider the following representation. We denote the conditional mean by

$$E(X_t \mid \mathbf{I}) = \mu_t = \mu + \theta_t$$

$$\lim_{t \to \infty} \theta_t = 0 \qquad \lim_{n \to \infty} \left| \sum_{t=1}^{n} \theta_t \right| < \infty.$$

† To see this one need only show $(1 - \alpha^{t^*})/(1 - \alpha^n) \geq t^*/n$. When the equalities at $t^* = 0, n$ and the convexity of the quantity on the left are noted, the result is quickly established.

Suppose that we begin the simulation with an arrival to an empty and idle system. Then we expect queue length and waiting time in the first several intervals to be below average so that $\theta_t \leq 0$ and $\mu \geq \mu_t$. Assuming that we can find an α in (0, 1) such that $\theta_t \leq \mu\alpha^t$, we have

$$\mu\left[1 - \frac{\alpha(1 - \alpha^n)}{n(1 - \alpha)}\right] \leq E(\overline{X}_n \mid \mathbf{I}) \leq \mu.$$

Neglecting α^n, we define a constant δ such that

$$\frac{\alpha}{n(1 - \alpha)} \leq \delta.$$

Here δ denotes an upper bound on the bias due to initial conditions. For $\alpha = 0.95$, we have $n \geq 19/\delta$ so that for $\delta = 0.01$, $n \geq 1900$; for $\delta = 0.05$, $n \geq 380$. More generally, these results suggest that if we adopt the relatively conservative position that $\alpha \leq 0.99$, a run length of $n = 2000$ guarantees $\delta \leq 0.05$.

Our example in this section has been based on a collection of observations $X_1, \ldots, X_n$ collected at periodic intervals in simulated time. Such a model is applicable to collecting observations on queue length, for example, but not on job waiting time. However, it is entirely possible to collect waiting times $W_1, \ldots, W_N$, where W_j is the waiting time of the jth job for which service was completed. If we specify that the simulation be run until N jobs have completed service then the sample mean waiting time is

$$\overline{W}_N = \frac{1}{N}\sum_{j=1}^{N} W_j. \tag{4.11}$$

The effect of initial conditions can be studied using the approach described above with job number j replacing time index t. The benefit of deleting the waiting times near the beginning of a run would be analogous.

There is one method that may prove useful in practice for assessing the dilution over time of bias due to initial conditions. Let

$$\omega_t = \begin{cases} 1 & X_t > \overline{X}_n, X_{t+1} < \overline{X}_n \quad \text{or} \quad X_t < \overline{X}_n, X_{t+1} > \overline{X}_n \\ 0 & \text{otherwise} \end{cases}$$

$$t = 1, \ldots, n - 1.$$

Then $\sum_{t=1}^{n-1} \omega_t$ tells us the number of times that the time series $X_1, \ldots, X_n$ crosses the mean. The larger this number is, the greater is our confidence in the fact that bias due to initial conditions has been resolved.

10.5 FINAL CONDITIONS

The desire to infer mean queue length and mean waiting time from the same simulation raises a design issue that rarely is considered by simulation investigators. At the beginning of Section 10.4 we spoke of specifying a run length of n time intervals. This fits well with collecting observations on queue length. At the end of the section we spoke of running a simulation until N jobs complete service. This fits well with collecting waiting times. Unfortunately the inherently stochastic nature of the experiment makes the specification of both n and N impossible. If we specify run length n then the number of completed jobs N is a random variable. If we specify N then the number of sampling intervals n is random.

To study the implication of this problem we first assume that the investigator selects n and collect $X_1, \ldots, X_n$ on queue length. For expository convenience we also assume that initial conditions play no role. Then $\bar{X}_n$ is an unbiased estimator of mean queue length. To estimate mean waiting time we may use (3.3.1), which yields

$$\bar{W}_n = \frac{\bar{Z}_n}{\bar{N}_n}$$

$$(5.1) \qquad \bar{Z}_n = \frac{1}{n}\sum_{t=1}^{n} Z_t \qquad Z_t = \max(X_t - m, 0)$$

$$\bar{N}_n = \frac{N}{n}.$$

Since n is fixed, N and consequently $\bar{N}_n$ are random variables. Therefore $\bar{W}_n$ is a ratio estimator. Let μ_W denote mean waiting time and let $E(\bar{N}_n \mid n) = \mu_N$. Then

$$(5.2) \qquad E(n\bar{Z}_n \mid n) = n\mu_N\mu_W + \theta \qquad 0 < \theta$$

where θ is the mean elapsed waiting time of jobs in the system when n time units have elapsed and

$$(5.3) \qquad E(\bar{Z}_n \mid n) = \mu_N\mu_W + \frac{\theta}{n}.$$

If the variations in $\bar{Z}_n$ and $\bar{N}_n$ about $\mu_N\mu_W$ and μ_W, respectively, are not too large, we may expand $\bar{W}_n$ in Taylor series

$$(5.4) \qquad \bar{W}_n = \mu_W + \frac{(\bar{Z}_n - \mu_N\mu_W)}{\mu_N} - \frac{\mu_N\mu_W(\bar{N}_n - \mu_N)}{\mu_N^2} - \frac{(\bar{Z}_n - \mu_N\mu_W)(\bar{N}_n - \mu_N)}{\mu_N^2} + \frac{\mu_N\mu_W(\bar{N}_n - \mu_N)^2}{\mu_N^3} + \cdots.$$

Let

(5.5)

$$\lim_{n\to\infty} n \operatorname{var}(\overline{Z}_n \mid n) = m_Z < \infty \qquad \lim_{n\to\infty} n \operatorname{var}(\overline{N}_n \mid n) = m_N < \infty$$

$$\lim_{n\to\infty} n \operatorname{cov}(\overline{Z}_n, \overline{N}_n \mid n) = m_{NZ} < \infty.$$

Then, to order $1/n$, $\overline{W}_n$ has bias

$$E(\overline{W}_n - \mu_W \mid n) = \left[\frac{\theta/\mu_N - m_{NZ}/\mu_N^2 + \mu_W m_N/\mu_N^2}{n}\right]. \tag{5.6}$$

The term $\theta/n\mu_N$ is due to the inclusion in $\overline{Z}_n$ of the waiting time of jobs that are in the system when the simulation terminates. The other two terms are specifically a consequence of using a ratio estimator.

Consider now the converse situation in which N is specified and n is random. Then it is not difficult to see that†

$$E(n\overline{Z}_n \mid N) = N\mu_W + \phi \qquad 0 < \phi \tag{5.7}$$

where ϕ is the mean elapsed waiting time of jobs that are in the system when the Nth job is completed. Therefore

$$E(\overline{W}_n - \mu_W \mid N) = \frac{\phi}{N} \tag{5.8}$$

so that a bias again appears in the sample mean waiting time.

The bias in (5.8) is academic if one decides to collect the N waiting times $W_1, \ldots, W_N$ when N is specified. Then (4.11) produces an estimate free of bias due to final conditions. However, these times are not sufficient to produce a useful estimator of mean queue length. If the queue lengths $X_1, \ldots, X_n$ for random n are also collected, $\overline{X}_n$ would produce a bias in mean queue length analogous to that in mean waiting time.

The presence of this bias due to final conditions is an unattractive feature of the methods of estimation that we have described. Fortunately some relief is possible when we estimate the parameters of interest using the *sample state times* as described in Section 3.3.2. We describe estimation procedures employing these data in Section 10.16.

In the rest of this chapter we include the conditioning n as in (5.2) and the conditioning N as in (5.7) only in situations where ambiguity may otherwise arise.

† Examining Z_n in terms of the H_i in (3.3.1) is helpful.

10.6 DATA COLLECTION ERRORS

Fortunately data collected in a computer simulation can be as precise as a user desires. Provided that the calculating formulae for summary statistics are correct and the numerical methods employed with the formulae preserve the number of significant digits needed for appropriate interpretation, these aspects of data collection should pose no special problem.

However, at least one potential source of error can occur in the data collection process and deserves note. This problem relates to the times at which observations should be collected. In Chapter 3, we noted that periodic data collection calls for explicit modeling. Sections 5.3 and 5.5 exemplify the necessary programming effort for GPSS/360 and SIMSCRIPT II, respectively. Now occasionally it is desirable to eliminate periodic data collection and alternatively to collect data when certain conditions are met in the simulation. For example, a user who plans to collect the waiting time of a job as it enters service may also elect to record queue length at that time. If initial and final conditions are ignored his estimator of mean waiting time is unbiased. However, the unweighted average of his queue length observations produces a biased estimator. This bias occurs because his estimate is based on queue length, given that a job is about to begin service. Since idle time plays no role the estimate is biased upward.

Although this error may be too obvious to some readers to deserve serious consideration, the fact remains that it and other, similar errors occur in practice. As mentioned in Section 3.3.2, an alternative to periodic sampling exists. This alternative, which we discuss in Section 10.16, avoids the error that results here.

10.7 VARIANCE CONSIDERATIONS

So far we have assessed design considerations with regard to the bias they induce in the sample mean. However, these considerations also affect its variance. For example, for initial conditions **I** and specified run length n one way to assess the characteristics of an estimator is to examine its mean-square error (MSE):

$$\begin{aligned} \text{mean-square error} &= \text{variance} + (\text{bias})^2 \\ &= \operatorname{var}(\bar{X}_n \mid \mathbf{I}) + [E(\bar{X}_n \mid \mathbf{I}) - \mu]^2. \end{aligned} \tag{7.1}$$

As mentioned earlier, the dominant bias term is usually of order $1/n$ so that the contribution of bias to the MSE is of order $1/n^2$. Since the dominant variance term is often of order $1/n$, it is clear that increasing n causes the bias contribution in (7.1) to decline in relative importance compared to the variance.

As mentioned in Section 10.4, deletion of observations at the beginning of a run leads to a reduced bias. We now must consider the effect of deletion on $\text{var}(\bar{X}_n \mid \mathbf{I})$. For the first-order autoregressive scheme with $\mathbf{I} = X_0$ we have [163]

(7.2)

$$\text{var}(\bar{X}_{t^*,n} \mid \mathbf{I}) = \frac{\sigma^2}{(n - t^*)(1 - \alpha)^2} \times \left\{1 - \frac{\alpha(1 - \alpha^{n-t^*})}{(n - t^*)(1 - \alpha^2)} [2 + \alpha^{2t^*+1}(1 - \alpha^{n-t^*})]\right\}.$$

Since $\bar{X}_n = \bar{X}_{0,n}$ it is clear that in large samples $\text{var}(\bar{X}_{t^*,n} \mid \mathbf{I}) > \text{var}(\bar{X}_n \mid \mathbf{I})$, which is an unappealing result.

This conflict involving reduced bias at the expense of increased variance is not generic to simulation work alone. It pervades much of statistical estimation theory. The analyst is usually forced to decide between a design practice which produces estimates that are on average unbiased but vary relatively widely around the true mean and a design practice that allows a systematic bias to occur in estimates that vary less widely. In the present case the design practices involved are deletion versus no deletion.

As also noted in Section 10.4, independent replications do not lead to reduced bias due to initial conditions. Moreover we note that for k replications

$$\text{MSE} = \frac{1}{k}\,\text{var}(\bar{X}_{n'} \mid \mathbf{I}) + [E(\bar{X}_{n'} - \mu \mid \mathbf{I})]^2 \tag{7.3}$$

so that as the number of replications increases for fixed n' the MSE converges to the squared bias. The simulation user who elects to employ the independent replications approach must choose k sufficiently large so that he can estimate $\text{var}(\bar{X}_{n'} \mid \mathbf{I})$ with acceptable accuracy and n' sufficiently large so that bias is relatively incidental.

10.8 VARIANCE ESTIMATION

In Chapter 6 we noted that a time series represents *one* observation on a stochastic sequence. Since we require several observations to compute a sample variance in most statistical analyses we would normally expect that the independent replications approach, which provides us with one time series or observation per replication, would be the natural method to employ when estimating the variance of $\bar{X}_n$. However, provided that certain ergodic properties hold, it is possible to estimate the variance of the sample mean from a single time series and have the resulting estimator be consistent in the statistical sense. The consistency property implies that as n increases the estimate converges in probability to the true variance of the sample mean.

We assume that we have a time series $X_1, \ldots, X_n$ free of the design considerations described in earlier sections. We also assume that the sample is drawn from a covariance stationary stochastic sequence with mean μ and autocovariance function† R. Then

(8.1)

$$\begin{aligned}
\operatorname{var}(\bar{X}_n) &= E[(\bar{X}_n - \mu)^2] = E\left[\frac{1}{n}\sum_{s=1}^{n}(X_s - \mu)\right]^2 \\
&= \frac{1}{n^2} E\left[\sum_{s,t=1}^{n}(X_s - \mu)(X_t - \mu)\right] = \frac{1}{n^2}\sum_{s,t=1}^{n} E[X_s - \mu)(X_t - \mu)] \\
&= \frac{1}{n^2}\sum_{s,t=1}^{n} R_{s-t}.
\end{aligned}$$

Notice that $\operatorname{var}(\bar{X}_n)$ requires knowledge of the n autocovariances (1 variance and $n - 1$ covariances) $R_0, R_1, \ldots, R_{n-1}$, since $R_s = R_{-s}$. This contrasts with (2.2), the unrestricted definition of $\operatorname{var}(\bar{X}_n)$, which requires n variances and $n(n - 1)/2$ covariances. Thus the formalism of covariance stationary sequences reduces the number of quantities to be known.

We can simplify $\operatorname{var}(\bar{X}_n)$ a bit more by first expanding the double summation in (8.1):

$$n^2 \operatorname{var}(\bar{X}_n)$$

$$(8.2) \quad = \begin{cases}
R_{1-n} + R_{2-n} + R_{3-n} + \cdots + R_{-1} + R_0 \\
\quad + R_{2-n} + R_{3-n} + \cdots + R_{-1} + R_0 + R_1 \\
\quad\quad + R_{3-n} + \cdots + R_{-1} + R_0 + R_1 + R_2 \\
\quad\quad\quad \ddots \qquad \vdots \quad \vdots \quad \vdots \\
\quad\quad\quad\quad + R_0 + R_1 + R_2 \\
\quad\quad\quad\quad\quad + R_3 + \cdots + R_{n-1}.
\end{cases}$$

† See Section 6.3.

Noting that R_s appears $n - |s|$ times in the summations which include R_{1-n} through R_{n-1}, we may write (8.2) as

$$\operatorname{var}(\bar{X}_n) = \frac{1}{n^2} \sum_{s=1-n}^{n-1} (n - |s|) R_s$$

$$(8.3) \qquad = \frac{1}{n} \sum_{s=1-n}^{n-1} \left(1 - \frac{|s|}{n}\right) R_s$$

$$= \frac{1}{n} \left[R_0 + 2 \sum_{s=1}^{n-1} \left(1 - \frac{s}{n}\right) R_s \right].$$

In Section 6.3 we noted that covariance stationarity requires each R_s to be bounded. Since $\operatorname{var}(\bar{X}_n)$ is a linear combination of n autocovariances, a question arises regarding the boundedness of this sum as n increases. Suppose that

$$(8.4) \qquad |R_s| \leq R_0 \alpha^{|s|} \qquad 0 \leq \alpha < 1.$$

Then one may show that

$$(8.5) \qquad \lim_{n \to \infty} n \operatorname{var}(\bar{X}_n) = \sum_{s=-\infty}^{\infty} R_s = 2\pi g(0),$$

$g(0)$ being the spectrum of $\{X_t\}$ at zero frequency as defined by (6.4.11). For convenience we define $m \equiv 2\pi g(0)$ so that, for large n, $\bar{X}_n$ has approximate variance m/n. For many problems we can express $\operatorname{var}(\bar{X}_n)$ as

$$(8.6) \qquad \operatorname{var}(\bar{X}_n) = \frac{m}{n} + O\left(\frac{1}{n}\right)$$

where $O(1/n)$ denotes terms of order higher than $1/n$. This notation facilitates the study of alternative estimators of m in the next several sections.

Before examining the alternative estimators it is instructive to examine the effects of a ratio estimator on variance. Consider $\bar{W}_n$ in (5.1) with n given. From the Taylor series expansion in (5.4), we can easily see that $\bar{W}_n$ has variance

$$(8.7) \qquad \operatorname{var}(\bar{W}_n \mid n) = \frac{1}{n} \left[\frac{m_Z}{\mu_N^2} - \frac{2 m_{NZ} \mu_W}{\mu_N^2} + \frac{m_N \mu_W^2}{\mu_N^2} \right] + O\left(\frac{1}{n}\right).$$

Consequently one clearly must acknowledge the variation due to N in (5.1) when computing the variance of $\bar{W}_n$.

10.9 VARIANCE ESTIMATOR BASED ON SUBSAMPLES IN A SINGLE TIME SERIES

Suppose that we divide the n observations in a single time series into k subsamples, each of length n', so that $n = kn'$. The sample mean of the jth subsample is†

$$\bar{X}_{(j)} = \frac{1}{n'}\sum_{s=1}^{n'} X_{n'(j-1)+s}. \tag{9.1}$$

We form the statistic

$$\begin{aligned} m_2 &= \frac{a}{k}\sum_{j=1}^{k}(\bar{X}_{(j)} - \bar{X}_n)^2 \\ &= a[\hat{\gamma} - (\bar{X}_n - \mu)^2] \\ \hat{\gamma} &= \frac{1}{k}\sum_{j=1}^{k}(\bar{X}_{(j)} - \mu)^2 \end{aligned} \tag{9.2}$$

where m_2 has expectation

$$E(m_2) = a\left[\frac{m}{n'} + o\left(\frac{1}{n'}\right) - \frac{m}{n} - o\left(\frac{1}{n}\right)\right] \tag{9.3}$$

since

$$\text{var}(\bar{X}_{(1)} = \text{var}(\bar{X}_{(2)}) = \cdots = \text{var}(\bar{X}_{(k)}). \tag{9.4}$$

As n increases $\text{var}(\bar{X}_n)$ converges to m/n, and as n' increases $\text{var}(\bar{X}_{(j)})$ converges to m/n'. For m_2 to be an unbiased estimator of m in large samples, we require

$$\lim_{n'\to\infty}\lim_{n\to\infty} E(m_2) = \lim_{n'\to\infty}\lim_{n\to\infty} am\left(\frac{1}{n'} - \frac{1}{n}\right) = m \tag{9.5}$$

$$a = \frac{nn'}{n - n'} = \frac{n}{k-1}. \tag{9.6}$$

In large samples m_2 has variance

$$\text{var}(m_2) \sim \frac{m^2(2+\kappa)}{k}, \tag{9.7}$$

κ being a fourth-order moment contribution.

† This approach for the analysis of simulation data is discussed in reference 159. Variations on this method can also be found there.

If we neglect κ and approximate the distribution of fm_2/m by that of $\chi^2(f)$

$$f \equiv \frac{2(\text{mean})^2}{\text{variance}} = k, \tag{9.8}$$

we then have a way of assessing the degree of variation in m_2. This assessment is especially helpful in Section 10.12, where we derive rough confidence intervals for the mean. The definition of f derives from the fact that the ratio of twice the squared mean and the variance gives the degrees of freedom for a chi-square variate.

We wish to emphasize that the reasonableness of this approach in practice depends on how large n' and k are. As n' increases, m_2 approaches m in expectation. If, however, n is fixed then k decreases, thereby increasing the variance of m_2. To derive a reasonable balance we suggest plotting m_2 for increasing n', and f for the correspondingly decreasing k. If m_2 appears to stabilize at an f value considered too low for the degree of statistical accuracy desired then we advise increasing n. If m_2 fails to stabilize at a given n we may increase n. If no improvement in stability is noted as n increases, the reason may be that the underlying model of covariance stationarity does not correctly characterize the data. A plot of the original data $X_1, \ldots, X_n$ may reveal such a departure from the stationarity assumption.

We note in passing that, if the process under study is a sequence of independent events, we underestimate degrees of freedom for m_2 as k instead of $n - 1$. Hence the determination of independence, when it exists, can improve the interpretation of our results considerably.

10.10 AN ESTIMATOR FROM SPECTRUM ANALYSIS

During the past two decades considerable development has occurred in the estimation of spectra of covariance stationary processes. Since $m = 2\pi g(0)$, it is natural to consider the estimation of m as the estimation of the spectrum g at zero frequency.† One of the first methods used to estimate $g(\lambda)$ consistently was

$$\tilde{g}(\lambda) = \frac{1}{2\pi} \sum_{s=-k+1}^{k-1} \left(1 - \frac{|s|}{k}\right) \hat{R}_s \cos \lambda s \qquad k < n, |\lambda| \leq \pi, \tag{10.1}$$

where $\hat{R}_s$ is given in Section 9.4.

† For a discussion of spectrum analysis, see references 161, 167, 168, and 170.

The corresponding estimate for $m = 2\pi g(0)$ is then easily seen to be

$$\tilde{m} = \sum_{s=-k+1}^{k-1} \left(1 - \frac{|s|}{k}\right) \hat{R}_s. \tag{10.2}$$

Since†

$$E(\hat{R}_s) = R_s - \frac{m}{n} + O\left(\frac{1}{n}\right),$$

m has expectation

$$E(\tilde{m}) = \sum_{s=-k+1}^{k-1} \left(1 - \frac{|s|}{k}\right) R_s - \frac{km}{n} + O\left(\frac{1}{n}\right). \tag{10.3}$$

Then m is asymptotically unbiased. Suppose that k is sufficiently large so that

$$\sum_{s=-k+1}^{k-1} \left(1 - \frac{|s|}{k}\right) R_s \sim m. \tag{10.4}$$

Then to order $1/n$

$$E(\tilde{m}) = m - \frac{km}{n}. \tag{10.5}$$

This observation suggests that in large but finite samples we should use $m_3 = \tilde{m}n/(n - k)$ as our estimator of m in order to reduce bias.

To order $1/n$ we have

$$\operatorname{var}(m_3) = \frac{4k}{3n} m^2 + \frac{\kappa}{n}. \tag{10.6}$$

Notice that as n and k increase the relative importance of the nonnormal contribution κ/n diminishes.

Assuming that $4k/3 \gg \kappa$, it is customary to approximate the distribution of m_3 by that of a multiple of chi square in large samples. Since the degrees of freedom f for a chi-square variate satisfy (9.8) we set

$$f = \frac{2[E(m_3)]^2}{\operatorname{var}(m_3)} = \frac{3n}{2k}. \tag{10.7}$$

To illustrate this estimating technique we use the queueing problem in Section 6.5 with

$$\omega_1 = 4.5, \qquad \omega_2 = 5, \qquad n = 4000. \tag{10.8}$$

† See, for example, reference 163, pp. 83–84.

We note from (6.5.5) that for the queue length process

$$m = 2\pi g(0) = 6840$$
(10.9)
$$\rho = \frac{\omega_1}{\omega_2} = 0.9$$

so that for $X_1, \ldots, X_n$ being the queue length at $t = 1, \ldots, n$

(10.10)
$$\operatorname{var}(\bar{X}_n) \sim \frac{6840}{n}.$$

This $\operatorname{var}(\bar{X}_n)$ actually overstates the variance since (6.5.5) applies to the continuous stochastic process on $(0, n)$ described in Section 6.5 and here we are considering discretely sampled points on $(0, n)$. The error, however, is not large in this example.† Figure 53 shows m_3 and k for $k = 150(50)500$. We

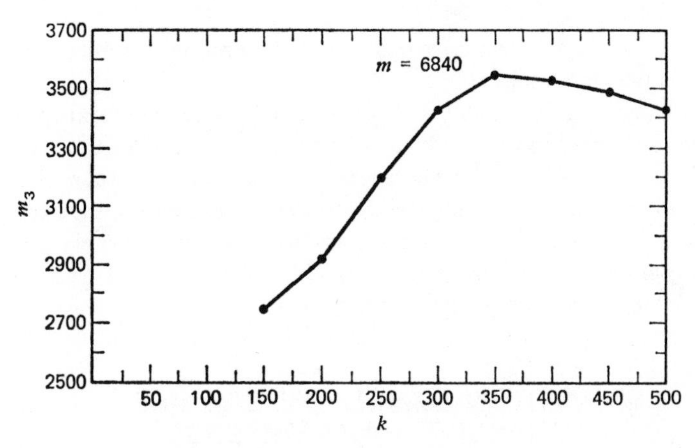

Fig. 53 k and m_3 for queueing problem in Section 6.5: $\omega_1 = 4.5$, $\omega_2 = 5$, $n = 4000$.

observe that, as k increases above 300, m_3 becomes more stable. Notice, however, that increasing k increases the variance so that we continue to expect some variation to occur.

In deriving an estimate of m in this way it is desirable to vary n as well as k, as the following example shows. For the same queueing model, the waiting time was recorded at unit intervals. Figure 54 shows m_3 for several values of k and n. Note that, for $n = 4000$, 8000, and 16,000, m_3 appears considerably more stable then for $n = 1000$ or 2000. This greater stability is

† See reference 164, p. 544.

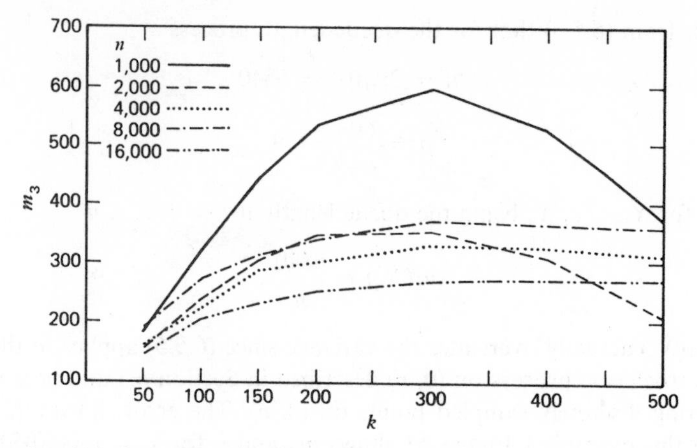

Fig. 54 k and m_3 for waiting time sequence: $\omega_1 = 4.5$, $\omega_2 = 5$.

attributable to the better approximation of the limiting formulae for large n and to the decrease in k/n.

The weighting function $1 - |s|/k$ is actually one of several forms commonly used when estimating m by a weighted sum of sample autocovariances. Other forms may be found in references 161, 168, and 169, where the estimation of m is part of the more general problem of estimating the spectrum g. The use of these alternative weighting functions is desirable since they produce better sampling properties than the one described here. We used $1 - |s|/k$ merely because of its simplicity.

10.11 ESTIMATION BASED ON SAMPLE AUTOGRESSIVE PARAMETERS

In Section 8.6 we found the autoregressive representation to be a convenient tool in facilitating the generation of autocorrelated events. Perhaps its most appealing feature was the fact that a small number of parameters, p autoregressive coefficients $b_1, \ldots, b_p$, and one residual variance σ^2 were all that was necessary to form the wide range of autocorrelation patterns.

Recent work [162] has indicated that the autoregressive representation can also contribute substantively to the present problem of estimating $\text{var}(\bar{X}_n)$. Consider the autoregressive representation in (6.6.1). It is easily seen from (6.6.14) that

$$m = 2\pi g(0) = \frac{\sigma^2}{[\sum_{s=0}^{p} b_s]^2}\,. \tag{11.1}$$

Since Section 9.4 provides us with a method of estimating $b_1, \ldots, b_p$ and σ^2, a convenient estimate of m is

$$(11.2) \qquad m_4 = \frac{\hat{\sigma}^2}{\hat{b}^2}$$

$$(11.3) \qquad \hat{b} = 1 + \sum_{s=1}^{p} \hat{b}_s.$$

From (9.4.18) and Table 28 we note that in large samples

$$(11.4) \qquad \lim_{n\to\infty} E(m_4 - m) = 0.$$

It can also be shown that

$$(11.5) \qquad \lim_{n\to\infty} n \operatorname{var}(m_4) = m^2 \left[2 + \kappa + \frac{4}{b} \sum_{s=0}^{p} (p - 2s) b_s \right].$$

Using (9.8) and neglecting κ, we have for the degrees of freedom

$$(11.6) \qquad f = \frac{nb}{(2p + 1)b - 4 \sum_{s=0}^{p} s b_s}.$$

Returning to our queueing example, we note in Table 29 that for the determined $p = 2$ we have $m_4 = 3009.62$, compared with the true $m = 6840$ and the estimate $m_3 = 3424$ for $k = 500$ in Section 10.10. Substituting the corresponding sample coefficients from Table 33 into (11.5), we estimate f to be 46.

The estimator m_4 has great appeal for us since we can usually compute it for a small number p of sample autocovariances, in contrast to the several hundred required in our example for m_3. The price we pay is to assume the existence of an autoregressive representation, an assumption we know incidentally to be invalid for the queue length process. Consequently we should regard the use of an autoregressive representation as an approximating method.

An additional advantage of m_4 over m_3 is the ability to determine the former in a self-contained computer program using $\hat{R}_0, \ldots, \hat{R}_q$ as input. As we discuss in Section 10.15, this feature enables us to compute m_4 during a simulation experiment and to use this information to determine how long to run the experiment [162]. By contrast the computation of an appropriate m_3 cannot be accomplished without an evaluation on our part as to what constitutes an adequate k. Both methods, nevertheless, are predicted on n being large.

If we choose to estimate the autoregressive parameters as suggested in Section 9.4 we gain the added convenience of testing the hypothesis that $\{X_t\}$

is an independent sequence ($p = 0$). This is especially desirable for it gives us a way of detecting when we may rely on the conventional t statistic approach to develop a confidence interval for μ, in contrast to an alternative approach designed to acknowledge dependence between events. Appendix B contains SIMSCRIPT II and FORTRAN subroutines for computing the autoregressive sample quantities of interest. A sample output is shown in Table 13.

By way of summary, Table 32 lists the four principal estimators we have considered together with their approximated degrees of freedom.

The question of estimating the variance of a ratio estimator as in (5.1) remains to be studied. Expression (8.7) indicates the dominant term in this variance. From (5.1) we note that we have a time series $Z_1, \ldots, Z_n$. Suppose that we also collect the time series $N_1, \ldots, N_n$ where N_t denotes the number of jobs completed in period t so that $N = \sum_{t=1}^{n} N_t$. Let

$$M_t = Z_t - N_t \qquad t = 1, \ldots, M$$

$$\overline{M}_n = \overline{Z}_n - \overline{N}_n.$$

Then in large samples

$$\operatorname{var}(\overline{M}_n \mid n) = \operatorname{var}(\overline{Z}_n \mid n) - 2\operatorname{cov}(\overline{Z}_n, \overline{N}_n \mid n) + \operatorname{var}(\overline{N}_n \mid n)$$

$$\sim \frac{m_M}{n} = \frac{m_Z - 2m_{NZ} + m_N}{n}.$$

We may proceed as follows:

1. Compute $\hat{m}_Z$ based on $Z_1, \ldots, Z_n$.
2. Compute $\hat{m}_N$ based on $N_1, \ldots, N_n$.
3. Compute $\hat{m}_M$ based on $M_1, \ldots, M_n$.
4. Compute $\hat{m}_{NZ} = (\hat{m}_Z + \hat{m}_N - \hat{m}_M)/2$.
5. Estimate $\operatorname{var}(\overline{W}_n \mid n)$ using (8.7) by substituting estimates for corresponding parameters.

10.12 CONFIDENCE INTERVALS FOR $\overline{X}_n$

Being able to estimate the variance of the sample mean for dependent events in a simulation experiment is a major step forward. It now remains for us to say something about the sampling distribution of $\overline{X}_n$ that enables us to use $\overline{X}_n$ and $\hat{m}/n$ ($\hat{m}$ being an estimate of m) to derive a confidence interval for the population mean μ.

In Section 10.1 we listed the "misuse of estimates" as contributing to the misinterpretation of simulation results. In practice the principal misuse

Table 32 Estimation of m

Section	Estimator	Approximated Degrees of Freedom	Definitions and Assumptions
10.8	$m_1 = \frac{n}{k(k-1)} \sum_{j=1}^{k} (\bar{X}_{j,n'} - \bar{X}_n)^2$	k	$k \equiv$ number of independent replications $n \equiv$ length of each replication $n \equiv n'k$ n large $\bar{X}_{j,n'} \equiv$ sample mean on replication j
10.9	$m_2 = \frac{n}{k(k-1)} \sum_{j=1}^{k} (\bar{X}_{(j)} - \bar{X}_n)^2$	k	$k \equiv$ number of divisions of one time series $n' \equiv$ length of each division $n \equiv n'k$ n', k large $\bar{X}_{(j)} \equiv$ sample mean on division j
10.10	$m_3 = \frac{n}{n-k} \left[\hat{R}_0 + 2 \sum_{s=1}^{k-1} \left(1 - \frac{s}{n}\right) \hat{R}_s \right]$	$\frac{3n}{2k}$	$n \equiv$ length of time series $k \equiv$ number of lags used to estimate spectrum ordinate $g(0)$ n, k large
10.11	$m_4 = \frac{\hat{\sigma}^2}{\hat{b}^2}$	$\frac{nb}{(2p+1)b - 4\sum_{s=1}^{p} sb_s}$ $b \equiv \sum_{s=0}^{p} b_s$	$n \equiv$ length of time series $p \equiv$ autoregressive order n large See Section 9.4 for $\hat{\sigma}^2$ and $\hat{b}^2$

is the basing of decisions on $\bar{X}_n$ without any consideration of how close to μ we can expect $\bar{X}_n$ to be. A confidence interval provides one way to assess the degree of closeness and hence to reduce the danger of misinterpretation. Although the methods described here are rough and can be considered suitable only when no additional specialized information is available to assess $\bar{X}_n$, the interpretation of simulation results can benefit from the presentation of corresponding confidence intervals.

The central limit theorem of classical statistics tells us that, if $X_1, \ldots, X_n$ are independent and identically distributed with finite mean and variance and meet other relatively weak conditions, then $\bar{X}_n$ has approximately the normal distribution when n is large. This is a remarkable fact that often allows us to derive a confidence interval for the mean when the parent distribution of the random variables $X_1, \ldots, X_n$ is unknown. It is natural, then, to inquire as to whether a modified central limit theorem exists when $X_1, \ldots, X_n$ are auto-correlated.

It is known that, if $\{X_t\}$ is a covariance stationary sequence with mean μ and continuous spectrum g, $X_t - \mu$ can be approximated arbitrarily closely by an infinite moving average†

$$X_t - \mu = \sum_{s=-\infty}^{\infty} a_s \varepsilon_{t-s} \tag{12.1}$$

$$\sum_{s=-\infty}^{\infty} |a_s| < \infty,$$

where $\{\varepsilon_t\}$ is an uncorrelated sequence of events with zero mean and variance σ^2. If $\{\varepsilon_t\}$ is also a sequence of independent, identically distributed events, then it is known that the distribution of $(\bar{X}_n - \mu)/\sqrt{m/n}$ converges to that of $\mathcal{N}(0, 1)$ as n increases [167]. This is a highly useful result which encourages us to assume independence for $\{\varepsilon_t\}$. Moreover an autoregressive process can be inverted to yield a corresponding moving average as in (12.1). An example of this was given in (8.6.10).

When we suspect that the assumption of independent disturbances is false we should regard the confidence interval that it provides to be, at best, a "ball-park" interval estimate of μ. We also note in passing that, if $X_1, \ldots, X_n$ jointly have a multivariate normal distribution, $\bar{X}_n$ has the normal distribution for all n, its variance being m/n for large n. Then we need not rely on asymptotic normality to provide a workable distribution theory.

Assuming $\bar{X}_n$ to be normal, we may make the confidence statement

$$\text{Prob}(\bar{X}_n - Q\sqrt{m/n} < \mu < \bar{X}_n + Q\sqrt{m/n}) \sim 1 - \alpha, \tag{12.2}$$

† See, for example, references 167 and 169.

where Q is the normal point corresponding to

$$(2\pi)^{-1/2} \int_{-\infty}^{Q} e^{-z^2/2}\, dz = 1 - \frac{\alpha}{2}. \tag{12.3}$$

We use the "approximately equal" sign ($\sim$) in (12.2) rather than the equal sign to emphasize that our results are not exact. Note in (12.1) that the larger n is, the shorter is the confidence interval around μ. Expressing (12.2) in terms of our original specification of a confidence interval in (2.1), we have

$$\begin{aligned} w_{\alpha,n} &\equiv \bar{X}_n - Q\sqrt{m/n} \\ v_{\alpha,n} &\equiv \bar{X}_n + Q\sqrt{m/n}. \end{aligned} \tag{12.4}$$

As first approximations we substitute m_3 and m_4 for m to compute a confidence interval for μ. For our queueing example we have

$$\begin{aligned} \alpha &= 0.10, \qquad Q = 1.64 \\ n &= 4000 \qquad \bar{X}_{4000} = 8.45 \\ m_3 = 3424,\ \kappa &= 500; \qquad m_4 = 3009,\ \ p = 2. \end{aligned} \tag{12.5}$$

Using the spectral estimate m_3 of m, we compute the rough confidence interval for μ of

$$w_{0.10,4000} = 6.93, \qquad v_{0.10,4000} = 9.97. \tag{12.6}$$

The autoregressive estimator m_4 yields the interval

$$w_{0.10,4000} = 7.03, \qquad v_{0.10,4000} = 9.87. \tag{12.7}$$

Since the several substitutions and approximations made in deriving these confidence intervals raise the question of how useful the intervals are, it is instructive to study the subject in some detail. The queueing model problem was used again with $1/\omega_1 = 0.25$ and $1/\omega_2 = 0.225$ to generate 100 replications, each with $n = 500$, 100 with $n = 1000$, and 50 with $n = 2000$. The normal intervals for $\alpha = 0.10$, using the known m computed in (10.10), are shown in column 3 of Table 33. Column 4 lists the expected number of replications in which the interval should contain $\bar{X}_n$ when $\bar{X}_n$ is normal. Column 5 gives our actual experience, where we note the apparent adequacy of the normal assumption for $\bar{X}_n$. Column 6 lists the number of replications whose confidence intervals based on m_3 instead of m contained the true mean queue length $\mu = 9$. Although the degradation is apparent it clearly diminishes with increasing n. The ratio of columns 6 and 4 in column 7 shows the relative improvement as n increases.

The results in Table 33 should serve as sobering for any reader who thinks that the substitution of $\hat{m}$ for m in (12.2) is inconsequential. It clearly

Table 33 Confidence and Probability Intervals for Queueing Problem[a]

$1/\omega_1 = 0.25, \quad 1/\omega_2 = 0.225, \quad m = 7695$

			Intervals Containing $\overline{X}_n$					
n (1)	Replications (2)	Normal Interval (3)	Expected (4)	Actual (5)	Intervals Containing μ Using m_4 (6)	(6)/(4) (7)	Modified Intervals Containing μ (8)	(8)/(4) (9)
500	100	2.56–15.48	90	92	69	0.77	74	0.82
1000	100	4.46–13.55	90	92	76	0.84	78	0.87
2000	50	5.78–12.22	45	42	39	0.87	42	0.93

[a] The estimate m_4 was used in columns 6 through 9.

has an effect, but one which diminishes as n increases. Since we know a priori that direct substitution of $\hat{m}$ for m leads to degradation it seems worth while to acknowledge this substitution explicitly. In earlier sections we approximated the distribution of $f\hat{m}/m$ by that of $\chi^2(f)$, f being defined in (11.6). By analogy to the case for independent events we suggest approximating the distribution of $(\bar{X} - \mu)/\sqrt{\hat{m}/n}$ by that of $t(f)$.

For our earlier example using m_4 we have $f = 46$ and

$$t_{46,0.95} \sim 1.68, \qquad t_{46,0.95}\sqrt{m_4/n} \sim 1.46 \tag{12.8}$$

so that

$$w_{0.10,4000} = 6.99, \qquad v_{0.10,4000} = 9.91. \tag{12.9}$$

Although the difference from (12.7) is insignificant in this case smaller values of f can lead to relatively important differences. Returning to Table 33, we find in column 8 the total number of modified intervals that contain† $\mu = 9$. The improvement is clear. The reader may now wish to restudy Table 13 in Chapter 5 in light of this exposition on confidence intervals.

The queueing model on which our examples have been based is known to be highly nonnormal. In particular the marginal p.d.f. of X_t is the geometric, a highly skewed distribution compared to the normal. Our choice of examples was deliberate, however, since highly skewed distributions often occur in queueing simulations and the statistical analysis must be capable of doing a reasonably acceptable job for most simulations encountered.

10.13 NONPARAMETRIC CONFIDENCE INTERVALS

Using the t distribution as described in Section 10.12 enables us to compensate for the substitution of $\hat{m}$ for m when $\bar{X}_n$ has the normal distribution. Since some investigators may consider the normal assumption questionable for their particular simulations, they may prefer to derive a confidence interval for μ based on a weaker assumption. This is especially true when a user believes that the p.d.f. of X_t is highly skewed so that the convergence to normality of $\bar{X}_n$ is slow. It is also true when the decisions to be based on the simulation results involve a considerable sum of money. In this case the investigator is often willing to spend more on his simulation to assure himself of the accuracy of the results. In this section we discuss weaker assumptions and their effects on required sample size.

† If one chooses to use m_3 instead of m_4 he can modify the confidence interval for μ by using $t(f)$, f then being defined as in (10.7).

The weakest assumption one can make in deriving a confidence interval is that $\bar{X}_n$ has finite mean and variance. Then Chebyshev's inequality gives us

$$\text{Prob}(|\bar{X}_n - \mu| < \theta\sqrt{m/n}) \geq 1 - \frac{1}{\theta^2}. \tag{13.1}$$

Normality plays no role here. If we choose a significance level of 0.1 then $1/\theta^2 = 0.1$, so that $\theta = \sqrt{10} = 3.16$, compared to $Q = 1.64$ for the normal assumption. Since sample size n is proportional to the square of these quantities we see that Chebyshev's inequality requires $(\theta/Q)^2 = 3.72$ more observations at the 0.1 level than the normal confidence interval does. The luxury of the weaker assumption is purchased at a high price.

We can impose an additional weak assumption that allows us to improve on the Chebyshev result. Suppose that $\bar{X}_n$ has a unimodal p.d.f. in addition to finite mean and variance. Then we may write [165]

$$\text{Prob}(|\bar{X}_n - \mu| < \theta\sqrt{m/n}) > \begin{cases} \dfrac{5\theta^2 - 3}{3(\theta^2 + 1)} & 0 \leq \theta \leq \sqrt{5/3} \\[2ex] \dfrac{9\theta^2 + 1}{9(\theta^2 + 1)} & \sqrt{5/3} \leq \theta. \end{cases} \tag{13.2}$$

For a significance level 0.1 the lower right-hand expression with $\theta = \sqrt{7.89} = 2.81$ applies. Compared to the normal result, the unimodal assumption requires $(\theta/Q)^2 = 2.93$ more observations at the 0.1 level. However, it requires $7.89/10 = 0.789$ as many observations as the Chebyshev result does.

Table 34 shows the required sample sizes for $\alpha \leq 0.05$ and $\alpha \leq 0.10$ for the normal, Chebyshev, and unimodal assumptions. It also shows the probabilities that occur if the normal sample sizes are used for the two values of α. We conclude that preference for the Chebyshev or unimodal assumption can be relatively expensive in terms of required sample size.

Other assumptions stronger than the unimodal can be made which require smaller sample size. Godwin [165] describes these assumptions and

Table 34 Sample Sizes for Normal, Chebyshev, and Unimodal Assumptions

Assumption	Prob. ≥ 0.90 $\alpha \leq 0.10$	Prob. ≥ 0.95 $\alpha \leq 0.05$	Prob. Using Normal n $\alpha = 0.10$	$\alpha = 0.05$
Normal	n	n	0.90	0.95
Chebyshev	$3.72n$	$2.61n$	≥ 0.63	≥ 0.74
Unimodal	$2.93n$	$2.05n$	> 0.76	> 0.82

their consequences, but the reader should be prepared for a major effort in comprehension. We emphasize once again that the use of the t distribution will increase the required sample size above that for the normal assumption.

10.14 THE AUTOCORRELATION FUNCTION AND SPECTRUM†

The autocorrelation function provides information about the extent of correlation between events in a series as a function of their time separation. The spectrum identifies cyclic phenomena and the extent of their regularity. These characterizations can be helpful in gaining insight into the process under study. For example, the discussion of the constant service problem in Section 6.4 establishes the value of this information for design purposes.

The conventional way to learn about the autocorrelation function is to form the estimates

$$\hat{\rho}_s = \frac{\hat{R}_s}{\hat{R}_0} \qquad s = 0, 1, \ldots \tag{14.1}$$

and to plot them against s. The graph is known as a *correlogram*. For large n this estimator has mean and variance to order $1/n$:

$$E(\hat{\rho}_s) \sim \rho_s - \frac{2\pi f(0)}{n}(1 - \rho_s) - \frac{4\pi}{n}\int_{-\pi}^{\pi} f^2(\omega)(\cos \omega s - \rho_s)\, d\omega \tag{14.2}$$

$$\operatorname{cov}(\hat{\rho}_s, \hat{\rho}_t) \sim \frac{1}{n}\sum_{v=-\infty}^{\infty} [\rho_v{}^2 + \rho_{v+s}\rho_{v-t} - 2(2\rho_s\rho_v\rho_{v+t} - \rho_s{}^2\rho_v{}^2)], \tag{14.3}$$

respectively. Here the function f is the spectral density function defined in (6.4.11). These properties are considered undesirable for a function estimator for three reasons. First, the bias in $\hat{\rho}_s$ for large s is $-2\pi f(0)/n$, which is independent of ρ_s. Second, $\hat{\rho}_s$ and $\hat{\rho}_t$ are correlated, implying that a sampling disturbance in one estimate also affects other estimates. Third, for large s we have

$$\operatorname{var}(\hat{\rho}_s) \sim \frac{1}{n}\sum_{v=-\infty}^{\infty} \rho_v{}^2, \tag{14.4}$$

which is independent of s. Since ρ_s approaches zero as s increases we see that the variance to mean ratio for $\hat{\rho}_s$ becomes large as s increases.

† In order not to obscure the central points of this section we assume that initial conditions do not play a role.

The estimator $\hat{\rho}_s$ is an unrestricted one. Given that we are willing to approximate the output process by an autoregressive representation, we can follow the procedure described in Section 6.6 to generate a correlogram. As shown in Fig. 39, this estimator has the attractive feature of producing a smooth picture for the autocorrelation function in contrast to (14.1). Also the ability to produce a correlogram based on only $p + 1$ autocovariances instead of one for every s cannot be underestimated. Although the sampling properties of this alternative estimator remain to be investigated fully, it appears to offer a useful way of learning about the correlation structure between events.

Unlike the autocorrelation function, unrestricted (no autoregressive assumption) spectrum estimates can be derived with relatively convenient sampling properties [161, 168]. Unfortunately the complex estimation procedure involved calls for a familiarity beyond that to be expected from an investigator interested in results rather than method. As an alternative we may follow the procedure in Section 6.6 with substituted estimates, which is based on the autoregressive representation. A comparison of the spectrum analysis estimator and the autoregressive estimator was given in Fig. 40. Expression (9.4.9) gives the variance of the autoregressive spectrum estimator.

At the end of a simulation experiment we would like to print useful sample quantities such as the estimates mean variance, variance of the sample mean, autocorrelation function, and spectrum. The estimated mean variance and variance of the sample mean are single numbers and present no difficulties; but the estimated spectrum is a function of frequency, and the estimated autocorrelation function is ordered on time. We now must decide at how many frequencies λ in $(0, \pi)$ we want to estimate and print the spectrum and at how many values of time $s = 0, 1, \ldots$ we want to estimate the autocorrelation function.

It is difficult to specify the number of points for the spectrum when we are initially unaware of its shape. We would like the points to be close enough to reveal narrow peaks, yet sufficiently limited in number to allow us to manage the volume of printed output. Suppose that we estimate the spectrum at the $T + 1$ frequency points $\lambda = 2\pi j/T$, $j = 0, \ldots, T$. Also suppose that we print the sets of quantities $[\lambda, g(\lambda)]$, four sets to a double-spaced line. If we assume 25 lines per page our estimated spectrum will require $(T + 1)/(4 \times 25) = (T + 1)/100$ pages. For example, if we decide on $T = 1000$, the printout consumes about 10 pages.

Presumably the sample autoregressive coefficients and the sample residual variance will also be printed, as in Table 13. Then we may regard the printed spectrum as a preliminary result that can be "filled in," if need be, after visual inspection. A graph of the spectrum (not necessarily all $T + 1$ points) can assist us in determining the adequacy of the initial printout. If successive plotted points differ substantially, it is worth while filling in points to obtain a more representative picture.

We emphasize that, if T is chosen too small, narrow peaks may be missed completely. A suggested approach is to print $T + 1 = 1001$ estimates (10 pages) and then to graph the points selectively. If change occurs rapidly in a frequency interval perhaps all printed points, and even additional ones in that interval, should be graphed. If change occurs slowly then the density of points to be graphed can be reduced in some cases, say from $\pi/1000$ to $\pi/100$. Greater reductions are also possible. Since the spectrum often changes by orders of magnitude over $(0, \pi)$ it is helpful to use scientific notation in expressing its printing format.

The plotting of the sample autocorrelation functions presents a problem of time length rather than density of points. We would like to find an s^* such that, for $s \geq s^*$, $\rho_s \sim 0$. Hence we would have a self-contained picture of ρ for $0 \leq s \leq s^*$. Using the recursive formula for the sample autocorrelation function, we may plot, say, $\bar{s}$ of $\hat{\rho}_s$; and if the curve has not converged by $\hat{\rho}_{\bar{s}}$ we need only use the sample autoregressive coefficients and $\hat{\rho}_{\bar{s}}, \ldots, \hat{\rho}_{\bar{s}-p+1}$ to generate additional $\hat{\rho}_s$. We hasten to mention, however, that the bias in $\hat{\rho}_s$ increases as s approaches n. Hence the user is encouraged to put little credence in the shape of $\hat{\rho}_s, \hat{\rho}_{s+1}, \ldots$ when s becomes a significant proportion of n.

10.15 DETERMINATION OF SAMPLE SIZE

In all simulations containing random phenomena the question of how long to run an experiment occurs. As might be expected, the answer depends on how statistically accurate we wish our results to be. If our interest is estimating the mean of a process then a confidence interval as described in Section 10.2 gives us a measure of the interval in which we expect to find μ with approximate probability $1 - \alpha$. As the sample size increases, this interval diminishes for fixed α and we gain a more precise picture of where μ is located.

Suppose that we require

$$\text{Prob}(|\bar{X}_n - \mu| < \Theta) = 1 - \alpha,$$

where the investigator specifies the maximum acceptable deviation Θ. From (12.2) it is clear that

$$\Theta = Q\sqrt{m/n^*}$$

so that

$$n^* = m\left(\frac{Q}{\Theta}\right)^2.$$

If we use an estimate $\hat{m}$ for m then

$$\Theta = t_{f,1-\alpha/2}\sqrt{\hat{m}/n^*}$$

$$n^* = \hat{m}\left(\frac{t_{f,1-\alpha/2}}{\Theta}\right)^2.$$

Suppose that $\hat{m}$ is based on an initial run of n observations. If $n^* > n$ we collect $n^* - n$ more observations. If $n^* \leq n$ we have met the criterion and the run may be terminated. Figure 55 illustrates this iterative scheme.

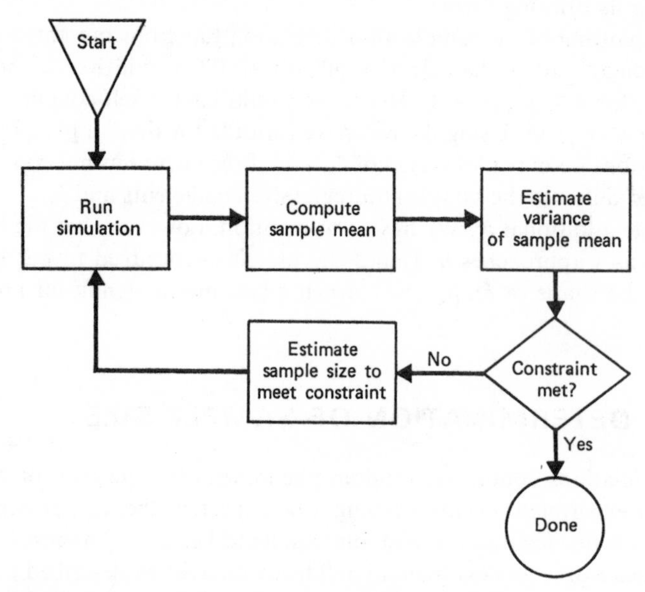

Fig. 55 Flowchart for specified statistical accuracy constraint.

If we are to follow the algorithm in Fig. 55 on a single run it is necessary to create a "data analysis" subroutine. In one approach reported in reference 5, in several investigations each involving 100 independent replications of a simulation of the first-come-first-served queueing problem the true mean lay in the resulting generated confidence intervals in at least 66 out of 100 cases. Since $1 - \alpha$ was set at 0.9 the expected number of inclusions was 90. Although the degradation was, of course, due to the many approximations and assumptions a success rate of $66/90 = 0.73$ is encouraging. Moreover slight modifications in the algorithm in reference 162 produce a minimum success rate of $75/90 = 0.83$.

The data analysis subroutines should contain an algorithm for estimating m. As reported in reference 162, the autoregressive approach offers many conveniences and it is this approach using Table 29 that we suggest. The subroutines in Appendix B can easily be modified to serve an interactive role. Figure 56 describes the major steps. The reader interested in using a more refined scheme is advised to consult reference 62.

The critical value of chi square in Fig. 56 may be approximated by

$$\chi^2_{q-r,1-\alpha} = \sum_{k=0}^{7} \frac{G_k(y)}{(q-r)^{k/2-1}},$$

where [166, 176]

$$\begin{aligned}
G_0(y) &= 1\\
G_1(y) &= \sqrt{2y}\\
G_2(y) &= 2(y^2 - 1)/3\\
G_3(y) &= (y^3 - 7y)/(9\sqrt{2})\\
G_4(y) &= -(6y^4 + 14y^2 - 32)/405\\
G_5(y) &= (9y^5 + 256y^3 - 433y)/(4860\sqrt{2})\\
G_6(y) &= (12y^6 - 243y^4 - 923y^2 + 1472)/25515\\
G_7(y) &= -(3753y^7 + 4353y^5 - 289517y^3 - 289717y)/(9185400\sqrt{2})\\
y &\equiv Q_{1-\alpha}.
\end{aligned}$$

To use the subroutine in a simulation based on SIMSCRIPT II, we can define an event named EVALUATION, which we initially schedule in M time units. The first operation called in this event routine is the data analysis subroutine, which returns with a new M. Since we are still in the event routine we schedule the next EVALUATION event in M time units and return to the timing routine. When $n \geq n^*$ the data analysis routine should be able to schedule an end-of-simulation event, which in turn should contain instructions for producing a complete statistical report. Programmers of simulations using FORTRAN, PL/1, and GASP II can also follow the same general directions. Variants of the procedure described above which may provide greater efficiency in special circumstances quickly come to mind.

10 16 THE SAMPLE STATE TIME APPROACH

We noted in Sections 3.3 that the time series approach is but one way of gathering data that can be used to infer population descriptors. For queueing-oriented problems the sample state time approach offers an alternative that has many attractive features. However, one traditional disadvantage has been the absence of any supporting statistical procedure to assess the adequacy

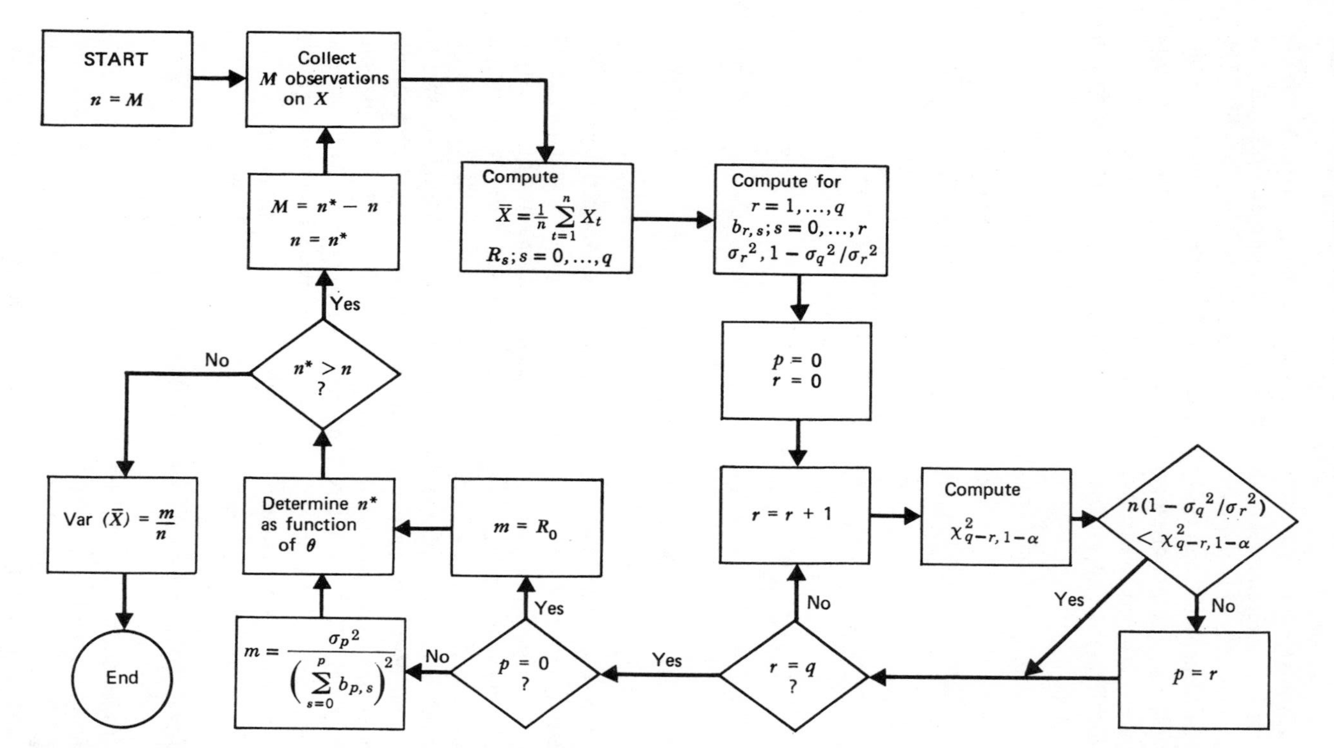

Fig. 56 Iterative scheme for estimating μ with specified accuracy (M, q, α, θ specified). *Note:* Assumes $n - q$ is large.

of the resulting estimates. In this section we describe a newly developed procedure intended to serve this purpose for certain types of queueing models.

We speak of a queueing system as being in state i when i jobs are waiting for or are in service. The null state $i = 0$ denotes the empty and idle condition and holds a unique position in many queueing analyses. In particular, once a system leaves this state, future behavior often is statistically independent of behavior before this exit. As an example, a queueing system with independent, identically distributed interarrival times and independent, identically distributed service times has this property. In the rest of this section we assume that this property of the null state holds.

During a simulation run, the system passes through many states. Let t_i denote the time spent in state i in a run of time length $n = \sum_{i=0}^{\infty} t_i$. Because of the stochastic elements in the simulation, t_i is inherently random and should properly be called a sample time. Then, conceptually, $\{t_i\}$ is a sequence of random variables, linear combinations of which enable us to estimate many relevant system performance measures. If we define

$$g_j = \sum_{i=0}^{\infty} c_{ij} t_i, \tag{16.1}$$

the sample *probability of finding the system in state k* is g_1/g_0, where

$$c_{i0} = \begin{cases} 1 & i = 0, \ldots, \infty \\ 0 & \text{elsewhere} \end{cases}$$
$$c_{i1} = \begin{cases} 1 & i = k \\ 0 & \text{elsewhere.} \end{cases} \tag{16.2}$$

The sample *probability of finding more than k jobs in the system* is g_2/g_0, where

$$c_{i2} = \begin{cases} 1 & i = k + 1, k + 2, \ldots, \infty \\ 0 & \text{elsewhere.} \end{cases} \tag{16.3}$$

The sample *probability of finding one or more of m servers* idle upon entry is g_3/g_0, where

$$c_{i3} = \begin{cases} 1 & i = 0, \ldots, m - 1 \\ 0 & \text{elsewhere.} \end{cases} \tag{16.4}$$

The sample *mean queue length* is g_4/g_0, where

$$c_{i4} = \begin{cases} i & i = 1, \ldots, \infty \\ 0 & \text{elsewhere.} \end{cases} \tag{16.5}$$

If the simulation begins with an exit from the empty and idle state and N jobs are completed during the run then for a system with m servers the sample *mean waiting time per job* is g_5/N, where

$$c_{i5} = \begin{cases} i - m & i = m + 1, \ldots, \infty \\ 0 & \text{elsewhere.} \end{cases} \tag{16.6}$$

In the single server queueing system the sample *mean busy period* is $n - t_0$.

As an experimental design consideration many simulation users specify that a simulation should run until N jobs have been completed. Then N is given, but $g_0 = n$, g_1, g_2, g_3, g_4, and g_5 are random variables. Conversely some users specify that the simulation run for n time units, so that $g_0 = n$ is a datum and N, g_1, g_2, g_3, g_4, and g_5 are random variables.† Regardless of the choice the $\{t_i\}$ are dependent sample data, a fact that has made the estimation of the *reliability* of the sample performance measures difficult. Moreover the specification of either N or n leaves open the possibility of bias due to both starting conditions and ending conditions. For example, how can one account for the waiting times of jobs present when the simulation begins and ends? The suggested solution is to pick large N or n to overcome the boundary effects, but the determination of large N or n is often vague.

Notice that every time the system leaves state zero, passage through different system states in the future is independent of the past behavior of the system. Let us define the time at which the system leaves the zero state as a *regeneration point*, and the interval between two successive regeneration points as an *epoch*. Let t_{ij} be the sample time spent in state i during epoch j. Then it should be clear that t_{ij} and $t_{kj'}$ are independent for $j \neq j'$ and i, $k = 0, \ldots, \infty$.

If we start the simulation with an arrival to an empty and idle system and run it for J epochs we may compute

$$\begin{aligned} y_{jl} &= \sum_{i=0}^{\infty} c_{il} t_{ij} \\ y_l &= \frac{1}{J} \sum_{j=1}^{J} y_{jl} \end{aligned} \tag{16.7}$$

so that $g_l = J y_l$ is a linear combination of independent, identically distributed random variables. Notice that, if we specify J, then the resulting number of completed jobs N and the total simulation time n are random variables. As a result we must regard the mean queue length estimator g_4/g_0, the mean waiting time estimator g_3/N, and the sample probability of finding an idle server on entry g_3/g_0 as *ratio* estimators.

† See Section 10.5 for the effect on the time series approach.

Let y_l have mean μ_l and variance σ_l^2, and let y_l and y_k have covariance σ_{kl}. The the distribution of $\sqrt{J}\,(y_l - \mu_l)/\sigma_l$ converges to $\mathcal{N}(0, 1)$ as J increases. However, the ratio requires more care when its sampling properties are discussed. Writing y_l/y_k in the form

$$\frac{y_l}{y_k} = \frac{\mu_l}{\mu_k}\frac{1 + (y_l - \mu_l)/\mu_l}{1 + (y_k - \mu_k)/\mu_k} \qquad l \neq k \tag{16.8}$$

and expanding the geometric power series that corresponds to the denominator,† we have to order $1/J$

$$\begin{aligned} E\left(\frac{y_l}{y_k}\right) &= \frac{\mu_l}{\mu_k}\frac{1 + \Theta_1}{J} \\ \text{var}\left(\frac{y_l}{y_k}\right) &= \left(\frac{\mu_l}{\mu_k}\right)^2 \frac{\Theta_2}{J} \\ \Theta_1 &\equiv \frac{\sigma_k^2}{\mu_k^2} - \frac{\sigma_{kl}}{\mu_k\mu_l} \\ \Theta_2 &\equiv \frac{\sigma_k^2}{\mu_k^2} - \frac{2\sigma_{kl}}{\mu_k\mu_l} + \frac{\sigma_l^2}{\mu_l^2}. \end{aligned} \tag{16.9}$$

Using queue length as an example, it is not difficult to see that the mean run time for J epochs is $J\mu_0$ and the mean job time spent in the system is $J\mu_4$, so that the mean queue length is $J\mu_4/J\mu_0 = \mu_4/\mu_0$.

Since reference 16.9 indicates that y_l/y_k has a bias whose dominant term is of order $1/J$, we first consider how to eliminate this term. Let

$$\begin{aligned} s_l^2 &= \frac{1}{J - 1}\sum_{j=1}^{J}(y_{jl} - y_l)^2 \\ s_{kl} &= \frac{1}{J - 1}\sum_{j=1}^{J}(y_{jk} - y_k)(y_{jl} - y_l). \end{aligned} \tag{16.10}$$

Then the expression $(y_l/y_k)(1 - \hat{\Theta}_1/J)$

$$\hat{\Theta} \equiv \frac{s_k^2}{y_k^2} - \frac{s_{kl}}{y_k y_l} \tag{16.11}$$

eliminates the bias to order $1/J$. Alternatively we may rewrite (16.9) as

$$\frac{E(y_l/y_k)}{(1 + \Theta_1/J)} = \frac{\mu_l}{\mu_k}. \tag{16.12}$$

† We assume that $|y_k - \mu_k| < \mu_k$, a result which holds in probability as n becomes large. See reference 175.

Then we may show that the estimator [175]

$$r_{k,l} = \frac{y_l}{y_k(1 + \hat{\Theta}_1/J)} \tag{16.13}$$

also eliminates the bias to order $1/J$.

Treating y_l/y_k as a normal variate, we have

$$\text{Prob}\left[\left|\frac{y_l}{y_k} - \left(\frac{\mu_l}{\mu_k}\right)\left(1 + \frac{\Theta_1}{J}\right)\right| < Q_{1-\alpha/2}\left(\frac{\mu_l}{\mu_k}\right)\sqrt{\hat{\Theta}_2/J}\right] \sim 1 - \alpha \tag{16.14}$$

where $Q_{1-\alpha/2}$ is the point corresponding to probability $1 - \alpha$ of the normal distribution. We can then rearrange terms to derive an *approximating* confidence interval for μ_l/μ_k in the form

$$\frac{y_l/y_k}{1 + \hat{\Theta}_1/J + Q_{1-\alpha/2}\sqrt{\hat{\Theta}_2/J}}, \qquad \frac{y_l/y_k}{1 + \hat{\Theta}_1/J - Q_{1-\alpha/2}\sqrt{\hat{\Theta}_2/J}} \tag{16.15}$$

$$\hat{\Theta}_2 \equiv \frac{s_k^2}{y_k^2} - \frac{2s_{kl}}{y_k y_l} + \frac{s_l^2}{y_l^2}.$$

Here we assume that J is large enough so that the error of substitution is not a principal issue.

For the waiting time we need to collect N_j, the number of jobs completed in epoch j. Then we may proceed to derive analogous results for $y_5/\bar{N}$, where

$$\bar{N} = \frac{1}{J}\sum_{j=1}^{N} N_j \tag{16.16}$$

The individual N_j are needed to compute the sample variance and covariance.

We observe that the design specification is now J instead of N or n. Moreover the initial conditions do not play the same role as before, since in each epoch their influence disappears once the system enters the zero state.

Consider the $M/M/1$ queueing problem of Section 6.5 with mean interarrival and service times 0.2222 and 0.20, respectively, and first-come-first-served queueing discipline. The above techniques were applied to the study of the mean queue length and mean waiting time, whose theoretical values were 9 and 1.8, respectively. One hundred independent replications of a SIMSCRIPT II.5 simulation of this problem were performed, each replication beginning with an arrival to the empty and idle system and running for $J = 1000$ epochs. On each run the mean queue length and mean waiting time were estimated and separate 90 per cent confidence intervals for mean queue length and mean waiting times were computed. Table 35 shows the results. These results are encouraging, for they suggest a mild degradation in performance due to the use of $\hat{\Theta}_1$ and $\hat{\Theta}_2$ for Θ_1 and Θ_2, respectively.

Table 35 Confidence Intervals for Queueing Problem (100 Replications)

	Mean		Number of Intervals That Include Population Mean		
	Sample	Theoretical	Sample	Theoretical	Sample Divided by Theoretical
Queue length	8.94	9	85	90	0.94
Waiting time	1.78	1.8	86	90	0.96

As an example of the summary statistics computable on each run we present those relevant to queue length on the forty-first replication:

$$y_4 = 23.66, \qquad y_0 = 2.43, \qquad r_{0,4} = 9.85$$

$$(16.17) \qquad s_4{}^2 = 31037.9, \qquad s_0{}^2 = 86.06, \qquad s_{04}{}^2 = 1554.66$$

$$\hat{\Theta}_1 = -12.47, \qquad \hat{\Theta}_2 = 15.96.$$

Having glossed over several practical issues, we now return to them and resolve those that can be resolved and qualify those that cannot.

Bias

As noted in Sections 10.4 and 10.5, bias is a real consideration at both the beginning and end of a simulation. If one analyzes the statistical data generated by the system in the present epoch perspective, the bias due to initial conditions plays no role. Since the probability distribution of the state times $\{t_i\}$ in an epoch is independent of and identical to the probability distribution of state times in any other epoch, behavior within an epoch represents steady state behavior. Consequently, no initial bias occurs, provided that each epoch begins with an arrival to an empty and idle state and every epoch ends just prior to an arrival to an empty and idle state.

Since we are using ratio estimators here, a bias due to the rule for terminating an epoch enters. However, the independence among epochs allows us to compute $\hat{\Theta}_1$ in (16.11) which is used in (16.13) to remove the bias to order $1/J$. In (5.1), where we deal with autocorrelated data, it is not clear that an analogous substitution is possible using (5.6) with estimates of m_{NZ}, m_N, μ_W, and μ_N. In addition Θ in (5.6) is unknown.

Normality

The accuracy of the approximating confidence interval in (16.15) depends heavily on the rate at which y_l/y_k converges to normality. Although it is generally safe to regard y_l and y_k as normal for $J > 30$, no analogous statement holds for their ratio. To order $1/J^{1/2}$ the skewness of y_l/y_k is [175]

$$\gamma_J = \frac{\Theta_2 - \Theta_1}{\sqrt{\Theta_2 J}}. \tag{16.18}$$

Substitution of sample parameters for population parameters in (16.18) provides us with an estimate of skewness. Naturally the closer this estimate is to zero the more comforting the result is. Dividing this estimate by $\sqrt{K}$ gives us an indication of the skewness reduction that results when data are collected on $(K - 1)J$ more epochs. For example, $K = 4$ reduces the skewness by half.

Since we customarily have neither the benefit of a theoretical result for comparison nor 100 replications for evaluating the normal approximation, computation of a sample skewness provides one means of assessing normality in a practical situation and also the reduction in skewness that added epochs produce. For the queue length data in (16.17) $\hat{\gamma}_{1000} = 0.23$, a value which indicates little skewness. Therefore additional epochs aimed at reducing skewness do not seem necessary.

Sample Parameter Calculation

To compute approximating confidence intervals for L performance measures we need at least JL memory locations to store the y_{jk} needed for (16.10). Alternatively, use of

$$s_k^2 = \frac{1}{J - 1}\left(\sum_{j=1}^{J} y_{jk}^2 - J y_k^2\right) \tag{16.19}$$

$$s_{kl} = \frac{1}{J - 1}\left(\sum_{j=1}^{J} y_{jk} y_{jl} - J y_k y_l\right) \qquad k \neq l \tag{16.20}$$

requires the collection of the indicated sums, which create a considerably smaller space need. However, it is now well known that sample variances and covariances calculated with (16.20) are subject to large numerical errors because of the finite length of storage locations. Double-precision arithmetic should reduce the problem, and the corresponding doubling of the space requirement should still not approach the amount required by (16.10).

Restarting a Simulation

Retention of the relevant sums mentioned above enables a user to run his simulation in segments. To begin a new segment he simply loads these sums into their appropriate locations in the program. Since we are starting with an arrival to an empty and idle state, we need no additional data.

Because our formulae are correct to order $1/J$, it behooves us to make the first run long enough so that our statistical results will be meaningful. Since the benchmark of $J > 30$ ensures a relatively smaller error in the normality assumption for numerators and denominators individually, we expect this to be the lower bound. A larger initial J would naturally be preferable.

Desired Statistical Reliability

The determination of the number of epochs to run is directly related to how statistically reliable we want our simulation results to be. For example, we may wish to produce results that satisfy the probability statement:

$$\text{Prob}\left(\left|r_{k,l} - \frac{\mu_l}{\mu_k}\right| < \frac{\phi\mu_l}{\mu_k}\right) \sim 1 - \alpha \tag{16.21}$$

where ϕ is a specified constant. Since $\text{var}(r_{k,l}) = (\mu_l/\mu_k)^2\Theta_2/J$ we have

$$\begin{aligned} \frac{\phi\mu_l}{\mu_k} &= Q_{1-\alpha/2}\frac{\mu_l}{\mu_k}\sqrt{\Theta_2/J^*} \\ J^* &= \left(\frac{Q_{1-\alpha/2}}{\phi}\right)^2 \Theta_2. \end{aligned} \tag{16.22}$$

Using $\hat{\Theta}_2 = 15.96$ from (16.17), we have for $\alpha = 0.10$

$$J^* = \frac{42.93}{\phi^2}.$$

For $\phi = 0.2$ we require $J^* = 1073$, which is close to the number of epochs already run. For $\phi = 0.1$ we need $J^* = 4293$, an additional 3293 epochs.

If an investigator specifies the desired statistical reliability for several performance measures, it will probably be the case that different J^*s are needed for each measure. Running the simulation for the largest $J^* - J$ is consistent with attempting to meet all reliability criteria.

The principal appeal of this approach is that it enables a user to collect one type of data, namely, sample state times, and to apply these data to learn about many population descriptors and their estimates. Although a time series on queue length also enables a user to estimate these parameters, as in Section 3.3, it does not provide a convenient way of resolving the bias that materializes, depending on whether run length n or number of completed jobs N is specified.

EXERCISES

1. Consider a sequence of independent, identically distributed random variables $X_1, \ldots, X_n$, each with continuous probability distribution function F. Let $F(0) = 0$ and $F(1) = 1$. A sampling experiment is planned to study the sampling properties of $\max_i (X_i)$. The investigator intends to run k independent replications, select the maximum on each replication, and use these sample maxima to estimate the population mean maximum and the cumulative distribution of this maximum. A colleague suggests that the investigator can derive more information from each sample in estimating the mean of the maximum by using a transformation of $\min_i(X_i)$, as well as $\max_i (X_i)$, provided that the probability density function f is symmetric.
   a. Which transformation does the colleague have in mind? A study of Exercise 3 in Chapter 8 may be helpful.
   b. What form would the new sample mean take?
   c. From variance considerations, determine whether or not the suggestion is helpful.
   d. Is there any benefit in using the added data to estimate the cumulative distribution of the sample maximum? Explain your answer.

2. In queueing theory there is a well-known result that, under rather general conditions, mean time spent in system = mean interarrival time × mean queue length. An investigator is interested in estimating M, the mean time spent in system, in each of a series of different simulation experiments. He must decide on one of two courses of action:

   1. run each simulation until N jobs are completed, collect the time each job spends in the system, compute the sample average $\bar{Y}_N$, and use this quantity as the estimate of M, and

   2. run each simulation until n time units have elapsed, collect observations on the queue length at unit intervals, compute the sample mean $\bar{X}_n$, and use $\lambda\bar{X}_n$ as the estimate of M, where λ is the mean interarrival time.

   In preliminary runs on the same experiment, action 1 produced an estimate σ_1^2/N for the variance of $\bar{Y}_N$, and action 2 produced an estimate σ_2^2/n for the variance of $\bar{X}_n$. Also the computer cost of action 1 was c_1 on the first run, and the computer cost of action 2 was c_2 on run 2. There is reason to believe that σ_1^2/σ_2^2 will be roughly constant for all experiments. Ignore bias considerations here.

   Given that the investigator wishes to obtain results with a specified variance, regardless of the method used, describe how he should make use of σ_1^2, σ_2^2, N, n, c_1, and c_2 to decide whether to adopt action 1 or action 2 in the remaining experiments.

3. Let X_t = queue length (number of jobs in the system) at time t, and let Z_t be the number of jobs being served at time t. An investigator wishes to estimate the mean number of jobs in the system. In many queueing simulations, μ_Z, the mean number of jobs in service, can be computed exactly from the arrival rate, service rate, and number of servers. The investigator is considering two alternative estimates for given n:

   1. $\bar{X}_n = \frac{1}{n}\sum_{t=1}^{n} X_t$, and

   2. $\tilde{X}_n = \frac{1}{n}\sum_{t=1}^{n} X_t - \frac{1}{n}\sum_{t=1}^{n} Z_t + \mu_Z$.

   Suggest a criterion for choosing between them. Assume that both are unbiased. Note that the two estimators can be used simultaneously.

4. Write an analysis routine to estimate the variance of the sample mean, using the spectrum analysis approach in Section 10.10. If the routine is written in SIMSCRIPT II, apply it to the analysis of queue length data in the example of Section 5.5; if written in FORTRAN, to the queue length data in the example of Section 5.3; and if written in SIMULA. to the waiting time data in the example of Section 5.7. Naturally in each case it is necessary to run the simulation to produce the necessary data.

5. Let $\{X_1, \ldots, X_n\}$ and $\{Y_1, \ldots, Y_n\}$ be sequences of independent normal variables. We estimate μ_y/μ_x by $\bar{Y}_n/\bar{X}_n$. One way to estimate a confidence interval for μ_y/μ_x is described in Section 10.16. An alternative method makes use of the fact that $\bar{Y}_n - v\bar{X}_n$ has mean $\mu_y - v\mu_x$ and variance $\sigma_y^2 - 2v\sigma_{2y} + v^2\sigma_x^2$. Let s_y^2, s_x^2, and s_{xy} denote the corresponding sample variances and covariance. Then $[\bar{Y}_n - v\bar{X}_n - (\mu_y - v\mu_x)]/\sqrt{(s_y^2 - 2vs_{xy} + v^2s_x^2)/n}$ has the t distribution with $n - 1$ degrees of freedom.

   By choosing v in an appropriate way derive a confidence interval for μ_y/μ_x, using the t distribution.

CHAPTER 11

THE DESIGN OF EXPERIMENTS

11.1 INTRODUCTION

Chapter 10 noted that a simulation user has the prerogative of specifying several design factors that influence the interpretation of his simulation output. Four of these factors are as follows:

1. initial conditions such as queue length and time of first arrival,
2. final conditions such as run simulated time length or number of completed simulated jobs,
3. data collection time points, and
4. balance between run length and number of replications.

There are at least two additional ways in which a user can affect his results. First, he can employ a *variance reduction technique* that enables him to obtain, for the same expenditure, results with greater statistical accuracy than would otherwise be possible. Second, he can adopt, for the input to the sequence of experiments that he has in mind, an *experimental layout* that provides him with improved statistical accuracy and facilitates his statistical analysis. These items, together with the above list, constitute a major portion of the *design of experiments*, a topic whose relevance to simulation work has long been acknowledged but rarely applied in practice. This chapter describes variance reduction and experimental layout issues as they relate to a simulation study. The principal emphasis here is on inference concerning the mean of a process of interest.

Occasionally an investigator can merge a priori information about the system under study with data collected during a simulation run to derive an estimate of the mean. Moreover this estimate often has smaller variance, for a given run length, then would obtain if the estimate were formed solely from the sample. In Sections 11.2 and 11.3 we describe examples of the use of prior information in an inventory problem.

Every stochastic simulation must adopt a *sampling plan* before it can produce the random variates needed to run the simulation. Each plan corresponds to a different way of using the generated uniform deviates to produce the needed variates from the diverse distributions that characterize the input. By a judicious pairing of sampling plans for two replications, we can often produce an estimate using data from both replications, whose variance is smaller than the variance that would result if the sampling plans were unrelated. The pairing of sampling plans is also a variance reduction technique. We discuss several plans in Sections 11.4 and 11.5.

The other component of the design of experiments concerns the choice of operating conditions or decision rules and parameter settings to be specified in a set of experiments intended to reveal changes in the output processes, if they exist. Suppose that the input to an experiment consists of j parameters called *factors*, and we are considering running experiments with each factor at k different levels. Each combination of levels is called a *treatment*, and it is easily seen that we have j^k experimental treatments to run. For example, with $j = 2$ and $k = 5$ there are 32 experiments. Clearly the number of experiments in the *full factorial* experiment can easily become prohibitive with a few factors and treatments.

To resolve this problem a variety of techniques can be used; some are based on close study of the system under examination, and others on the statistical purpose of the analysis. Occasionally these techniques are also called sampling plans. To distinguish them from variance reduction methods, we refer to them here as comprising the *experimental layout*.

If we are going to discuss experimental layout, it behooves us to first describe the ultimate purpose of a simulation study. From a statistical viewpoint there are principally two purposes. One is the *comparison* of experimental results under alternative operating conditions. The other is the *detection* and *estimation* of the functional relationship that exists between the quantitative input factors and the experimental results. These topics in themselves deserve lengthy exposition. For present purposes we describe the problems as they may occur in a simulation, outline the traditional statistical framework for solution, discuss variations in the statistical framework that a simulation may introduce, and provide references to more definitive statistical accounts. It is hoped that putting problems into perspective in this way will encourage the reader to concern himself with their solution.

One particularly important area in which the comparison of experiments is a central consideration is the *validation* of a simulation model. By validation we mean a study of how well the behavior of a model accords with that of the true system. As Section 11.8 makes clear, validation can imply different types of agreement. Therefore test procedures are needed to accommodate the search for these different forms of agreement. Section 11.8 suggests several alternative procedures.

11.2 PRIOR INFORMATION

In addition to having the knowledge to model a system for simulation, a user would like to bring this knowledge to bear on the estimation of output parameters. Presumably doing so enables him to estimate these quantities more accurately. Although the incorporation of prior knowledge into estimation techniques is a substantive topic in the Monte Carlo literature [194], the extension of the methods suggested there to discrete event simulation is not obvious. No doubt, the complexity of many discrete event simulation models in contrast to the more mathematically precise Monte Carlo experiments contributes to this difficulty. In this section we describe a procedure for incorporating prior information and illustrate its use with an example from Carter and Ignall [186].

Let X and Y be observations drawn from distributions with means μ_X and μ_Y, respectively. Assume that we know the ratio $r = \mu_Y/\mu_X$ and wish to estimate μ_Y. One estimate is

$$\hat{\mu}_Y = Y;$$

an alternative one is

$$\tilde{\mu}_Y = rX.$$

Both are unbiased. However, if

$$\frac{\text{var}(Y)}{\text{var}(X)} > r^2$$

we prefer $\tilde{\mu}_Y$ to $\hat{\mu}_Y$. Moreover we prefer the estimate

$$\ddot{\mu} = \frac{rX + Y}{2}$$

if

$$\text{var}(Y) > r^2 \text{var}(X) > \frac{r^2 \text{var}(X) + 2r \text{cov}(X, Y) + \text{var}(Y)}{4}.$$

Although this discussion may appear academic to a potential simulation user, he often may be in a position to use this technique when his prior information includes certain probability distributions.

Suppose that we have two times series $V_1, \ldots, V_n$ and $W_1, \ldots, W_n$, and wish to estimate μ_V. We assume that V_i and W_i are discrete. Then theory tells us that

$$\mu_V = \sum_{j=-\infty}^{\infty} j \text{ Prob}(V = j).$$

For $t = 1, \ldots, n$ let

$$X_{jt} = \begin{cases} 1 & W_t = j \\ 0 & W_t \neq j \end{cases}$$

$$Y_{jt} = \begin{cases} 1 & V_t = j \\ 0 & V_t \neq j \end{cases}$$

$$Y_{jkt} = \begin{cases} 1 & V_t = j, W_t = k \\ 0 & \text{elsewhere,} \end{cases}$$

$$X_j = \sum_{t=1}^{n} X_{jt}, \qquad Y_j = \sum_{t=1}^{n} Y_{jt}, \qquad \tilde{Y}_{jk} = \sum_{t=1}^{n} Y_{jkt}.$$

As an estimate of $\text{Prob}(V = j)$ we have Y_j/n and hence the sample mean

$$\hat{\mu}_V = \frac{1}{n} \sum_{j=-\infty}^{\infty} jY_j. \tag{2.1}$$

This is easily seen to be the sample mean $\overline{V}_n$. Noting that

$$Y_j = \sum_{k=-\infty}^{\infty} \tilde{Y}_{jk}, \qquad X_k = \sum_{j=-\infty}^{\infty} \tilde{Y}_{jk},$$

we can write (2.1) as

$$\hat{\mu}_V = \frac{1}{n} \sum_{j=-\infty}^{\infty} j \sum_{k=-\infty}^{\infty} \tilde{Y}_{jk} = \sum_{k=-\infty}^{\infty} \left(\sum_{j=-\infty}^{\infty} j \frac{\tilde{Y}_{jk}}{X_k} \right) \frac{X_k}{n}. \tag{2.2}$$

The quantities $\tilde{Y}_{jk}/X_k$ and X_k/n are estimates of $\text{Prob}(V = j \mid W = k)$ and $\text{Prob}(W = k)$, respectively.

Suppose that our prior knowledge includes $\text{Prob}(V = j \mid W = k)$, which corresponds to the ratio μ_Y/μ_X in our earlier example. By substituting this probability for Y_{jk}/X_k we can form the alternative estimates:

$$\begin{aligned} \tilde{\mu}_V &= \frac{1}{n} \sum_{k=-\infty}^{\infty} \left[\sum_{j=-\infty}^{\infty} j \, \text{Prob}(V = j \mid W = k) \right] X_k \\ &= \frac{1}{n} \sum_{t=1}^{n} Z_t \end{aligned} \tag{2.3}$$

$$Z_t = \sum_{j=-\infty}^{\infty} j \, \text{Prob}(V = j \mid W = W_t) \tag{2.4}$$

and

$$\ddot{\mu}_V = \frac{1}{n} \sum_{t=1}^{n} (V_t + Z_t). \tag{2.5}$$

The attractive feature here is that we can compute $\hat{\mu}_V$, $\tilde{\mu}_V$, and $\ddot{\mu}_V$ on the same replication, estimate their respective variances using, for example, the autoregressive subroutine in Appendix B, and then select the estimator with the smallest sample variance. Alternatively we can use the spectrum analysis approach of Section 10.9. Analogously, a user who prefers to work with independent replications can compute sample variances for the grand sample means obtained over all replications and select the grand estimate with least sample variance.

The use of (2.4) requires care. If V and W have finite ranges then the computation and storing the sums in (2.4) in a table for later use is a judicious step. If an analytical expression exists for (2.4) that is even a greater convenience.

To illustrate the desirability of $\tilde{\mu}_V$ over $\hat{\mu}_V$ we describe an application of this technique by Carter and Ignall [186] to an inventory system. They have also applied this technique to a simulation of fire department operations in New York City [185].

Consider a period model (Section 3.7) of an inventory system† with
$D_t \equiv$ demand during period t, $I_t \equiv$ inventory level at the end of period t,
$\phi_t \equiv$ quantity ordered at beginning of period t.

Demand is generated from a known distribution. The order quantity is determined according to the rule:

$$\phi_t = \begin{cases} 0 & I_{t-1} > s \\ S - I_{t-1} & I_{t-1} \le s \end{cases} \qquad s < S.$$

Orders are delivered instantaneously so that

$$I_t = I_{t-1} + \phi_t - D_t.$$

Let h = unit holding cost, b = unit back-order cost, K = setup cost for ordering, so that operating cost in period t is

$$\begin{aligned} C_t &= hI_t^+ + b(-I_t)^+ + K\delta_t \\ x^+ &= \max(x, 0) \end{aligned} \tag{2.6}$$

$$\delta_t = \begin{cases} 1 & \phi_t > 0 \\ 0 & \phi_t \le 0. \end{cases}$$

For an n period run we have the sample mean operating cost per period

$$\bar{C}_n = \frac{1}{n} \sum_{t=1}^{n} C_t. \tag{2.7}$$

† We use the notation of Carter and Ignall throughout this section.

The objective is to select s and S to produce a minimum operating cost per period.

Looking at the unit costs, we note that b large relative to h implies that, under an optimal choice of s and S, back orders ($I_t < 0$) occur infrequently, although the mean back-order cost $b(-I_t)^+$ is a significant share of the mean operating cost per period. Under these circumstances a good estimate $\overline{C}_n$ requires us to choose n large enough so that sufficient back-order situations occur to obtain an accurate estimate of mean back-order cost. It is worth while therefore to search for an alternative estimate of mean back orders that overcomes this undesirable step. Carter and Ignall offer the following alternative, which makes use of prior information on demand.

Let

$$A_t \equiv I_{t-1} + \phi_t$$

$$\pi(j) = \text{probability of } j \text{ demands in a period.}$$

Here A_t is the stock position after an order arrives but before demand occurs. We note that

$$\text{(2.8)} \qquad \text{Prob}(I_t = k) = \sum_{j=-\infty}^{\infty} \text{Prob}(I_t = k \mid A_t = j)\,\text{Prob}(A_t = j).$$

However, from the definitions of A_t and ϕ_t it is clear that

$$\text{Prob}(A_t = j) = 0 \qquad j = -\infty, \ldots, s;\ S + 1, \ldots, \infty.$$

Also

$$\text{Prob}(I_t = k \mid A_t = j) = \text{Prob}(D_t = j - k) = \pi(j - k).$$

Then (2.6) becomes

$$\text{Prob}(I_t = k) = \sum_{j=s+1}^{S} \pi(j - k)\,\text{Prob}(A_t = j).$$

The expected number of back orders in period t is

$$E(-I_t)^+ = -\sum_{k=-1}^{-\infty} k\,\text{Prob}(I_t = k)$$

$$= -\sum_{k=-1}^{-\infty} k \sum_{j=s+1}^{S} \pi(j - k)\,\text{Prob}(A_t = j)$$

$$\text{(2.9)} \qquad = \sum_{j=s+1}^{S} \left[\sum_{r=j+1}^{\infty} (r - j)\pi(r)\right] \text{Prob}(A_t = j).$$

Let

$$Y_{jt} = \begin{cases} 1 & A_t = j \\ 0 & A_t \neq j. \end{cases} \tag{2.10}$$

Then $1/n \sum_{t=1}^{n} Y_{jt}$ is an estimate of $\text{Prob}(A_t = j)$, and

$$B = \frac{b}{n} \sum_{t=1}^{n} \sum_{j=s+1}^{S} Y_{jt} \left[E(D_t) - j + \sum_{k=0}^{j} (j - k)\pi(k) \right] \tag{2.11}$$

is an estimate of mean back-order cost. The bracketed quantity replaces the bracketed quantity in (2.9). Then Carter and Ignall suggest estimating mean operating cost by

$$\bar{C}'_n = \frac{h}{n} \sum_{t=1}^{n} I_t^+ + B + \frac{K}{n} \sum_{t=1}^{n} \delta_t. \tag{2.12}$$

Noting the definition of Y_{jt} in (2.10), we can rewrite (2.12) as

(2.13)

$$\bar{C}'_n = bE(D_t) + \frac{1}{n} \sum_{t=1}^{n} \left\{ hI_t^+ - b\left[A_t - \sum_{k=0}^{A_t} (A_t - k)\pi(k) \right] + K\delta_t \right\}.$$

The rationale for considering this alternative estimator is that substitution of factual information into the estimator should produce $\text{var}(\bar{C}'_n) < \text{var}(\bar{C}_n)$. To check this conjecture, Carter and Ignall ran two sets of 99 independent replications with

$$n = 100, \qquad h = 1, \qquad K = 64$$

$$\pi(k) = \pi(73 - k) = \frac{2k - 41}{512} \qquad k = 21, \ldots, 36$$

$$E(D_t) = 36.$$

On the first set,

$$b = 9, \qquad s = 30, \qquad S = 79;$$

on the second,

$$b = 99, \qquad s = 41, \qquad S = 88.$$

On each replication $\bar{C}_n$ and $\bar{C}'_n$ were computed. These statistics were then used to compute estimates of $\text{var}(\bar{C}_n)$ and $\text{var}(\bar{C}'_n)$, which are shown in Table 36. The benefit of $\bar{C}'_{100}$ is easily appreciated.

Table 36 Variance Comparisons for Two Estimators

	$\widehat{\text{var}}\,(\bar{C}_{100})$	$\widehat{\text{var}}\,(\bar{C}'_{100})$
$b = 9,\quad s = 30,\quad S = 79$	3.42	0.88
$b = 99,\quad s = 41,\quad S = 88$	14.07	1.60

We wish to note that (2.6) and (2.12) would enable a user to generate a time series from which he could estimate $\text{var}(\bar{C}_n)$ and $\text{var}(\bar{C}'_n)$ on a single run either by spectrum analysis (Section 10.10) or by the approximating autoregressive method (Section 10.11).

11.3 IMPORTANCE SAMPLING

Examining the cost function (2.6) suggests an alternative variance reduction technique. Notice that, for given I_{t-1}, C_t is solely a function of D_t. Since demands are independent it follows that $\{C_t \mid I_{t-1}\}$ is a sequence of independent observations. Suppose that we specify $I_{t-1} = j$ and generate k_j independent demands $D_{1j}, \ldots, D_{k_j j}$. Then the cost on replication i is

$$(3.1)\qquad C_{ij} = \begin{cases} h(S - D_{ij})^+ + b(D_{ij} - S)^+ + K & j \le s \\ h(j - D_{ij})^+ + b(D_{ij} - j)^+ & j > s. \end{cases}$$

Suppose that C_{ij} have variance $\sigma_j^{\,2}$. Then

$$\bar{C}_{k_j} = \frac{1}{k_j} \sum_{i=1}^{k_j} C_{ij}$$

is an estimate of the conditional cost, given $I_{t-1} = j$, and has variance $\sigma_j^{\,2}/k_j$. Then

$$(3.2)\qquad \tilde{C}_n = \sum_{j=-\infty}^{\infty} \bar{C}_{k_j}\,\text{Prob}(I_{t-1} = j)$$

is an unbiased estimate of mean cost and has variance

$$\text{var}(\tilde{C}_n) = \sum_{j=-\infty}^{\infty} \frac{\sigma_j^{\,2}}{k_j}\,[\text{Prob}(I_{t-1} = j)]^2.$$

Since our principal concern is to obtain accurate cost data for $I_{t-1} < s$, we can simply take more replications for $j < s$, thereby reducing the variances of the $\bar{C}_{k_j}$ for $j < s$ to a greater relative extent. This approach is called *importance sampling* because it concentrates the sampling effort on the js that are most likely to contribute to back-order cost.

Since $\text{Prob}(I_{t-1} = j)$ is unknown we need an estimate of it. Suppose that we run the inventory simulation straightforwardly and generate $\bar{C}_n$ as in (2.7) and $I_1, \ldots, I_n$. This replication is to be independent of the sampling mentioned above. Let

$$W_{jt} = \begin{cases} 1 & I_i = j \\ 0 & I_t \neq j \end{cases}$$

so that

$$\bar{W}_{jn} = \frac{1}{n} \sum_{t=1} W_{jt}$$

is an estimate of $\text{Prob}(I_{t-1} = j)$. Substitution into (3.2) gives us

$$\tilde{\tilde{C}}_n = \sum_{j=-\infty}^{\infty} \bar{C}_{k_j} \bar{W}_{jn} \tag{3.3}$$

Since

$$E(\tilde{\tilde{C}}_n) = \sum_{j=-\infty}^{\infty} E(\bar{C}_{k_j}) E(\bar{W}_{jn})$$

$\tilde{\tilde{C}}_n$ is an unbiased estimator of mean cost. If we let

$$X_t = \sum_{j=-\infty}^{\infty} W_{jt} \bar{C}_{k_j} = \bar{C}_{k_{I_t}}$$

then

$$\tilde{\tilde{C}}_n = \bar{X}_n = \frac{1}{n} \sum_{t=1}^{n} X_t.$$

As before, the quantity $(\tilde{\tilde{C}}_n + \bar{C}_n)/2$ may provide an improved estimate of mean cost, and the time series $X_1 + C_1, \ldots, X_n + C_n$ yields data for estimating its variance.

The appeal of this variance reduction technique is that the form of the cost function (2.7) is the only prior information required. In comparison to the technique described in Section 11.2, however, this approach requires more sampling. Since it is highly unlikely that $\text{Prob}(I_{t-1} = j)$ will be known in advance, users of this variance reduction technique must content themselves with the added sampling.

In practice it is advisable to perform the n period run to generate the sample probability mass function in advance of the sampling for each j. Doing so establishes limits on j which replace the infinite limits in (3.3) and facilitate the subsequent sampling.

11.4 ANTITHETIC SAMPLING

Whereas the two techniques just described concentrated on the manipulation of information given about the system under study, antithetic and stratified (next section) sampling procedures concentrate on the manipulation of the uniform deviate stream to induce variance reduction. Their reliance on prior information is considerably less; however, it pays to use available information to avoid serious errors that can arise.

Suppose that we wish to estimate a parameter Ω in the process of interest. If S_j is an unbiased estimate of Ω on the jth replication then our grand estimate over k such replications is

$$\bar{S}_k = \frac{1}{k}\sum_{j=1}^{k} S_j \tag{4.1}$$

with variance

$$\operatorname{var}(\bar{S}_k) = \frac{1}{k}\left[\sum_{j=1}^{k} \operatorname{var}(S_j) + 2\sum_{i\neq j}^{k} \operatorname{cov}(S_i, S_j)\right]. \tag{4.2}$$

For independent replications the covariance terms vanish. However, if we can make the sum of the covariances negative we can produce a smaller variance than independent replications offer.

Inducing negative correlation (covariance) between the results of different replications is one strategy for obtaining more statistically reliable answers than independent replications can offer. In the case of antithetic variates and stratified sampling the objective is to induce negative correlation between corresponding elements of the random number streams used to generate input variates on different replications. The intuitive feeling is that some of the negative correlation between inputs finds its way to the corresponding output, and hence a variance reduction is realized. The method of *control variates* operates in a different way, as described in Section 11.6.

Other variance reduction techniques exist and are described in references 187, 194, and 202. We prefer the three mentioned above, however, because of the relative simplicity they offer in practice and the fact that every replication produces a useful result, so that we need not complete all replications to make use of the results from any one. Most other techniques described in the references suffer from the absence of either one or the other of these desirable properties.

In one variant of the antithetic variate method we use the random deviate sequence $U_1, \ldots, U_n$ to generate events in an input sequence on one replication, and $1 - U_1, \ldots, 1 - U_n$ to generate the corresponding input sequence in a second replication. We consequently expect the two sequences to be negatively correlated. For example, if the input sequences are

$$\left.\begin{aligned} W_j &= aU_j + b \\ W_j' &= a(1 - U_j) + b \end{aligned}\right\} \quad j = 1, \ldots, n, \tag{4.3}$$

respectively, then Table 21 gives

$$\begin{aligned} E(W_j) &= E(W_j') = \frac{a}{2} + b \\ \operatorname{var}(W_j) &= \operatorname{var}(W_j') = \frac{a^2}{12}. \end{aligned} \tag{4.4}$$

Moreover,

$$\begin{aligned} &E\{[W_j - E(W_j)][W_j' - E(W_j')]\} \\ &\qquad = E\left[aU_j + b - \frac{a}{2} - b\right]\left[a(1 - U_j) + b - \frac{a}{2} - b\right] \\ &\qquad = -a^2 \operatorname{var}(U_j) = \frac{-a^2}{12} \end{aligned} \tag{4.5}$$

so that

$$\operatorname{corr}(W_j, W_j') = -1. \tag{4.6}$$

This means that corresponding elements in $\{W_j\}$ and $\{W_j'\}$ have perfect negative correlation.

Since we usually transform uniform variates to variates from other distributions it is instructive to study the effect on correlation. Suppose

$$\left.\begin{aligned} W_j &= -\beta \log U_j \\ W_j' &= -\beta \log(1 - U_j) \end{aligned}\right\} \quad j = 1, \ldots, n. \tag{4.7}$$

For example, $\{W_j\}$ may be the interarrival times in our single server queueing model on one replication, and $\{W'_j\}$ may be the interarrival times on a second replication. We have from Table 21

$$E(W_j) = E(W_j') = \beta, \qquad \operatorname{var}(W_j) = \operatorname{var}(W_j') = \beta^2. \tag{4.8}$$

Moreover

$$\begin{aligned} \operatorname{cov}(W_j, W_j') &= E[(W_j - \beta)(W_j' - \beta)] \\ &= \beta^2 E[\log U_j - 1][\log(1 - U_j) - 1] \\ &= \beta^2 \left[\int_0^1 \log u \log(1 - u)\, du - 1\right] \\ &= \beta^2 \left(1 - \frac{\pi^2}{6}\right) = \beta^2 \times (-0.644934). \end{aligned} \tag{4.9}$$

so that

$$\text{corr}(W_j, W_j') \sim -0.645. \tag{4.10}$$

In most simulations and Monte Carlo studies input variables from specified distributions are exposed to a variety of transformations including translation in time. Although such transformation attenuates the influence of the induced correlation there is always the hope that some negative influence will reach the outputs and produce an overall smaller variance.

Page [204] has provided one theoretical insight into the single server queueing problem with interarrival times from $\mathscr{E}(\beta_1)$ and service times from $\mathscr{E}(\beta_2)$. Let X_i and Y_i be the sample interarrival and service times of job i. Then

$$d = \sum_{i=1}^{K} (X_i - Y_i) \tag{4.11}$$

is the sample idle time that elapses with K completed jobs. Clearly d has variance $K(\beta_1^2 + \beta_2^2)$. Using a prime to indicate a second independent replication, we note that

$$\text{var}\left(\frac{d + d'}{2}\right) = \frac{K(\beta_1^2 + \beta_2^2)}{2}. \tag{4.12}$$

If, however, we elect to use (4.7) on corresponding interarrival time streams and on corresponding service time streams then we produce a sample idle time d'', where

$$\text{var}\left(\frac{d + d''}{2}\right) = \frac{K(\beta_1^2 + \beta_2^2)(2 - \pi^2/6)}{2}. \tag{4.13}$$

Hence the variance reduction due to use of the antithetic sampling plan is

$$\frac{\text{var}[(d + d')/2]}{\text{var}[(d + d'')/2]} = \frac{1}{2 - \pi^2/6} = 2.82. \tag{4.14}$$

Notice that this variance reduction is independent of the parameters β_1 and β_2. In applying this method to the estimation of mean waiting time in the single server problem, Page [204, p. 304] reported variance reductions ranging from 1.4 to 2.1 for traffic intensities varying from 0.75 to 0.90.

An alternative antithetic sampling plan suggested for queueing simulations is to use the random number stream that generates the interarrival times on the first replication to generate the service times on the second replication, and to similarly use the random number stream for service times on the first replication to generate interarrival times on the second. Naturally we would like some way of assessing this technique in comparison to the one mentioned earlier.

Using the concept of sample idle time again, we note that (4.12) continues to hold for independent replications, but for d''' generated under the above plan

$$\frac{\text{var}[(d + d')/2]}{\text{var}[(d + d''')/2]} = \frac{1}{1 - 2r/(1 + r^2)}$$

$$r = \frac{\beta_2}{\beta_1}.$$

In contrast to the use of (4.7) the present technique theoretically improves variance reduction as the activity level r increases. However, Page's study of this technique for estimating mean waiting time revealed a variance reduction of 1.6 for $r = 0.9$, a result which suggests that this particular problem may blunt the variance reduction potential of this technique.

Antithetic sampling can be applied to the generation of deviates from other techniques as well. For example, if $X \sim \mathcal{N}(0, 1)$ then $Y = -X$ is also from $\mathcal{N}(0, 1)$ and X and Y have perfect negative correlation. The reader should note, however, that indiscriminate use of antithetic sampling can lead to undesired results. Suppose that the normal deviate X is transformed to $V = X^2$ for use in the simulation and Y is transformed to $W = Y^2$. Then V and W are perfectly *positively* correlated, an undesired result. Reference 192 describes a technique for inducing negative correlation between V and W in this context.

Because of the many conditional decision rules in a simulation we cannot always expect a one-to-one correspondence between the stochastic variates in a replication and those in the original run unless added programming steps are undertaken. Although the impact of this lack of correspondence is difficult to assess we intuitively expect it to dilute the benefits of variance reduction. Burt and Garman [183] discuss a method of ensuring a one-to-one correspondence in the study of PERT networks. Favorable results from using antithetic variates for variance reduction in PERT networks have also been reported by Burt, Gaver, and Perlas [184].

11.5 STRATIFIED SAMPLING

Suppose that we plan to run two replications, the first with $U_1, \ldots, U_n$ and the second with $U_1', \ldots, U_n'$, where

$$U_j' = \begin{cases} U_j + \frac{1}{2} & 0 \leq U_j < \frac{1}{2} \\ U_j - \frac{1}{2} & \frac{1}{2} \leq U_j < 1. \end{cases} \tag{5.1}$$

Here U_j and U'_j are each from $\mathscr{U}(0, 1)$ with means $\frac{1}{2}$ and variances $\frac{1}{12}$, but they are dependent. In particular,

$$E(U_jU'_j) = \int_0^{1/2} u(u + \tfrac{1}{2})\, du + \int_{1/2}^1 u(u - \tfrac{1}{2})\, du = \tfrac{5}{24} \tag{5.2}$$

$$\text{cov}(U_j, U'_j) = E(U_jU'_j) - \tfrac{1}{4} = -\tfrac{1}{24} \tag{5.3}$$

$$\text{corr}(U_j, U'_j) = -0.5. \tag{5.4}$$

Suppose that we generate exponential variates

$$X = -\beta \log U$$

$$X' = -\beta \times \begin{cases} \log(\frac{1}{2} + U) & 0 \le U \le \frac{1}{2} \\ \log(U - \frac{1}{2}) & \frac{1}{2} < U \le 1. \end{cases} \tag{5.5}$$

Then†

$$\begin{aligned} \text{corr}(X, X') &= \int_0^1 [\log u - 1][\log(u + \tfrac{1}{2}) - 1]\, du \\ &\quad + \int_{1/2}^1 [\log u - 1][\log(u - \tfrac{1}{2}) - 1]\, du \sim -0.303. \end{aligned} \tag{5.6}$$

If we consider the more generalized stratified sampling scheme

$$U' \equiv \Theta + U \pmod 1 \qquad 0 \le \Theta < 1 \tag{5.7}$$

we have

$$X' = -\beta \times \begin{cases} \log(\Theta + U) & 0 \le U \le 1 - \Theta \\ \log(\Theta - 1 + U) & 1 - \Theta < U \le 1 \end{cases} \tag{5.8}$$

So that

$$\begin{aligned} \text{corr}(X, X' \mid \Theta) &= \int_0^{1-\Theta} \log u \log(\Theta + u)\, du \\ &\quad + \int_{1-\Theta}^1 \log u \log(\Theta - 1 + u)\, du - 1. \end{aligned} \tag{5.9}$$

In a review paper on variance reduction technique Gaver [193] suggested that one might profitably combine antithetic variates with stratified

†See reference 191.

sampling to gain the benefits of both. A study of this merger in reference 191 revealed that, for

$$
\begin{aligned}
W_1 &= U \\
W_2 &= 1 - U \\
W_3 &\equiv \Theta + U \quad (\text{mod } 1) \\
W_4 &\equiv \Theta + 1 - U \quad (\text{mod } 1)
\end{aligned}
\tag{5.10}
$$

$$X_j = -\beta \log W_j \qquad j = 1, \ldots, 4, \tag{5.11}$$

$4 \operatorname{var}(\overline{X})/\beta^2$, for $\overline{X} = \frac{1}{4}\sum_{j=1}^{4} X_j$, attains a minimum of about 0.002 for $\Theta = 0.75$. When this sampling plan was applied to the interarrival times in the first-come-first-served queueing model a variance reduction for sample mean queue length of 3.3 was reported.

11.6 CONTROL VARIATES

The method of control variates can induce a variance reduction on a single replication without tampering with the sampling mechanism. Suppose that we approximate a simulation model by a simpler mathematical model that is analytically tractable. Assume that we can solve the simpler model to derive the value of the parameter corresponding to the parameter of interest in the simulation. For example, our simulation may be a complex queueing model in which we are interested in the mean queue length, and our gross model is the first-come-first-served single server problem in Section 6.5, where mean queue length can be determined analytically.

To use the control variate approach we build a simulation model of the simpler model, as well, and run each model with the same random deviates $U_1, \ldots, U_N$. If $\overline{X}_n$ is the estimate of the parameter of interest in the complex simulation and $\overline{Y}_n$ is the estimate of the corresponding parameter in the simpler simulation, we use

$$\overline{Z}_n = E(\overline{Y}_n) + \overline{X}_n - \overline{Y}_n, \tag{6.1}$$

$E(\overline{Y}_n)$ being known analytically, as a modified estimate of the parameter under study. If $\overline{X}_n$ and $\overline{Y}_n$ are unbiased estimates then so is Z. Here Z has variance

$$\operatorname{var}(\overline{Z}_n) = \operatorname{var}(\overline{X}_n) + \operatorname{var}(\overline{Y}_n) - 2\operatorname{cov}(\overline{X}_n, \overline{Y}_n) \tag{6.2}$$

so that

$$\text{var}(\overline{Z}_n) < \text{var}(\overline{X}_n), \tag{6.3}$$

provided that

$$\begin{aligned} \text{cov}(\overline{X}_n, \overline{Y}_n) &> \frac{\text{var}(\overline{Y}_n)}{2} \\ \text{corr}(\overline{X}_n, \overline{Y}_n) &> \tfrac{1}{2}\sqrt{\text{var}(\overline{Y}_n)/\text{var}(\overline{X}_n)}. \end{aligned} \tag{6.4}$$

Now we expect the likelihood that (6.4) is satisfied to increase as the simple model improves in its resemblance to the complex model.

To assess the benefit of the control variate technique on a single replication we can generate two times series on the process of interest. The first times series is $\{X_t; t = 1, \ldots, n\}$; the second, $\{Z_t = E(\overline{Y}_n) + X_t - Y_t; t = 1, \ldots, n\}$. These provide us with the information needed to estimate $\text{var}(\overline{Z}_n)$ and $\text{var}(\overline{X}_n)$ in (6.3). If the inequality is satisfied then the control variate approach is beneficial. Otherwise we use $\overline{X}_n$.

11.7 COMPARISON OF EXPERIMENTS

In most discrete event simulations there exists a series of operating rules that control the states of entities in the system under study and are ultimately responsible for the events that occur during a run. Each of these rules may assume one of several possible forms. For example, the rule that selects jobs for service in a queueing model may be (1) take jobs in the order in which they arrive, (2) take the last job that arrived, (3) take the job requiring least service, and so on.

We may think of each operating rule as a *factor* and of each variant of a particular rule as a *treatment*. Suppose that there are only one rule (factor) and two variants (treatments). Then a test of the difference of the two experimental results, each corresponding to one treatment, reveals whether or not the form that the rule takes affects the output. This is the simplest *comparison of experiments*, but one that often occurs in practice.

Suppose that we perform two experiments, one with one variant of a particular operating rule and the other with an alternative variant. We collect observations $X_1, \ldots, X_{n_1}$ at unit intervals on the first experiment and $Y_1, \ldots, Y_{n_2}$ at unit intervals on the second. To estimate the means μ_X and μ_Y we use $\overline{X}_{n_1}$ and $\overline{Y}_{n_2}$, respectively, with corresponding variances m_X/n_1 and m_Y/n_2. We wish to test the hypothesis that $\mu_X = \mu_Y$.

Assuming that the random numbers generating the events in each experiment are independent, we may treat $\bar{X}_{n_1} - \bar{Y}_{n_2}$ as being $\mathcal{N}(0, m_X/n_1 + m_Y/n_2)$ in large samples, when the hypothesis is true. Under the hypothesis we expect the interval

$$\bar{X}_{n_1} - \bar{Y}_{n_2} - Q_{1-\alpha/2}\sqrt{m_X/n_1 + m_Y/n_2},$$
$$\bar{X}_{n_1} - \bar{Y}_{n_2} + Q_{1-\alpha/2}\sqrt{m_X/n_1 + m_Y/n_2},$$

$Q_{1-\alpha/2}$ being the normal point corresponding to $1 - \alpha/2$ probability, to contain zero with probability $1 - \alpha$. If it does we accept the hypothesis; otherwise we reject it. The assumption of normality in this case is more plausible for $\bar{X}_{n_1} - \bar{Y}_{n_2}$ than for $\bar{X}_{n_1}$ and $\bar{Y}_{n_2}$ separately since, if $\bar{X}_{n_1}$ and $\bar{Y}_{n_2}$ both have skewness with the same sign, the skewness of $\bar{X}_{n_1} - \bar{Y}_{n_2}$ is reduced.†

Suppose that we set $n_1 = n_2 = n$. Then we may create the time series

$$Z_t = X_t - Y_t \qquad t = 1, \ldots, n, \tag{7.1}$$

compute $\bar{Z}_n$, and estimate m_Z as in Chapter 10. If the resulting confidence interval for

$$E(\bar{Z}_n) = \mu_Z \tag{7.2}$$

includes zero, we accept the hypothesis $\mu_X = \mu_Y$; otherwise we reject it.

Choosing equal n_1 and n_2 also enables us to employ a relatively simple variance reduction sampling plan. If we use the same random number stream in each experiment we expect

$$\operatorname{cov}(\bar{X}_n, \bar{Y}_n) \geq 0 \tag{7.3}$$

so that $m_Z \leq m_X + m_Y$. This means that $\bar{Z}_n$ has smaller variance than in the uncorrelated case and we have a more precise estimate of $\mu_X - \mu_Y$ than earlier. Since

$$\operatorname{var}(\bar{Z}_n) = \frac{m_Z}{n} = \frac{1}{n}[m_X + m_Y - 2\operatorname{cov}(\bar{X}_n, {}_n\bar{Y})] \tag{7.4}$$

variance reduction can be measured by $m_Z/(m_X + m_Y)$, which gives the ratio of the variances in the correlated and uncorrelated cases.

This method of variance reduction is attractive because of its simplicity and ease of implementation. In view of the additional simplicity of antithetic variates a user might well consider running several antithetic replications for a set of operational conditions and then using the same random numbers for

† See Table 31 for the definition of skewness.

another set of operating conditions. Here the objective would be to derive the benefits of the two variance reduction techniques simultaneously.†

Let $\hat{\Theta}_{ij}$ be our estimate of Θ_j on replication i of experiment j. Suppose that all estimates have variance m/n. We wish to test the hypothesis $\Theta_1 = \Theta_2$ using the statistic

$$\tilde{\Theta} = \hat{\Theta}_{11} + \hat{\Theta}_{21} - \hat{\Theta}_{12} - \hat{\Theta}_{22}. \tag{7.5}$$

By considering the random number streams used, it can be seen that

$$\begin{aligned} \text{cov}(\hat{\Theta}_{11}, \hat{\Theta}_{12}) &= \text{cov}(\hat{\Theta}_{21}, \hat{\Theta}_{22}) \\ \text{cov}(\hat{\Theta}_{11}, \hat{\Theta}_{22}) &= \text{cov}(\hat{\Theta}_{21}, \hat{\Theta}_{12}) \end{aligned} \tag{7.6}$$

so that

$$\begin{aligned} \text{var}(\tilde{\Theta}) = \frac{4m}{n} &+ 2\,\text{cov}(\hat{\Theta}_{11}, \hat{\Theta}_{21}) + 2\,\text{cov}(\hat{\Theta}_{12}, \hat{\Theta}_{22}) \\ &- 4\,\text{cov}(\hat{\Theta}_{11}, \hat{\Theta}_{12}) - 4\,\text{cov}(\hat{\Theta}_{21}, \hat{\Theta}_{12}). \end{aligned} \tag{7.7}$$

Here $\text{cov}(\hat{\Theta}_{11}, \hat{\Theta}_{21}) \leq 0$ as is $\text{cov}(\hat{\Theta}_{12}, \hat{\Theta}_{22})$ because of the antithetic character of the replications. Also $\text{cov}(\hat{\Theta}_{11}, \hat{\Theta}_{12}) \geq 0$ because of the use of the same random number stream. However, we would expect $\text{cov}(\hat{\Theta}_{21}, \hat{\Theta}_{12}) \leq 0$ since $\hat{\Theta}_{21}$ and $\hat{\Theta}_{12}$ are produced using antithetic streams. This last covariance has a negating effect on variance reduction. Because of this conflict the user is well advised to work with one variance reduction method or the other until more is learned about how to combine variance techniques effectively for comparing experiments.

Although the convenience that equal n_1 and n_2 offer cannot be underestimated the equality brings with it at least two major shortcomings. We generally would like to estimate parameters with a specified reliability or variance. Equal sample sizes make this impossible, however, unless $m_X = m_Y$. Moreover most testing in the analysis of variance is predicted on the assumption of equal variances. Violation of this assumption can seriously undermine the validity of conclusions based on such testing.

Suppose that we require mean estimates to have variance σ^2. The corresponding sample sizes are

$$n_1 = \frac{m_X}{\sigma^2}, \qquad n_2 = \frac{m_Y}{\sigma^2}. \tag{7.8}$$

† See reference 193.

Section 10.15 describes one way of meeting this requirement with estimates of m_X and m_Y. Then our test statistic is

$$\frac{\bar{X}_{n_1} - \bar{Y}_{n_2}}{\sqrt{\hat{m}_X/n_1 + \hat{m}_Y/n_2}} \tag{7.9}$$

whose distribution we approximate by Student's t distribution with $f_1 + f_2$ degrees of freedom, f_1 being determined for $\hat{m}_X$ and f_2 for $\hat{m}_Y$ as in Table 32.

Failure to acknowledge the degrees of freedom leads to an underestimate of the test interval for (7.9) and consequently to a rejection of the hypothesis more often than theory suggests. In addition to testing for difference due to two variants of a particular operating rule we can, of course, perform similar testing for two numerical values of an input parameter.

Since σ^2 determines n_1 and n_2 it is instructive to consider a way of selecting σ^2. Suppose that we wish to detect mean absolute differences greater than c with probability α under the null hypothesis. Then $c = Q_{1-\alpha/2}\sqrt{2\sigma^2}$, so that $\sigma^2 = (c/Q_{1-\alpha/2})^2/2$. Although this procedure necessitates specifications of c this quantity seems within reach.

There is at least one point regarding statistical efficiency that deserves note in passing. Neither the case $n_1 = n_2$ nor the case $m_X/n_1 = m_Y/n_2$ produces a minimum variance test statistic when $m_X \neq m_Y$. A minimum variance would obtain for $n = n_1 + n_2$ total observations when $n_1/n_2 = \sqrt{m_X/m_Y}$. This result, which is discussed in reference 190, is derived by minimizing $m_X/n_1 + m_Y/n_2$ subject to the constraint $n = n_1 + n_2$.

From a cost point of view the results of minimizing n_1 and n_2 to produce a specified total variance $2\sigma^2$ may also differ. If c_1 and c_2 are the costs of obtaining observations on experiments 1 and 2, respectively, then the cost minimizing solution is $n_1/n_2 = \sqrt{m_X c_2/(m_Y c_1)}$. This follows from minimizing $c_1 n_1 + c_2 n_2$ subject to the constraint $2\sigma^2 = m_X/n_1 + m_Y/n_2$ [190].

Although the desirability of efficient or cost minimizing solutions is clear these results are presently of only academic interest because we do not have a way of estimating both m_X and m_Y before conducting the experiment. For this reason we must presently content ourselves with earlier mentioned methods of selecting n_1 and n_2.

11.8 VALIDATION

Before an investigator claims that his simulation model is a useful tool for studying behavior under new hypothetical conditions, he is well advised to check its consistency with the true system, as it exists before any change is made. The success of this validation establishes a basis for confidence in the results that the model generates under new conditions. After all, if a model cannot reproduce system behavior without change, we hardly expect it to produce truly representative results with change.

The problem of how to validate a simulation model arises in every simulation study in which some semblance of a system exists. The space devoted to validation in Naylor's book, *Computer Simulation Experiments with Models of Economic Systems* [203], indicates both the relative importance of the topic and the difficulty of establishing universally applicable criteria for accepting a simulation model as a valid representation. In this section we offer one view of validation that should be regarded as a minimal requirement for reconciling a model with a system.

We assume that the investigator has both input data and output data on the system under study. The input data are observations on the factors that drive the system, whereas the output data are observations on system performance. Now it seems legitimate to expect that a given input history to the model should produce an output similar to that produced when the input stimulated the true system. Hence one way of checking the model is to expose it to the given input history and then compare its output with that of the true system. In a queueing environment, the input history may consist of a stream of jobs with specified interarrival times, service times, and priorities. The output would then consist of sample mean waiting time, queue length, mean idle time, and so on. In an economic model the input may consist of a sequence of values for the exogenous variables of the system.

In response to the input, let X_t and Y_t denote queue length at time t in a queueing system and in the corresponding model, respectively. Then for n periods we have

$$Z_t = X_t - Y_t$$

and we test $\bar{Z}_n$ for significance. Since both outputs are based on the same input, we expect X_t and Y_t to be positively correlated and hence $\bar{Z}_n$ should have smaller variance than would occur if independent inputs were used.

A more rigorous check of consistency would be to test the hypothesis that $Z_1, \ldots, Z_n$ has a zero mean and also is a set of independent observations. This could be done using the periodogram test described in Section 7.5. The justification for testing for independence is that we expect the model value Y_t to differ from the system value X_t by only a nonsystematic error due to an approximation, namely, the model, for the true system.

It is the writer's opinion that in many cases application of this more rigorous test may be more than is needed for obtaining a useful model. In particular, we encouraged a model builder in Chapter 2 to pay special attention to the selection of a performance measure. Presumably this choice is the measure which is used for decision making. Since this performance measure is often an average, say, mean operating cost in our early inventory example, or mean waiting time or mean queue length in queueing examples, it seems reasonable to check the consistency of such a measure in both model and system. In simple terms, an acceptable model should produce a sample value for the performance measure that would lead to the same decision that one would make on the basis of the corresponding sample performance measure for the system. Hence the actual structure of the time series may be incidental to the principal objective.

If $\bar{Z}_n$ is found to be a significant deviation it behooves the investigator to learn more about system and model behaviors in order to reduce the statistically established difference. Examination of the spectra of the X and Y time series may reveal differences that lead to model modification. Chapter 6 described the value of studying the spectrum for verification purposes, and it is not hard to see that such study also offers benefits in reconciling system and model behaviors.

We now consider the situation in which an input data stream is unavailable to the investigator. Assuming that the output $X_1, \ldots, X_n$ is available, we can simply generate a series $Y_1, \ldots, Y_n$ from the model, compute $\bar{Z}_n$, and test it for significance. However, we expect $\bar{Z}_n$ to have larger variance than in the earlier case. As a consequence, the test is less discriminating than before and is more tolerant of deviations between system and model behavior.

Although the topic of validation deserves a book unto itself, the writer inclines to the view that minimal steps such as estimating input parameters from actual data, as in Chapter 9, and the testing of deviations, as described here, can be highly beneficial in reconciling a model to the corresponding system. In addition, the many techniques described in reference 184 are naturally helpful in any validation effort.

11.9 MORE THAN TWO TREATMENTS

When two experiments are to be compared the testing and detection occur at the same time; that is, if the test reveals a significant difference then the relevant parameters in the two experiments differ. For three or more experiments, however, testing precedes detection. If, for example, we compare three experiments we may apply the standard *analysis of variance* techniques to determine whether $\mu_1 = \mu_2 = \mu_3$, the subscript indicating the experiment number. The test is binary; it either accepts the hypothesis $\mu_1 = \mu_2 = \mu_3$ or rejects it. Rejection implies that at least one difference exists. Either

$$
\begin{aligned}
&\mu_1 = \mu_2 \neq \mu_3 \\
&\mu_1 = \mu_3 \neq \mu_2 \\
&\mu_1 \neq \mu_2 = \mu_3 \\
&\mu_1 \neq \mu_2 \neq \mu_3.
\end{aligned}
\tag{9.1}
$$

Suppose that we wish to test k variants of a particular operating rule. Let W_j be the sample mean for rule j with variance m_j/n_j such that

$$n_j = \frac{\sigma^2}{m_j}. \tag{9.2}$$

The n_j can be estimated by an iterative technique as in Section 10.15. All replications are assumed independent. For the hypothesis $\mu_1 = \mu_2 = \cdots = \mu_k$ the test statistic is

$$\Theta_k = \frac{f}{k-1} \frac{\sum_{j=1}^{k} (W_j - \overline{W})^2}{\sum_{j=1}^{k} \hat{m}_j/n_j} \tag{9.3}$$

where

$$\overline{W} = \frac{1}{k} \sum_{j=1}^{k} W_j, \qquad f = \sum_{j=1}^{k} f_j.$$

Under the null hypothesis we approximate the distribution of Θ_k by the F distribution with $k - 1$ and f degrees of freedom. The f_j are computed as in Table 32.

Rejection of the hypothesis implies that at least one difference exists. It is suggested that the *method of multiple comparisons* be used to detect which differences are real. The account in reference 205 states the basic steps in the method. References 181, 199, and 206 offer more comprehensive accounts.

11.10 2^k FACTORIAL EXPERIMENTS

Section 11.9 described the one factor layout with k treatments. In simulation work it is not at all unusual for an investigator to study the effects of several operating rules, each having several variants. To illustrate this point we consider three operating rules A, B, and C, where A has variants A_1 and A_2, B has variants B_1 and B_2, and C has variants C_1 and C_2. We refer to this arrangement as a 2^3 *factorial* experiment.

We are interested in determining output changes due to a change in A alone, due to a change in B alone, and due to a change in C alone. These three potential output changes are called *main effects* and are denoted by A, B, and C, respectively. There are also second-order effects due to joint changes in A and B, A and C, and B and C. For example, AB refers to the change in output caused by changing rule A from A_1 to A_2 and rule B from B_1 to B_2. These *interactions* are denoted by AB, AC, and BC, respectively. A third-order interaction due to simultaneous changes in all three rules is denoted by ABC.

One convenient way of determining the test statistics for the three main effects and four interactions is to assign -1 to A_1, B_1, C_1 and $+1$ to A_2, B_2, C_2. Then for each effect we determine the signs in Table 37 by multiplying the corresponding A, B, and C. The quantity W_{ijk} corresponds to the sample output at treatment A_i, B_j, and C_k.

Table 37 Statistics for the 2^3 Factorial Experiment

	Signs										
Output	A	B	C	Mean	A	B	AB	C	AC	BC	ABC
W_{111}	-1	-1	-1	1	-1	-1	1	-1	1	1	-1
W_{211}	1	-1	-1	1	1	-1	-1	-1	-1	1	1
W_{121}	-1	1	-1	1	-1	1	-1	-1	1	-1	1
W_{221}	1	1	-1	1	1	1	1	-1	-1	-1	-1
W_{112}	-1	-1	1	1	-1	-1	1	1	-1	-1	1
W_{212}	1	-1	1	1	1	-1	-1	1	1	-1	-1
W_{122}	-1	1	1	1	-1	1	-1	1	-1	1	-1
W_{222}	1	1	1	1	1	1	1	1	1	1	1

To derive the numerator of the test statistic for the main effect A we multiply each element on column 1 by its corresponding element in column 4. This quantity is

(10.1)

$$W_A = -W_{111} + W_{211} - W_{121} + W_{221} - W_{112} + W_{212} - W_{122} + W_{222}$$

which, under the hypothesis of no difference due to A, has mean zero and variance $\sum_{i,j,k=1}^{2} m_{ijk}/n_{ijk}$. Here m_{ijk} and n_{ijk} correspond to W_{ijk}. We choose n_{ijk} so that all W_{ijk} have the same variance σ^2. Then

$$n_{ijk} = \frac{m_{ijk}}{\sigma^2} \qquad i, j, k = 1, 2. \tag{10.2}$$

The technique of Section 10.15 provides a method of obtaining roughly equal variances.

As an estimate of $\sum_{ijk=1}^{2} m_{ijk}/n_{ijk}$ we use $\sum_{i,j,k=1}^{2} \hat{m}_{ijk}/n_{ijk}$ with degrees of freedom $\sum_{i,j,k=1}^{2} f_{ijk}$ as determined in Table 32. The test statistic for the main effect is then

$$\Theta_A = \frac{W_A}{\sqrt{\sum_{i,j,k=1}^{2} \hat{m}_{ijk}/n_{ijk}}} \tag{10.3}$$

whose distribution we approximate by that of a Student t variate with $\sum_{i,j,k=1}^{2} f_{ijk}$ degrees of freedom.

To test for the interaction effect due to changing A and B together, the numerator of the test statistic is derived by summation of the products of corresponding elements of columns 1 and 6:

(10.4)

$$W_{AB} = W_{111} - W_{211} - W_{121} + W_{221} + W_{112} - W_{212} - W_{122} + W_{222}.$$

The test statistic is

$$\Theta_{AB} = \frac{W_{AB}}{\sqrt{\sum_{i,j,k=1}^{2} \hat{m}_{ijk}/n_{ijk}}} \qquad (10.5)$$

whose distribution we approximate by that of a t variate with $\sum_{i,j,k=1}^{2} f_{ijk}$ degrees of freedom. Test statistics for the remaining two main effects and three interactions can be derived in a similar way.

One important property of the above procedure is that under the null hypothesis all seven test statistics are independent. This means that we test separately for each effect. To see this independence, say, between B and ABC we note that, since the W_{ijk} are independent,

$$\begin{aligned} E(W_B W_{ABC}) &= E(W_{111}^2) - E(W_{211}^2) + E(W_{121}^2) - E(W_{221}^2) \\ &\quad - E(W_{112}^2) + E(W_{212}^2) - E(W_{122}^2) + E(W_{222}^2) \qquad (10.6) \\ &= 0. \end{aligned}$$

Three natural extensions of the above procedure exist. First, we can increase the statistical reliability of the test by adding independent replications. Provided there is an independent replication for each treatment combination, the adjustment to the test statistic is minimal, being described in most standard references. Here testing and identification of the difference continue to occur in the same step.

The second extension concerns adding factors. This has the effect of increasing the set of factor combinations needed to test for main effects and interactions. It is easily seen that this number grows rapidly even for the two treatment case. However, testing and identification still occur in the same step.

The third extension involves adding treatments. This situation, which is also covered in the references, requires two steps for testing and identification. If a test reveals that variants of a rule induce a difference one uses the method of multiple comparisons to identify which variant (or variants) is responsible for the change.

11.11 FRACTIONAL FACTORIAL DESIGNS

Because of cost and time limitations we occasionally wish to investigate an experimental model with less than the full factorial design. A question then arises regarding the effect that this reduction in treatment combinations has on our ability to test for sources of variation. One type of partial layout is called a *fractional factorial design.* Suppose that we carry out experiments corresponding to the first four treatment combinations in Table 37. The numerator of the test statistic for the A effect is

$$V_A = -W_{111} + W_{211} - W_{121} - W_{221}. \tag{11.1}$$

But

$$V_{BC} = -W_{111} + W_{211} - W_{121} - W_{221}, \tag{11.2}$$

which is identical to the numerator for the test statistic for BC. This means that variation due to A is *confounded* with variation due to BC. In a similar manner variations due to B and C are confounded with variations due to AC and AB, respectively. We are therefore unable to test for main effects in this fractional factorial experiment *unless* we can safely assume that the interactions BC, AC, and AB are relatively inconsequential in comparison to the main effects A, B, and C, respectively. In some experiments this assumption is acceptable; in others it is not. Hence the experimenter who wishes to use a fractional factorial design is encouraged a priori to determine the potential effect of such a plan on response changes that he wishes to investigate. References 177, 178, 188, and 208 offer a variety of design layouts and the consequent effects that can be studied.

One additional observation regarding the choice of treatment combinations in fractional factorial experiments is important. The fractional design mentioned above preserves the independence property among test statistics. If, however, we were to use row 6 instead of row 4 in Table 37 we would lose this property. We see therefore that not all fractional layouts offer the same convenience.

Because of cost we may have to choose between adding to the completeness of a fractional factorial experiment and running replications of specific treatment combinations that have already been used. Since a fractional factorial design often induces a confounding of effects it is generally desirable to complete the factorial design before beginning replications in order to permit testing of all sources of variation. Preliminary test results may then suggest to the investigator which treatments are worth replicating to gain greater precision for further testing.

The concepts of fractional factorial design were developed for use in agriculture and industry, where experiments could consume a lengthy interval and often required a high setup cost. This meant that, if experiments were to be carried out sequentially, considerable time would pass and large setup costs would be encountered in the course of completing a full factorial experiment. By contrast, the calendar time for running a treatment on a simulation experiment is negligible and the central processing unit (CPU) time on the computer represents most of the cost. The fact that calendar time is relatively unimportant in making simulation runs suggests that the full factorial experiment be run whenever possible, thereby avoiding the possibility of making a poor assumption about interactions, an assumption which may lead to misinterpretation of results. Of course the budget for CPU time must be able to accommodate these runs.

11.12 RESPONSE SURFACES

Comparing results is one possible objective in the statistical analysis of simulations. Another is to infer the functional relationship between the input parameters and the output. With an analytical form in hand the investigator can estimate the output due to sets of inputs that differ from those originally used to fit the relationship. A common purpose here is to determine the combinations of input parameters that produce a minimum or a maximum in the output.

Over the past twenty years publications on response surfaces have developed into an integral part of the literature of applied statistics. References 177–180 provide detailed descriptions of a variety of problems with which an investigator may wish to familiarize himself. In this section we give a limited description of considerations in several relatively simple problems.

One Input Factor

In studying a system we generally compute the output $Y = \bar{X}_n$ for a number of different settings or treatments of the input variables $\mathbf{z} = (z_1, z_2, \ldots)$. Let experiment i denote the exercise of the system using the ith treatment. For the ith experiment w ehave Y_i with variance

$$\text{var}(Y_i) \sim \frac{m_i}{n_i}. \tag{12.1}$$

A number of questions now arise concerning the relationship between input and response. These questions include the following:

How reliable do we wish our estimates to be?

What is the functional form relating input and output?

How does one estimate the parameters of this form?

With regard to reliability it is convenient to specify equal variances σ^2 for all Y_i so that

$$\sigma^2 = \frac{m_i}{n_i}. \tag{12.2}$$

One way of ensuring (12.2) is to use the algorithm described in Section 10.15. Generally this means different run lengths n_i, but this requirement should not be an issue of significance.

Sometimes the functional form relating an input $\mathbf{z}$ to a response Y follows as a consequence of the theory relating cause and effect. When this is so no identification problem exists, provided the theory is correct. More often, however, theory merely suggests a causal relationship without specifying its form. Here we infer the form from the sample data, carefully qualifying the extent to which we regard the form as valid.

Suppose that there is one input $\mathbf{z}$. We initially perform k experiments and collect a set of k points† $(z_i, Y_i; i = 1, 2, \ldots, k)$. The sample range of $\mathbf{z}$ is assumed to include all values for which we intend to predict Y. The functional form is therefore to be used for *interpolation*, not *extrapolation.*

Column 1 of Table 38 lists functions that fit a variety of data. A graphical analysis may be used to determine how well a particular function fits the data. For a function to be considered acceptable the corresponding variables in column 2 should have an approximately linear relationship on arithmetic graph paper. A computationally simpler way is to plot the variables in column 3 using the corresponding scales indicated in column 4. A linear relationship again implies correct specification.

Suppose that system response y is a proportion restricted to the interval (0, 1) for all z in $(0, \infty)$. Then forms 12 through 15 may often represent the data. If form 12 is correct

$$y = 1 - ae^{bz} \qquad b < 0. \tag{12.3}$$

For graphical purposes we note that

$$\log(1 - y) = \log a + bz \tag{12.4}$$

linearizes the relationship. Here y is a monotonic function of z. If y increases sharply and then decreases slowly form 13,

$$y = 1 - aze^{bz} \qquad b < 0, \tag{12.5}$$

may be appropriate. Linearization yields

$$\log\left(\frac{1 - y}{z}\right) = \log a + bz. \tag{12.6}$$

† Here we drop the subscript n_i on the sample mean Y_i.

Table 38 Graphical Techniques

Form	(1) y	(2) $f(z)$	$f(y)$	(3) Plot		(4) Scale[a]
1	$a + bz$	z,	y	z,	y	A, A
2	$a + b/z$	z,	zy	z,	y	R, A
3	$1/(a + bz)$	z,	$1/y$	z,	y	A, R
4	$z/(a + bz)$	z,	z/y	z,	z/y	A, A
5	$z/(a + bz^2)$	z^2,	z/y	z^2,	z/y	A, A
6	az^b	$\log z$,	$\log y$	z,	y	L, L
7	az^{bz}	$z \log z$	$\log y$	$z \log z$,	y	A, L
8	ae^{bz}	z,	$\log y$	z,	y	A, L
9	aze^{bz}	z,	$\log (y/z)$	z,	y/z	A, L
10	aze^{bz^2}	z^2,	$\log (y/z)$	z^2,	y/z	A, L
11	$az^2e^{bz^2}$	z^2,	$\log (y/z^2)$	z^2,	y/z^2	A, L
12	$1 - ae^{bz}$	z,	$\log (1 - y)$	z,	$1 - y$	A, L
13	$1 - aze^{bz}$	z,	$\log [(1 - y)/z]$	z,	$(1 - y)/z$	A, L
14	$1 - aze^{bz^2}$	z^2,	$\log [(1 - y)/z]$	z^2,	$(1 - y)/z$	A, L
15	$1 - az^2e^{bz^2}$	z^2,	$\log [(1 - y)/z^2]$	z^2,	$(1 - y)/z^2$	A, L

Source: [196].
[a] A = arithmetic, L = logarithmic, R = reciprocal.

The most beneficial aspect of linearization is that it enables us to apply the linear least-squares methods of regression analysis to the estimation of a and b, and as a consequence we wish to emphasize the value of graphical techniques for suggesting the appropriate form and for checking how well a particular regression curve fits. Very often, results in the neighborhoods of small and large z are crucial. Hence it is important that the functional forms fit well in these intervals. Failure to recognize a poor fit can easily cause misinterpretation of results.

As in the comparison of experiments the ability to choose the treatments to run is an option that can help us significantly in estimating a functional relationship between input and output. Let z^* and z^{**} be the minimum and the maximum values of z to be considered. Then it is convenient to work with the standardized variate

$$w = \frac{2z - z^* - z^{**}}{z^{**} - z^*} \tag{12.7}$$

so that $-1 \leq w \leq 1$. We consider the linear relationship

$$Y = \beta_1 + \beta_2 w + \varepsilon \tag{12.8}$$

where

$$E(\varepsilon) = 0, \qquad E(\varepsilon^2) = \sigma^2. \tag{12.9}$$

Our objective is to estimate β_1 and β_2. To recover the relationship to z we need only substitute (12.7) into (12.8) so that

$$Y = \beta_1 + \left(\frac{2\beta_2}{z^{**} - z^*}\right) z - \frac{\beta_2(z^{**} + z^*)}{z^{**} - z^*} + \varepsilon$$

$$a = \beta_1 - \frac{\beta_2(z^{**} + z^*)}{z^{**} - z^*} \tag{12.10}$$

$$b = \frac{2\beta_2}{z^{**} - z^*}.$$

Letting the subscript be the observation number and assuming the elements of sequence $\varepsilon_1, \ldots, \varepsilon_k$ to be independent random variables, we obtain the least-squares estimates of β_1 and β_2 as

$$\hat{\beta}_1 = \frac{\sum w_i^2 \sum Y_i - \sum w_i \sum w_i Y_i}{k \sum w_i^2 - (\sum w_i)^2} \tag{12.11}$$

$$\beta_2 = \frac{k \sum w_i Y_i - \sum w_i \sum Y_i}{k \sum w_i^2 - (\sum w_i)^2}$$

with covariance matrix

$$\mathbf{cov}(\hat{\beta}) = \frac{\sigma^2}{k \sum w_i^2 - (\sum w_i)^2} \begin{pmatrix} \sum w_i^2 & -\sum w_i \\ -\sum w_i & k \end{pmatrix}, \tag{12.12}$$

k being the number of observations.

We note in (12.12) that, if k is even and we pair our observations so that $w_i = -w_{i+1}$, $i = 1, 3, \ldots, k-1$, then $\hat{\beta}_1$ and $\hat{\beta}_2$ are independent since $\sum w_i = 0$. One value of (12.11) becomes apparent when we wish to compute a confidence region for the regression line. It is easily seen that $\hat{\beta}_1 + \hat{\beta}_2 w$ is an unbiased estimator of $\beta_1 + \beta_2 w$ with variance $\sigma^2(1/k + w^2/\sum_{i=1}^k w_i^2)$.

If we, in fact, know that (12.8) holds we will follow the design layout of running pairs of replications at $w = \pm 1$, which produces minimum variance estimates. Since we seldom know the exact form of a relationship we want to choose our treatment levels to serve two purposes. One is to derive statistically efficient estimates; the other, to assist us in identifying an appropriate functional form. One compromise is to string out treatment levels over $-1 \leq w \leq 1$ with the density of points greater near -1 and 1. This procedure allows for identification while preserving some efficiency. The pairing of observations is also recommended.

Few of the forms in Table 38 allow us to find a finite maximum system response for an input factor w. One alternative approach to solving this problem is to assume that Y may be represented as a low-order polynomial in w. As an example, we consider a third-order representation

$$Y = \beta_0 + \beta_1 w + \beta_2 w^2 + \beta_3 w^3 + \varepsilon \qquad -1 \le w_i \le 1 \tag{12.13}$$

where β_j is the jth-order effect. This expression has maxima or minima for

$$\frac{dy}{dw} = \beta_1 + 2\beta_2 w + 3\beta_3 w^2 = 0$$

$$w_1 = \frac{-2\beta_2 - \sqrt{4\beta_2 - 6\beta_1\beta_3}}{3\beta_3} \tag{12.14}$$

$$w_2 = \frac{-2\beta_2 + \sqrt{4\beta_2 - 6\beta_1\beta_3}}{3\beta_3},$$

provided that $4\beta_2 > 6\beta_1\beta_3$. Otherwise y is monotonic in w. An alternative use for (12.13) is to specify a $\tilde{y}$ and then solve $\sum_{j=0}^{3} \beta_j w^j = \tilde{y}$ for w in $(-1, 1)$. This enables us to determine the input needed for a desired response.

As in the linear case we can profit by a judicious choice of treatments. We note that only four distinct treatments are needed to estimate $\beta_0, \ldots, \beta_3$. Since the covariance matrix of $\hat{\beta}_0, \ldots, \hat{\beta}_3$ is

$$\mathbf{cov}(\hat{\beta}) = \sigma^2 \begin{bmatrix} k & \sum w_i & \sum w_i^2 & \sum w_i^3 \\ \sum w_i & \sum w_i^2 & \sum w_i^3 & \sum w_i^4 \\ \sum w_i^2 & \sum w_i^3 & \sum w_i^4 & \sum w_i^5 \\ \sum w_i^3 & \sum w_i^4 & \sum w_i^5 & \sum w_i^6 \end{bmatrix}^{-1} \tag{12.15}$$

it can be shown [201] that for $k = 4$ the treatment combination $w_1 = -1$, $w_2 = 1$, $w_3 = -0.4472$, and $w_4 = 0.4472$ minimizes the *generalized variance* $|\mathbf{cov}(\hat{\beta})|$. If more observations are desired the minimization continues to hold if sets of independent replications are performed at the four settings specified above. In terms of the original z_i in (12.7) we have

$$z_i = \frac{z^* + z^{**} + (z^{**} - z^*)w_i}{2}. \tag{12.16}$$

Reference 201, p. 160, lists the designs that minimize the generalized variance $|\mathbf{cov}(\hat{\beta})|$ for schemes of orders one through six.

As a measure of the error variance we can average the sample variances on the individual experiments and denote this average by $\hat{\sigma}^2$. For a polynomial of order j we may compute the least-squares estimator of σ^2, provided that the number of treatments exceeds k. Its form is

$$\tilde{\sigma}^2 = \frac{1}{k - j} \sum_{i=1}^{k} \left(Y_i - \sum_{l=0}^{j} \hat{\beta}_l w_i^l\right)^2. \tag{12.17}$$

A comparison of $\tilde{\sigma}^2$ with $\hat{\sigma}^2$ gives us a measure of the adequacy of a jth-order polynomial. If $\hat{\sigma}^2 < \tilde{\sigma}^2$ then there is reason to believe that the polynomial does not include sufficient terms to explain the variance of Y. As the number of terms increase, we would expect to reach a polynomial representation for which $\hat{\sigma}^2 \sim \tilde{\sigma}^2$.

Two Inputs

The analysis of more than one factor again creates the problem of sorting out main effects and interactions. Here we consider a two factor problem in which we wish to estimate response by

$$Y = \beta_0 + \beta_1 v + \beta_2 w + \beta_{11} v^2 + \beta_{12} vw + \beta_{22} w^2 + \varepsilon. \tag{12.18}$$

We have one total effect β_0, two main effects β_1 and β_2, two second-order effects β_{11} and β_{22}, and one interaction β_{12}. There are six effects, requiring at least six treatment combinations. Since we would prefer to identify the need for a second-order representation with as few observations as possible we can work with an augmented version of a 2^2 factorial design. For $\beta_{11} = \beta_{12} = \beta_{22} = 0$ the variance minimizing design for v and w is $(-1, -1)$, $(-1, 1)$, $(1, -1)$, $(1, 1)$.

Table 39 shows this design. The entries in each column are the coefficients of the βs in (12.18) for the treatment in each:

$$\begin{aligned} E(Y_1) &= \beta_0 - \beta_1 - \beta_2 + \beta_{11} + \beta_{12} + \beta_{22} \\ E(Y_2) &= \beta_0 - \beta_1 + \beta_2 + \beta_{11} - \beta_{12} + \beta_{22} \\ E(Y_3) &= \beta_0 + \beta_1 - \beta_2 + \beta_{11} - \beta_{12} + \beta_{22} \\ E(Y_4) &= \beta_0 + \beta_1 + \beta_2 + \beta_{11} + \beta_{12} + \beta_{22}. \end{aligned} \tag{12.19}$$

Substituting Y_i for $E(Y_i)$ in (12.19) and combining rows appropriately gives us the unbiased estimates:

$$\begin{aligned} 4(\hat{\beta}_0 + \hat{\beta}_{11} + \hat{\beta}_{22}) &= Y_1 + Y_2 + Y_3 + Y_4 \\ 4\hat{\beta}_1 &= -Y_1 - Y_2 + Y_3 + Y_4 \\ 4\hat{\beta}_2 &= -Y_1 + Y_2 - Y_3 + Y_4 \\ 4\hat{\beta}_{12} &= Y_1 - Y_2 - Y_3 + Y_4. \end{aligned} \tag{12.20}$$

Table 39 Treatments for Second-Order Polynomial with Two Factors

Treatment v,	w	Response	β_0	β_1	β_2	β_{11}	β_{12}	β_{22}
−1,	−1	Y_1	1	−1	−1	1	1	1
−1,	1	Y_2	1	−1	1	1	−1	1
1,	−1	Y_3	1	1	−1	1	−1	1
1,	1	Y_2	1	1	1	1	1	1

These can be shown to be the LSE of the corresponding population quantities. Each of the quantities $\hat{\beta}_1$, $\hat{\beta}_2$, and $\hat{\beta}_{12}$ has variance $\sigma^2/4$, so that, for example, we can test β_{12} for significance.

Unfortunately β_0, β_{11}, and β_{22} are confounded, so that we cannot estimate them separately. Adding the treatment $v = 0$, $w = 0$ to Table 39, however, gives

(12.21) $E(Y_1 + Y_2 + Y_3 + Y_4 + Y_5) = 5\beta_0 + 4(\beta_{11} + \beta_{22}).$

With (12.21) this yields $\hat{\beta}_0 = Y_5$.

More important, however, is the observation that

$$E\left(5\sum_{j=1}^{4} Y_j - 4\sum_{j=1}^{5} Y_j\right) = 4(\beta_{11} + \beta_{22})$$

so that

$$\hat{\beta}_{11} + \hat{\beta}_{22} = \frac{Y_1 + Y_2 + Y_3 + Y_4 - 4Y_5}{4}$$

gives us an unbiased estimate of $\beta_{11} + \beta_{22}$, with variance $5\sigma^2/4$. This enables us to test the statistic $(\hat{\beta}_{11} + \hat{\beta}_{22} - \beta_{11} - \beta_{22})/\sqrt{5\hat{\sigma}^2/4}$ to determine whether or not the quadratic terms in v and w are needed. If they are then only one more distinct treatment combination is required to estimate β_{11} and β_{22} separately.

11.13 THE VALUE OF FORESIGHT

In the introduction to this chapter we remarked that studying k input factors at each of two treatment levels straightforwardly requires 2^k experiments. For T treatments we require T^k experiments. Clearly this number can quickly become unmanageable. There are at least two ways of resolving this difficulty in many experiments. The first approach requires careful analysis of the problem under study to see whether factors can be eliminated without loss of generality. We give two examples of this approach. The other method relies on the theory of experimental design to select the treatment combinations that best facilitate estimating and testing the quantities of interest. We discuss this approach with regard to the second of our examples.

Weibull Example

The paper by Thoman, Bain, and Antle [207] studies the small sample properties of the maximum likelihood estimators of α and β in the Weibull distribution and offers an example of the rewards of forethought.† The purpose of the study is to find percentage points of the distributions of $\hat{\alpha}$ and $\hat{\beta}$, the MLEs of α and β, for $5 \leq n \leq 120$. There are three *factors*, α, β, and n. Suppose that we wish to investigate the sampling properties for three values of each factor. Straightforward sampling would require $3^3 = 27$ separate experiments. For ten values of each factor we would require $3^{10} = 59{,}049$ experiments, a number far beyond our capabilities.

Let $X_1, \ldots, X_n$ be a set of independent observations each with the p.d.f.

$$f(x) = \frac{\alpha}{\beta^\alpha} x^{\alpha-1} e^{-(x/\beta)^\alpha}. \tag{13.1}$$

The corresponding MLE equations are

$$\frac{n}{\hat{\alpha}} - n \frac{\sum_{j=1}^n X_j^{\hat{\alpha}} \log X_j}{\sum_{j=1}^n X_j^{\hat{\alpha}}} + \sum_{j=1}^n \log X_j = 0 \tag{13.2}$$

$$\hat{\beta} = \left(\frac{1}{n} \sum_{j=1}^n X_j^{\hat{\alpha}}\right)^{1/\hat{\alpha}}. \tag{13.3}$$

Thoman, Bain, and Antle make several highly useful observations. Let

$$Y_j = \left(\frac{X_j}{\beta}\right)^\alpha. \tag{13.4}$$

Then Y_i has the exponential distribution $\mathscr{E}(1)$. Also we may write (13.2) in terms of

$$\begin{aligned} &\frac{n}{\hat{\alpha}} - n \frac{\sum_{j=1}^n \beta^{\hat{\alpha}} (Y_j)^{\hat{\alpha}/\alpha} \log(\beta Y_j^{1/\alpha})}{\sum_{j=1}^n \beta^{\hat{\alpha}} Y_j^{\hat{\alpha}/\alpha}} + \sum_{j=1}^n \log[\beta Y_j^{1/\alpha}] = 0 \\ &\frac{n}{(\hat{\alpha}/\alpha)} - n \frac{\sum_{j=1}^n Y_j^{\hat{\alpha}/\alpha} \log Y_j}{\sum_{j=1}^n Y_j^{\hat{\alpha}/\alpha}} + \sum_{j=1}^n \log Y_j = 0. \end{aligned} \tag{13.5}$$

† See Section 8.2.8 for the Weibull distribution.

Therefore the statistic $\hat{\alpha}/\alpha$ is a function of exponential variates from $\mathscr{E}(1)$ and is independent of α and β.

Working with (13.3), we can write

$$\hat{\beta} = \left[\frac{1}{n}\sum_{j=1}^{n}(\beta Y_j^{1/\alpha})^{\hat{\alpha}}\right]^{1/\hat{\alpha}} \tag{13.6}$$

$$\frac{\hat{\beta}}{\beta} = \left[\frac{1}{n}\sum_{j=1}^{n} Y_j^{\hat{\alpha}/\alpha}\right]^{1/\hat{\alpha}} \tag{13.7}$$

$$\frac{1}{\hat{\alpha}}\log\frac{\hat{\beta}}{\beta} = \log\left(\frac{1}{n}\sum_{j=1}^{n} Y_j^{\hat{\alpha}/\alpha}\right) \tag{13.8}$$

so that $(1/\hat{\alpha})\log(\hat{\beta}/\beta)$ is a function of the Y_js, which are from $\mathscr{E}(1)$, and is independent of α and β. This means that all one has to do to generate the distributions $\hat{\alpha}/\alpha$ and $1/\hat{\alpha}\log(\hat{\beta}/\beta)$ for a given n is the following:

1. on each of k independent replications sample n observations from $\mathscr{E}(1)$,
2. solve for $\hat{\alpha}/\alpha$ and $(1/\hat{\alpha})\log(\hat{\beta}/\beta)$ in (13.5) and (13.8), respectively,
3. compute the sample quantiles corresponding to the percentage points of interest, say 0.02, 0.05, 0.10, 0.25, 0.40, 0.50, and 0.60, based on the k sample statistics, and
4. repeat for each value of n.

We see that there are now only as many experiments as there are different ns, a considerable improvement over the three-factor layout discussed earlier. Setting up the sampling study in terms of factors and treatments quickly establishes the magnitude of the efforts being undertaken. In turn the sheer magnitude thus revealed often motivates an investigator to look for ways of eliminating factors, as in the Weibull case.

Design Layouts in Response Surface Methodology

As indicated in Section 11.12, fitting a function to the relationship between input and output is no small task when there are many input factors. One approach to this problem is to use a fractional factorial design to determine the treatments to be considered in fitting a polynomial in the input factors. Reference 178 presents a concise description of how one goes about selecting appropriate treatments and constructing the relationship. The emphasis is on iteratively exploring the response surface with successively more complex relationships until an adequate representation is made. The techniques in reference 178 can be especially useful.

Here we wish to stress the importance of thinking deeply about the problem at hand before applying the techniques unqualifiedly. To make our point we cite an example due to Ignall [198]. Consider an inventory system with the following parameters:†

$$\begin{aligned}
D &= \text{daily demand} \\
LT &= \text{production lead time} \\
f(D) &= \text{p.d.f. of daily demand} \\
g(LT) &= \text{p.d.f. of production lead time} \\
EOQ &= \text{optimum production order} = \text{economic order quantity} \\
ROP &= \text{inventory level at which an order is placed} = \text{reorder point} \\
c_1 &= \text{carrying cost} \\
c_2 &= \text{setup cost} \\
c_3 &= \text{shortage cost.}
\end{aligned}$$

The objective of the investigation is to study total cost TC as a function of the seven factors D, LT, EOQ, ROP, c_1, c_2, and c_3.

Looking at the problem from the perspective of a factorial design, we see that for k levels of each factor we have a full factorial experiment of k^7 runs. For $k = 2$, $2^7 = 128$. Hunter and Naylor [197] recommend a fractional factorial design to investigate such problems. Since this approach leads to confounding, as discussed in Section 11.12, we would have to assume that certain second-order interactions were negligible before we could estimate the main effects. With seven factors many assumptions are called for and this is clearly undesirable. However, Ignall [198] has made several observations that can reduce the factor space considerably. He notes that expected total cost has the form

$$TC = c_2 p + c_1 E(I_t) - c_3 E(S_t)$$

where p = fraction of days on which orders are placed, $E(I_t)$ = mean inventory level ($I_t \geq 0$), $E(S_t)$ = mean stockout level ($S_t \leq 0$).

Clearly, if one specifies EOQ, ROP, $f(D)$, and $g(LT)$, runs an experiment, and estimates p, $E(I_t)$, and $E(S_t)$, he can combine these estimates with any c_1, c_2, and c_3 to find TC. Hence three factors are easily eliminated. Then for the two treatments we would have $2^4 = 16$ combinations or 1/8 as many as the seven factors require.

Ignall goes on to point out how the factor space can be reduced to EOQ and ROP. Unfortunately, not all simulation models are amenable to reduction of this kind. However, the time spent in looking for simplifications is well worth the effort when we consider the assumption that we must make to utilize a fractional factorial design and the plethora of treatments to be considered in the full factorial design.

† See reference 197. The notation here follows that in references 197 and 198.

EXERCISES

1. Exercise 3 of Chapter 8 concerns the generation of the maximum of n independent, uniformly distributed random variables on (0, 1). In a second replication we wish to use the method of antithetic variates to induce negative correlation between corresponding maxima.
   a. Derive the expression for the antithetic maximum Z' corresponding to a maximum Z, based on U, in the original experiment.
   b. Derive corr(Z, Z').
   c. Plot corr(Z, Z') for $n = 2(1)10$.

2. The input to a simulation is a binary sequence

$$X_i = \begin{cases} 1 & 0 \leq U_i < p \\ 0 & p \leq U_i < 1. \end{cases}$$

   We wish to create a second binary sequence $\{Y_i\}$ as input to a second run. Moreover we want X_i and Y_i to be negatively correlated for variance reduction purposes.
   a. What is the largest negative correlation we can expect to achieve regardless of the method of variance reduction chosen?
   b. Suppose that $\frac{1}{2} \leq p < 1$ and we generate Y_i by

$$Y_i = \begin{cases} 1 & 0 \leq 1 - U_i < p \\ 0 & p \leq 1 - U_i < 1 \end{cases}$$

   Find the correlation between X_i and Y_i.

3. Consider the single server queueing problem with interarrival and service times from $\mathcal{N}(\mu_1, \sigma_1)$ and $\mathcal{N}(\mu_2, \sigma_2)$, respectively. Following the suggestion in Section 11.4, we wish to reduce the variance of the sample idle time by using the stream of normal deviates that generates interarrival times on the first replication to generate service times on the second replication, and using the stream of normal deviates that generate service times on the first replication to generate interarrival times on the second. A normal deviate has zero mean and unit variance.
   a. Determine the variance reduction.
   b. Contrast the result obtained in Exercise 3a with the result in Section 11.4 for exponential variates.
   c. Suggest a variance reduction technique that dominates the above approach.

4. We wish to test the hypothesis that the first-come-first-served selection

rule and the shortest service time selection rule lead to the same mean queue length in the single server problem.

a. Modify the SIMSCRIPT II programs in Section 5.5 and in Exercise 3 of Chapter 5 to allow corresponding interarrival times to be produced from the same pseudorandom number on both runs and also to allow corresponding service times to be produced from the same pseudorandom number on both runs.
b. Collect data as in (7.1) and use these data with the analyses routine in Appendix B to compute relevant statistics.
c. Test the hypothesis of no difference at $\alpha = 0.10$.
d. Estimate the extent of variance reduction due to the matching of pseudorandom numbers.

5. Follow the steps in Exercise 4 for the GPSS programs in Section 5.3 and in Exercise 4 of Chapter 5.

6. Apply a variance reduction technique to the estimation of π in Exercise 8 of Chapter 7.

7. Consider a two-factor experiment with two treatments per factor. Denote the factors as A and B and the treatments for each factor as -1 and 1.
   a. Write down the design matrix (see Table 37) for deriving test statistics for the total effect (estimation of the grand mean over all experiments), for the main effect A, for the main effect B, and for the interaction AB.
   b. The above design matrix calls for four replications. Let X_i be the observation on replication i. Assume that each observation has unit variance. Also assume that the experimenter has the ability to induce a correlation:

$$\text{corr}(X_i, X_j) = \rho \qquad i, j = 1, \ldots, 4; i \neq j.$$

   The question he must resolve is whether to make ρ positive or negative in order to reduce the variances of the test statistics below the corresponding variances that independent replications produce. By working with the covariance matrix of $\mathbf{X} = (X_1, \ldots, X_4)$ and the design matrix of Exercise 7a, determine whether positive or negative correlation produces smaller variances.

CHAPTER 12

QUESTIONS AND PROCEDURES

The multifarious considerations that enter into a thoughtful discrete event digital simulation may well have left the reader a bit doubtful as to how to proceed. Although no one can provide an infallible procedure for success in every case, it is possible to identify certain questions that a simulation user should ask himself in the course of his study. This chapter presents a hierarchical list of such questions.

TO SIMULATE OR NOT TO SIMULATE

A. Can problem be solved analytically?

1. Yes. Do so.
2. No. Is it amenable to discrete event digital simulation techniques?
   a. Yes. Define system in terms of
      i. entities,
      ii. attributes,
      iii. class relationships among entities,
      iv. dynamic relationships,
      v. input stimuli,
      vi. relevant performance criteria.
   b. No. Search for other method.

CHOICE OF LANGUAGE AND MODELING APPROACH

B. Does investigator have access to a simulation programming language (SPL)?

1. Yes.
   a. Model system using approach consistent with selected SPL.
   b. Program model.
2. No.
   a. Select a modeling approach.
   b. Model system using selected approach.
   c. Write timing routine to handle scheduling.
   d. Program model.

INPUT DATA ANALYSIS

C. Are real world data available on input?

1. Yes.
   a. Plot data.
   b. Identify plausible distributional forms.
   c. Estimate parameters.
   d. Test adequacy of distributional forms.
2. No.
   a. Make best guess regarding distributional forms.
   b. Make best guess regarding parameter values for each distribution.

PSEUDORANDOM NUMBER GENERATOR SELECTION

D. What pseudorandom number generator do you propose to use?

1. Does generator have any theoretically known defects?
   a. Yes. Select a generator without defect.
   b. No. Continue.
2. Do statistical test results in the literature support the chosen generator?
   a. Yes.
      i. If possible, assign separate stream to each source of variation.
      ii. Allow for recording of initial and final pseudorandom numbers for all streams.
   b. No. Test generator.

EXPERIMENTAL DESIGN CONSIDERATIONS

E. Is the objective of the study a comparison of different operating rules?

1. Yes.

    a. Use same pseudorandom streams on corresponding replications.
    b. Determine appropriate test statistic and testing procedure.
    c. Establish degree of accuracy required.

2. No. Is objective to measure response to change in quantitative input factors?

    a. Yes.
        i. Develop plan for the sequence of treatment sets to be studied.
        ii. Establish degree of accuracy required.
        iii. Incorporate variance reduction techniques, if possible.

DATA ANALYSIS TECHNIQUE SELECTION AND PRESENTATION

F. Choose analysis technique for estimating descriptors, their variances, and their estimated confidence intervals (Chapter 10).

G. Include in simulation termination routine a summary report program that provides for a clear, concise, properly labeled presentation of all summary data.

No one can claim the above list to be exhaustive of all issues that deserve attention in a simulation. However, following these suggestions lends a feeling of thoughtfulness and confidence to the whole simulation exercise. We wish to emphasize that the inclusion in the report of summary statistics that are used to form final statistics is often a beneficial practice, since the simulation user may wish to employ these data for computing other statistics after the simulation is over.

APPENDIX A

Table A1 Maximum Likelihood Estimates of $\hat{\alpha}$, $\Delta(\hat{\alpha})$, and $\psi(\hat{\alpha})/\Delta(\hat{\alpha})$ for the Gamma Distribution

$M = \log a - c,\ \Delta(\hat{\alpha}) = \hat{\alpha}\Psi'(\hat{\alpha}) - 1$

$1/M$	$\hat{\alpha}$	$\Delta(\hat{\alpha})$	$\Psi'(\alpha)/\Delta(\hat{\alpha})$
.020	.0187	52.46	54.45
.030	.0275	35.43	37.42
.040	.0360	26.86	28.84
.050	.0442	21.68	23.65
.060	.0523	18.20	20.17
.070	.0602	15.70	17.67
.080	.0679	13.82	15.78
.090	.0756	12.34	14.30
.100	.0831	11.16	13.11
.200	.1532	5.731	7.664
.300	.2178	3.861	5.579
.400	.2790	2.906	4.185
.500	.3381	2.326	4.228
.600	.3955	1.935	3.834
.700	.4517	1.654	3.551
.800	.5070	1.443	3.338
.900	.5615	1.278	3.173
1.000	.6155	1.146	3.041
1.100	.6690	1.038	2.934
1.200	.7220	.9481	2.845
1.300	.7748	.8720	2.770
1.400	.8272	.8068	2.706
1.500	.8794	.7505	2.652
1.600	.9314	.7012	2.604
1.700	.9832	.6579	2.562
1.800	1.034	.6194	2.526
1.900	1.086	.5851	2.493
2.000	1.137	.5543	2.464
2.100	1.188	.5264	2.438
2.200	1.240	.5012	2.415
2.300	1.291	.4782	2.393
2.400	1.342	.4572	2.374
2.500	1.393	.4379	2.356
2.600	1.444	.4202	2.340
2.700	1.494	.4038	2.325
2.800	1.545	.3886	2.311
2.900	1.596	.3744	2.299
3.000	1.646	.3613	2.287
3.200	1.748	.3375	2.266
3.400	1.849	.3167	2.248
3.600	1.950	.2982	2.232
3.800	2.051	.2817	2.217
4.000	2.151	.2669	2.205
4.200	2.252	.2536	2.194
4.400	2.353	.2416	2.183
4.600	2.453	.2306	2.174
4.800	2.554	.2205	2.166
5.000	2.654	.2113	2.158

Source: [139a]. See p. 246 for the definitions of a and c.

Table A1 (continued)

$M = \log a - c, \Delta(\hat{\alpha}) = \hat{\alpha}\Psi'(\hat{\alpha}) - 1$

$1/M$	$\hat{\alpha}$	$\Delta(\hat{\alpha})$	$\Psi'(\alpha)/\Delta(\hat{\alpha})$
5.200	2.755	.2028	2.151
5.400	2.855	.1950	2.145
5.600	2.956	.1878	2.139
5.800	3.056	.1810	2.134
6.000	3.156	.1747	2.129
6.200	3.257	.1689	2.124
6.400	3.357	.1634	2.119
6.600	3.457	.1583	2.115
6.800	3.558	.1534	2.112
7.000	3.658	.1489	2.108
7.300	3.808	.1426	2.103
7.600	3.958	.1367	2.099
7.900	4.109	.1314	2.094
8.200	4.259	.1264	2.091
8.500	4.409	.1218	2.087
8.800	4.560	.1175	2.084
9.100	4.710	.1135	2.081
9.400	4.860	.1098	2.078
9.700	5.010	.1063	2.075
10.000	5.160	.1030	2.073
10.300	5.311	.1000	2.070
10.600	5.461	.0971	2.068
10.900	5.611	.0944	2.066
11.200	5.761	.0918	2.064
11.500	5.911	.0893	2.062
11.800	6.061	.0870	2.061
12.100	6.211	.0848	2.059
12.400	6.362	.0827	2.058
12.700	6.512	.0807	2.056
13.000	6.662	.0788	2.055
13.300	6.812	.0770	2.053
13.600	6.962	.0752	2.052
13.900	7.112	.0736	2.051
14.200	7.262	.0720	2.050
14.500	7.412	.0705	2.049
14.800	7.562	.0690	2.048
15.100	7.712	.0676	2.047
15.400	7.862	.0663	2.046
15.700	8.013	.0650	2.045
16.000	8.163	.0637	2.044
16.300	8.313	.0626	2.043
16.600	8.463	.0614	2.042
16.900	8.613	.0603	2.041
17.200	8.763	.0592	2.041
17.500	8.913	.0582	2.040
17.800	9.063	.0572	2.039
18.100	9.213	.0562	2.038
18.400	9.363	.0553	2.038
18.700	9.513	.0544	2.037
19.000	9.663	.0535	2.036
19.300	9.813	.0527	2.036
19.600	9.963	.0519	2.035
20.000	10.16	.0508	2.034

Table A2 Percentage Points l_γ for the Maximum Likelihood Estimator $\hat{\alpha}/\alpha$ for the Weibull Distribution

$$\text{Prob}\,(\hat{\alpha}/\alpha < l_\gamma) = \gamma$$

n \ γ	0.02	0.05	0.10	0.25	0.40	0.50	0.60
5	0.604	0.683	0.766	0.951	1.116	1.238	1.378
6	0.623	0.697	0.778	0.937	1.080	1.188	1.304
7	0.639	0.709	0.785	0.930	1.059	1.155	1.256
8	0.653	0.720	0.792	0.926	1.045	1.131	1.223
9	0.665	0.729	0.797	0.925	1.035	1.114	1.198
10	0.676	0.738	0.802	0.924	1.028	1.101	1.179
11	0.686	0.745	0.807	0.924	1.022	1.090	1.163
12	0.695	0.752	0.811	0.924	1.017	1.082	1.151
13	0.703	0.759	0.815	0.924	1.014	1.075	1.140
14	0.710	0.764	0.819	0.925	1.011	1.069	1.132
15	0.716	0.770	0.823	0.925	1.008	1.064	1.124
16	0.723	0.775	0.826	0.926	1.006	1.059	1.117
17	0.728	0.779	0.829	0.927	1.004	1.056	1.111
18	0.734	0.784	0.832	0.927	1.003	1.052	1.106
19	0.739	0.788	0.835	0.928	1.001	1.049	1.101
20	0.743	0.791	0.838	0.929	1.000	1.047	1.097
22	0.752	0.798	0.843	0.930	0.998	1.042	1.090
24	0.759	0.805	0.848	0.932	0.997	1.038	1.084
26	0.766	0.810	0.852	0.933	0.995	1.035	1.079
28	0.772	0.815	0.856	0.934	0.994	1.033	1.074
30	0.778	0.820	0.860	0.935	0.993	1.030	1.070
32	0.783	0.824	0.863	0.937	0.993	1.028	1.067
34	0.788	0.828	0.866	0.938	0.992	1.027	1.064
36	0.793	0.832	0.869	0.939	0.992	1.025	1.061
38	0.797	0.835	0.872	0.940	0.991	1.024	1.059
40	0.801	0.839	0.875	0.940	0.991	1.023	1.056
42	0.804	0.842	0.877	0.941	0.990	1.022	1.054
44	0.808	0.845	0.880	0.942	0.990	1.021	1.052
46	0.811	0.847	0.882	0.943	0.990	1.020	1.051
48	0.814	0.850	0.884	0.944	0.990	1.019	1.049
50	0.817	0.852	0.886	0.944	0.989	1.018	1.048
52	0.820	0.854	0.888	0.945	0.989	1.017	1.046
54	0.822	0.857	0.890	0.946	0.989	1.017	1.045
56	0.825	0.859	0.891	0.946	0.989	1.016	1.044
58	0.827	0.861	0.893	0.947	0.989	1.015	1.043
60	0.830	0.863	0.894	0.948	0.989	1.015	1.041
62	0.832	0.864	0.896	0.948	0.989	1.014	1.040
64	0.834	0.866	0.897	0.949	0.989	1.014	1.040
66	0.836	0.868	0.899	0.949	0.988	1.014	1.039
68	0.838	0.869	0.900	0.950	0.988	1.013	1.038
70	0.840	0.871	0.901	0.950	0.988	1.013	1.037
72	0.841	0.872	0.903	0.951	0.988	1.012	1.036
74	0.843	0.874	0.904	0.951	0.988	1.012	1.036
76	0.845	0.875	0.905	0.952	0.988	1.012	1.035
78	0.846	0.876	0.906	0.952	0.988	1.011	1.034
80	0.848	0.878	0.907	0.952	0.988	1.011	1.034
85	0.852	0.881	0.910	0.953	0.988	1.011	1.032
90	0.855	0.883	0.912	0.954	0.988	1.010	1.031
95	0.858	0.886	0.914	0.955	0.988	1.009	1.030
100	0.861	0.888	0.916	0.956	0.988	1.009	1.029
110	0.866	0.893	0.920	0.958	0.988	1.008	1.027
120	0.871	0.897	0.923	0.959	0.988	1.007	1.025

Table A2 (continued)

n \ γ	0.70	0.75	0.80	0.85	0.90	0.95	0.98
5	1.557	1.671	1.812	2.001	2.277	2.779	3.518
6	1.453	1.543	1.662	1.812	2.030	2.436	3.067
7	1.386	1.461	1.561	1.688	1.861	2.183	2.640
8	1.338	1.404	1.491	1.602	1.747	2.015	2.377
9	1.303	1.361	1.439	1.538	1.665	1.896	2.199
10	1.275	1.328	1.399	1.489	1.602	1.807	2.070
11	1.253	1.302	1.367	1.450	1.553	1.738	1.972
12	1.234	1.281	1.341	1.418	1.513	1.682	1.894
13	1.219	1.263	1.319	1.391	1.480	1.636	1.830
14	1.206	1.248	1.300	1.369	1.452	1.597	1.777
15	1.195	1.234	1.284	1.349	1.427	1.564	1.732
16	1.185	1.223	1.270	1.332	1.406	1.535	1.693
17	1.176	1.213	1.258	1.317	1.388	1.510	1.660
18	1.168	1.204	1.247	1.303	1.371	1.487	1.630
19	1.162	1.196	1.237	1.291	1.356	1.467	1.603
20	1.155	1.188	1.228	1.281	1.343	1.449	1.579
22	1.144	1.176	1.213	1.262	1.320	1.418	1.538
24	1.135	1.165	1.200	1.246	1.301	1.392	1.504
26	1.128	1.156	1.189	1.232	1.284	1.370	1.475
28	1.121	1.148	1.180	1.220	1.269	1.351	1.450
30	1.115	1.141	1.171	1.210	1.257	1.334	1.429
32	1.110	1.135	1.164	1.201	1.246	1.319	1.409
34	1.105	1.129	1.157	1.193	1.236	1.306	1.392
36	1.101	1.125	1.151	1.186	1.227	1.294	1.377
38	1.097	1.120	1.146	1.179	1.219	1.283	1.363
40	1.094	1.116	1.141	1.173	1.211	1.273	1.351
42	1.091	1.112	1.137	1.167	1.204	1.265	1.339
44	1.088	1.109	1.132	1.162	1.198	1.256	1.329
46	1.085	1.106	1.129	1.158	1.192	1.249	1.319
48	1.083	1.103	1.125	1.153	1.187	1.242	1.310
50	1.081	1.100	1.122	1.149	1.182	1.235	1.301
52	1.078	1.098	1.119	1.145	1.177	1.229	1.294
54	1.076	1.095	1.116	1.142	1.173	1.224	1.286
56	1.075	1.093	1.113	1.139	1.169	1.218	1.280
58	1.073	1.091	1.111	1.135	1.165	1.213	1.273
60	1.071	1.089	1.108	1.133	1.162	1.208	1.267
62	1.070	1.087	1.106	1.130	1.158	1.204	1.262
64	1.068	1.086	1.104	1.127	1.155	1.200	1.256
66	1.067	1.084	1.102	1.125	1.152	1.196	1.251
68	1.066	1.083	1.100	1.122	1.149	1.192	1.246
70	1.064	1.081	1.098	1.120	1.146	1.188	1.242
72	1.063	1.080	1.097	1.118	1.144	1.185	1.237
74	1.062	1.078	1.095	1.116	1.141	1.182	1.233
76	1.061	1.077	1.093	1.114	1.139	1.179	1.229
78	1.060	1.076	1.092	1.112	1.136	1.176	1.225
80	1.059	1.075	1.090	1.110	1.134	1.173	1.222
85	1.057	1.072	1.087	1.106	1.129	1.166	1.213
90	1.055	1.069	1.084	1.102	1.124	1.160	1.206
95	1.053	1.067	1.081	1.099	1.120	1.155	1.199
100	1.051	1.065	1.079	1.096	1.116	1.150	1.192
110	1.048	1.061	1.074	1.090	1.110	1.141	1.181
120	1.046	1.058	1.070	1.086	1.104	1.133	1.171

Source: [156].

Table A3 Percentage Points t_γ for the Maximum Likelihood Estimator $\hat{\alpha} \log (\hat{\beta}/\beta)$ for the Weibull Distribution

$$\text{Prob}\ [\hat{\alpha} \log (\hat{\beta}/\beta) < t_\gamma] = \gamma$$

n \ γ	0.02	0.05	0.10	0.25	0.40	0.50	0.60
5	-1.631	-1.247	-0.888	-0.444	-0.241	-0.056	0.085
6	-1.396	-1.007	-0.740	-0.385	-0.194	-0.045	0.079
7	-1.196	-0.874	-0.652	-0.344	-0.168	-0.038	0.074
8	-1.056	-0.784	-0.591	-0.313	-0.150	-0.032	0.070
9	-0.954	-0.717	-0.544	-0.289	-0.137	-0.029	0.067
10	-0.876	-0.665	-0.507	-0.269	-0.126	-0.026	0.065
11	-0.813	-0.622	-0.477	-0.253	-0.118	-0.023	0.062
12	-0.762	-0.587	-0.451	-0.239	-0.111	-0.021	0.061
13	-0.719	-0.557	-0.429	-0.228	-0.106	-0.019	0.059
14	-0.683	-0.532	-0.410	-0.217	-0.100	-0.018	0.057
15	-0.651	-0.509	-0.393	-0.208	-0.096	-0.016	0.056
16	-0.624	-0.489	-0.379	-0.200	-0.092	-0.015	0.054
17	-0.599	-0.471	-0.365	-0.193	-0.089	-0.014	0.053
18	-0.578	-0.455	-0.353	-0.187	-0.085	-0.013	0.052
19	-0.558	-0.441	-0.342	-0.181	-0.083	-0.013	0.051
20	-0.540	-0.428	-0.332	-0.175	-0.080	-0.012	0.050
22	-0.509	-0.404	-0.314	-0.166	-0.075	-0.011	0.048
24	-0.483	-0.384	-0.299	-0.158	-0.071	-0.009	0.047
26	-0.460	-0.367	-0.286	-0.150	-0.068	-0.009	0.046
28	-0.441	-0.352	-0.274	-0.144	-0.065	-0.008	0.044
30	-0.423	-0.338	-0.264	-0.139	-0.062	-0.007	0.043
32	-0.408	-0.326	-0.254	-0.134	-0.059	-0.006	0.042
34	-0.394	-0.315	-0.246	-0.129	-0.057	-0.006	0.041
36	-0.382	-0.305	-0.238	-0.125	-0.055	-0.005	0.040
38	-0.370	-0.296	-0.231	-0.121	-0.053	-0.005	0.040
40	-0.360	-0.288	-0.224	-0.118	-0.052	-0.004	0.039
42	-0.350	-0.280	-0.218	-0.115	-0.050	-0.004	0.038
44	-0.341	-0.273	-0.213	-0.112	-0.048	-0.004	0.037
46	-0.333	-0.266	-0.208	-0.109	-0.047	-0.003	0.037
48	-0.325	-0.260	-0.203	-0.106	-0.046	-0.003	0.036
50	-0.318	-0.254	-0.198	-0.104	-0.045	-0.003	0.036
52	-0.312	-0.249	-0.194	-0.102	-0.043	-0.003	0.035
54	-0.305	-0.244	-0.190	-0.100	-0.042	-0.002	0.035
56	-0.299	-0.239	-0.186	-0.098	-0.041	-0.002	0.034
58	-0.294	-0.234	-0.183	-0.096	-0.040	-0.002	0.034
60	-0.289	-0.230	-0.179	-0.094	-0.039	-0.002	0.033
62	-0.284	-0.226	-0.176	-0.092	-0.039	-0.002	0.033
64	-0.279	-0.222	-0.173	-0.091	-0.038	-0.001	0.032
66	-0.274	-0.218	-0.170	-0.089	-0.037	-0.001	0.032
68	-0.270	-0.215	-0.167	-0.088	-0.036	-0.001	0.032
70	-0.266	-0.211	-0.165	-0.086	-0.035	-0.001	0.031
72	-0.262	-0.208	-0.162	-0.085	-0.035	-0.001	0.031
74	-0.259	-0.205	-0.160	-0.084	-0.034	-0.001	0.031
76	-0.255	-0.202	-0.158	-0.083	-0.033	-0.001	0.030
78	-0.252	-0.199	-0.155	-0.081	-0.033	-0.001	0.030
80	-0.248	-0.197	-0.153	-0.080	-0.032	-0.000	0.030
85	-0.241	-0.190	-0.148	-0.078	-0.031	-0.000	0.029
90	-0.234	-0.184	-0.144	-0.075	-0.030	0.000	0.028
95	-0.227	-0.179	-0.139	-0.073	-0.028	0.000	0.028
100	-0.221	-0.174	-0.136	-0.071	-0.027	0.000	0.027
110	-0.211	-0.165	-0.129	-0.067	-0.025	0.001	0.026
120	-0.202	-0.158	-0.123	-0.064	-0.024	0.001	0.025

Table A3 (continued)

n \ γ	0.70	0.75	0.80	0.85	0.90	0.95	0.98
5	0.254	0.349	0.452	0.587	0.772	1.107	1.582
6	0.221	0.302	0.404	0.516	0.666	0.939	1.291
7	0.200	0.272	0.362	0.465	0.598	0.829	1.120
8	0.185	0.251	0.331	0.427	0.547	0.751	1.003
9	0.174	0.235	0.307	0.397	0.507	0.691	0.917
10	0.165	0.222	0.288	0.372	0.475	0.644	0.851
11	0.157	0.211	0.273	0.351	0.448	0.605	0.797
12	0.150	0.202	0.260	0.334	0.425	0.572	0.752
13	0.145	0.194	0.249	0.319	0.406	0.544	0.714
14	0.140	0.187	0.239	0.306	0.389	0.520	0.681
15	0.135	0.180	0.230	0.294	0.374	0.499	0.653
16	0.131	0.175	0.223	0.284	0.360	0.480	0.627
17	0.128	0.170	0.216	0.274	0.348	0.463	0.605
18	0.124	0.165	0.209	0.266	0.338	0.447	0.584
19	0.121	0.161	0.204	0.258	0.328	0.433	0.566
20	0.118	0.157	0.199	0.251	0.318	0.421	0.549
22	0.113	0.150	0.189	0.239	0.302	0.398	0.519
24	0.109	0.144	0.181	0.228	0.288	0.379	0.494
26	0.105	0.138	0.174	0.219	0.276	0.362	0.472
28	0.102	0.134	0.168	0.210	0.265	0.347	0.453
30	0.098	0.129	0.163	0.203	0.256	0.334	0.435
32	0.095	0.125	0.158	0.197	0.247	0.323	0.420
34	0.093	0.122	0.153	0.191	0.239	0.312	0.406
36	0.090	0.118	0.149	0.185	0.232	0.302	0.393
38	0.088	0.115	0.145	0.180	0.226	0.293	0.382
40	0.086	0.113	0.142	0.175	0.220	0.285	0.371
42	0.084	0.110	0.139	0.171	0.214	0.278	0.361
44	0.082	0.108	0.136	0.167	0.209	0.271	0.352
46	0.080	0.105	0.133	0.164	0.204	0.264	0.344
48	0.079	0.103	0.130	0.160	0.199	0.258	0.336
50	0.077	0.101	0.128	0.157	0.195	0.253	0.328
52	0.076	0.099	0.126	0.154	0.191	0.247	0.321
54	0.074	0.097	0.123	0.151	0.187	0.243	0.315
56	0.073	0.096	0.121	0.148	0.184	0.238	0.309
58	0.072	0.094	0.119	0.146	0.181	0.233	0.303
60	0.071	0.092	0.117	0.143	0.177	0.229	0.297
62	0.070	0.091	0.116	0.141	0.174	0.225	0.292
64	0.068	0.089	0.114	0.139	0.171	0.221	0.287
66	0.067	0.088	0.112	0.137	0.169	0.218	0.282
68	0.066	0.087	0.111	0.135	0.166	0.214	0.278
70	0.065	0.085	0.109	0.133	0.164	0.211	0.274
72	0.064	0.084	0.108	0.131	0.161	0.208	0.269
74	0.064	0.083	0.107	0.129	0.159	0.205	0.266
76	0.063	0.082	0.105	0.128	0.157	0.202	0.262
78	0.062	0.081	0.104	0.126	0.155	0.199	0.258
80	0.061	0.080	0.103	0.125	0.153	0.197	0.255
85	0.059	0.077	0.100	0.121	0.148	0.190	0.246
90	0.057	0.075	0.097	0.118	0.143	0.185	0.239
95	0.056	0.073	0.095	0.115	0.139	0.179	0.232
100	0.054	0.071	0.093	0.112	0.136	0.175	0.226
110	0.051	0.067	0.089	0.107	0.129	0.166	0.215
120	0.049	0.064	0.085	0.103	0.123	0.159	0.205

Source: [156].

Table A4 Unbiasing Factors for the Weibull Maximum Likelihood Estimator of α

$$E\left[B(n)\hat{\alpha}\right] = \alpha$$

n	B(n)	n	B(n)	n	B(n)
5	0.669	26	0.947	58	0.977
6	0.752	28	0.951	60	0.978
7	0.792	30	0.955	62	0.979
8	0.820	32	0.958	64	0.980
9	0.842	34	0.960	66	0.980
10	0.859	36	0.962	68	0.981
11	0.872	38	0.964	70	0.981
12	0.883	40	0.966	72	0.982
13	0.893	42	0.968	74	0.982
14	0.901	44	0.970	76	0.983
15	0.908	46	0.971	78	0.983
16	0.914	48	0.972	80	0.984
18	0.923	50	0.973	85	0.985
20	0.931	52	0.974	90	0.986
22	0.938	54	0.975	100	0.987
24	0.943	56	0.976	120	0.990

Source: [156].

For $n < 30$ use of the asymptotic variance of $\hat{\alpha}$, $0.608\alpha^2/n$, in place of the true variance results in at least a 20 per cent underestimate in variance.* For $n \geq 66$, however, use of the asymptotic variance and normal distribution understates the confidence interval on $\hat{\alpha}/\alpha$ by less than 0.04 at the $\gamma = 0.04$ significance level.† For $\hat{\beta}$, use of the asymptotic variance and normal distribution for $n \geq 85$ results in an understatement of the confidence level for $\hat{\beta}/\beta$ for $0.6 < \hat{\alpha} < 1.6$ at the $\gamma = 0.04$ significance level.‡

* See Fig. 29 in reference 156, p. 457.
† See reference 156, p. 453.
‡ See reference 156, p. 456.

Table A5 Complete Sample Maximum Likelihood Estimates of Parameters of the Beta Distribution

Range of G_1 is 0.01, 0.1(0.1)0.9. Range of G_2 is 0.01, 0.1(0.1)0.9. ($G_1 + G_2 \le 1.0$). In each cell the first entry is $\hat{\alpha}$ and the second entry is $\hat{\beta}$. ($G_1 = e^c$ and $G_2 = e^d$; see Table 23).

G_1 / G_2	.01	.1	.2	.3	.4	.5	.6	.7	.8	.9	.98
.01	.112 .112	.192 .135	.254 .147	.318 .157	.395 .168	.495 .179	.639 .192	.877 .210	1.376 .237	3.162 .299	42.128 .850
.1		.245 .245	.337 .278	.441 .310	.576 .345	.770 .389	1.093 .451	1.756 .560	3.864 .846	* *	
.2			.395 .395	.537 .456	.735 .531	1.057 .640	1.701 .834	3.669 1.367	* *		
.3				.647 .647	.947 .804	1.532 1.086	3.280 1.869	* *			
.4					1.320 1.320	2.832 2.358	* *				
.5	BY SYMMETRY**					* *					
.6					* *						
.7				* *				NOT POSSIBLE ($G_1 + G_2 > 1$)			
.8			* *								
.9		* *									
.98											

* $G_1 + G_2 = 1$

** For example, the estimates of α and β when $G_2 = 0.2$ and $G_1 = 0.1$ are respectively the estimates of β and α when $G_1 = 0.2$ and $G_2 = 0.1$, namely 0.278 and 0.337.

Source: [153a].

Table A6 Modified Critical Values of the Kolmogorov–Smirnov Test for Normality with the Maximum Likelihood Estimates Used for Parameters

Sample Size N	Level of Significance for $D = \text{Max}\ \lvert F^*(X) - S_N(X) \rvert$				
	.20	.15	.10	.05	.01
4	.300	.319	.352	.381	.417
5	.285	.299	.315	.337	.405
6	.265	.277	.294	.319	.364
7	.247	.258	.276	.300	.348
8	.233	.244	.261	.285	.331
9	.223	.233	.249	.271	.311
10	.215	.224	.239	.258	.294
11	.206	.217	.230	.249	.284
12	.199	.212	.223	.242	.275
13	.190	.202	.214	.234	.268
14	.183	.194	.207	.227	.261
15	.177	.187	.201	.220	.257
16	.173	.182	.195	.213	.250
17	.169	.177	.189	.206	.245
18	.166	.173	.184	.200	.239
19	.163	.169	.179	.195	.235
20	.160	.166	.174	.190	.231
25	.142	.147	.158	.173	.200
30	.131	.136	.144	.161	.187
Over 30	$\frac{.736}{\sqrt{N}}$	$\frac{.768}{\sqrt{N}}$	$\frac{.805}{\sqrt{N}}$	$\frac{.886}{\sqrt{N}}$	$\frac{1.031}{\sqrt{N}}$

Source: [147].

Table A7 Modified Critical Values of the Kolmogorov–Smirnov Test of the Exponential Distribution with the Maximum Likelihood Estimate used for the Parameter

Sample Size N	Level of Significance for $D = \text{Max}\,|F^*(X) - S_N(X)|$				
	.20	.15	.10	.05	.01
3	.451	.479	.511	.551	.600
4	.396	.422	.449	.487	.548
5	.359	.382	.406	.442	.504
6	.331	.351	.375	.408	.470
7	.309	.327	.350	.382	.442
8	.291	.308	.329	.360	.419
9	.277	.291	.311	.341	.399
10	.263	.277	.295	.325	.380
11	.251	.264	.283	.311	.365
12	.241	.254	.271	.298	.351
13	.232	.245	.261	.287	.338
14	.224	.237	.252	.277	.326
15	.217	.229	.244	.269	.315
16	.211	.222	.236	.261	.306
17	.204	.215	.229	.253	.297
18	.199	.210	.223	.246	.289
19	.193	.204	.218	.239	.283
20	.188	.199	.212	.234	.278
25	.170	.180	.191	.210	.247
30	.155	.164	.174	.192	.226
Over 30	$\frac{.86}{\sqrt{N}}$	$\frac{.91}{\sqrt{N}}$	$\frac{.96}{\sqrt{N}}$	$\frac{1.06}{\sqrt{N}}$	$\frac{1.25}{\sqrt{N}}$

Source: [148].

APPENDIX B*

This appendix contains SIMSCRIPT II and FORTRAN IV routines for analyzing data as described in Sections 9.4 and 10.11, and as shown in output form in Section 5.5.

SIMSCRIPT II ANALYSIS ROUTINE

When calling the routine, the number of elements in a prespecified *global* array X(·) must be given. The quantity N assumes this value. The input consists of Q, the highest-order autoregressive scheme to be considered, and ALPHA, the significance level for the mean. We suggest Q < 50. ALPHA must be 0.025, 0.05, or 0.10.

* See Sections 9.4 and 10.11.

SIMSCRIPT II Data Analysis Routine

```
ROUTINE ANALYSIS (N)
NORMALLY MODE IS REAL
DEFINE N AS A REAL VARIABLE
DEFINE B AS A 2-DIM ARRAY
DEFINE D, G, H, R, S, V, W AS 1-DIM ARRAYS
READ Q, ALPHA
LET Q1 = Q + 1
RESERVE D AS 4,G AS 6,H AS 4,R, S, V, W AS Q1
LET D(1)=2.24   LET D(2)=1.96   LET D(4)=1.64
LET G(1)=1   LET G(2)=2.771808   LET G(3)=1.894306
LET G(4)=-.486382   LET G(5)=-.272398   LET G(6)=.194832
LET I = ALPHA/.025   LET Y = D(I)
LET H(1) = (Y**3+Y)/4
LET H(2) = (5*Y**5+16*Y**3+3*Y)/96
LET H(3) = (3*Y**7+19*Y**5+17*Y**3-15*Y)/384
LET H(4) = (79*Y**9+776*Y**7+1482*Y**5-1920*Y**3-945*Y)/92160
FOR T = 1 TO N, COMPUTE XBAR AS THE MEAN OF X(T)
FOR I = 1 TO Q1, DO
   FOR T = 1 TO N-I+1, COMPUTE R(I) AS THE MEAN OF (X(T)-XBAR)*(X(T+I-1)-XBAR)
LOOP
RESERVE B(*,*) AS Q+1 BY *
RESERVE B(1,*) AS 1
LET B(1,1) = 1
LET S(1) = R(1)
FOR I = 1 TO Q, DO
   RESERVE B(I+1,*) AS I+1
   FOR J=1 TO I, DO
      COMPUTE V(I) AS THE SUM OF B(I,J)*R(J)
      COMPUTE W(I) AS THE SUM OF B(I,J)*R(I-J+2)
      LOOP
   LET K=I+1
   LET B(K,K)=-W(I)/V(I)
   LET B(K,1)=1
   IF I>1,
      FOR J=2 TO I,LET B(K,J)=B(I,J)+B(K,K)*B(I,K-J+1)
   ELSE
      FOR J = 1 TO K, COMPUTE  S(K)  AS THE SUM OF B(K,J)*R(J)
   LOOP
FOR I = 1 TO Q, DO
   LET Q2 = Q- I + 1
   FOR K = 1 TO 6,COMPUTE CHI.SQUARE AS THE SUM OF Q2**((3-K)/2)*G(K)
   LET STAT = N*(1-S(Q1)/S(I))
   IF STAT<CHI.SQUARE,
      GO TO 'ORDER'
   ELSE
   IF Q2 = 1, PRINT 1 LINE THUS
  NO AUTOREGRESSIVE ORDER FOUND
      RETURN
   ELSE
LOOP
'ORDER'
   FOR J=1 TO I, COMPUTE BSUM AS THE SUM OF B(I,J)
   LET   VXBAR =S(I)/(N*BSUM*BSUM)
   FOR J=1 TO I, COMPUTE LZ AS THE SUM OF (I-1-2*(J-1))*B(I,J)
LET DF =  N/(1+2*LZ/BSUM)-1
IF DF<1, LET DF = 1
ELSE
FOR J = 1 TO 4, COMPUTE QU AS THE SUM OF H(J)/DF**J
ADD Y TO QU
START NEW PAGE
PRINT 28 LINES WITH XBAR,ALPHA/2,XBAR-QU*SQRT.F(VXBAR),ALPHA/2,
   XBAR+QU*SQRT.F(VXBAR),R(1),VXBAR,N,DF,QU,I-1,S(I) THUS
```

SIMSCRIPT II Data Analysis Routine (continued)

```
STATISTICAL RESULTS OF SIMULATION EXPERIMENT USING AUTOREGRESSIVE APPROACH

                                          SAMPLE MEAN=............
                    **.*** LOWER CONFIDENCE POINT=............
                    **.*** UPPER CONFIDENCE POINT=............
                       SAMPLE POPULATION VARIANCE=............
                    SAMPLE VARIANCE OF SAMPLE MEAN=............
                                    FINAL SAMPLE SIZE=******
                      EQUIVALENT DEGREES OF FREEDOM=******
                           COMPUTED CRITICAL T VALUE=***.**
                                AUTOREGRESSIVE ORDER=**
                            SAMPLE RESIDUAL VARIANCE=............
SKIP 3 LINES
IF DF = 1, PRINT 2 LINES THUS
   COMPUTED EQUIVALENT DEGREES OF FREEDOM < 1.
  SET = 1 FOR COMPUTATION OF CONFIDENCE LEVEL
   SKIP 3 LINES
ELSE
IF I>1, PRINT 1 LINE THUS
                                             SAMPLE AUTOREGRESSIVE COEFFICIENTS
   SKIP 1 LINE
   FOR J = 1 TO I, WRITE J-1,B(I,J) AS "B(",I 2,")=",E(11,4),S 5
SKIP 1 LINE
PRINT 1 LINE WITH STAT, Q2 AND CHI.SQUARE THUS
  TEST STATISTIC = *****.**   D.F. = ****   CRITICAL 0.025 VALUE = ***.**
SKIP 5 LINES
   PRINT 1 LINE THUS
                                               SAMPLE AUTOCORRELATIONS
   SKIP 1 LINE
   FOR J = 1 TO Q1, WRITE R(J)/R(1) AS E(11,4),S 5
   ELSE
START NEW OUTPUT LINE
SKIP 7 LINES
PRINT 2 LINES THUS
   AUTOREGRESSIVE ORDER DETERMINED AS IN FISHMAN, G.S., CONCEPTS AND METHODS
   IN DISCRETE EVENT DIGITAL COMPUTER SIMULATION
RETURN
END
```

FORTRAN IV ANALYSIS ROUTINE

This routine is similar to the SIMSCRIPT II in all but two respects. First, when calling this routine the user must specify the data array as well as N. The *local* array X then holds the data to be analyzed. Second, Q, the highest order autoregressive scheme cannot exceed 25. However, a user can modify the program to accept a larger Q.

FORTRAN IV Data Analysis Routine

```
      SUBROUTINE ANALYS(X,N)
      DIMENSION X(N)
      DIMENSION B(26,26)
      DIMENSION D(4),G(6),H(4),R(26),S(26),V(26),W(26)
      INTEGER DF
      INTEGER Q1,Q2,Q
      READ (5,1000) Q,ALPHA
 1000 FORMAT (I5,F10.4)
      Q1=Q+1
      D(1)=2.24
      D(2)=1.96
      D(4)=1.64
      G(1)=1.
      G(2)=2.771808
      G(3)=1.894306
      G(4)=-.486382
      G(5)=-.272398
      G(6)=.194832
      I=ALPHA*40.0001
      Y=D(I)
      Y3=Y**3
      Y5=Y3*Y*Y
      Y7=Y5*Y*Y
      Y9=Y7*Y*Y
      H(1)=(Y3+Y)*.25
      H(2)=(5.*Y5+16.*Y3+3.*Y)/96.
      H(3)=(3.*Y7+19.*Y5+17.*Y3-15.*Y)/384.
      H(4)=(79.*Y9+776.*Y7+1482.*Y5-1920.*Y3-945.*Y)/92160.
      XBAR=0.
      DO 2 IT=1,N
2     XBAR=XBAR+X(IT)
      XBAR=XBAR/FLOAT(N)
      DO 3 I=1,Q1
      NI1=N-I+1
      R(I)=0.
      DO 4 IT=1,NI1
4     R(I)=R(I)+(X(IT)-XBAR)*(X(IT+I-1)-XBAR)
3     R(I)=R(I)/FLOAT(NI1)
      B(1,1)=1
      S(1)=R(1)
      DO 5 I=1,Q
      V(I)=0.
      W(I)=0.
      DO 6 J=1,I
      V(I)=V(I)+B(I,J)*F(J)
6     W(I)=W(I)+B(I,J)*R(I-J+2)
      K=I+1
      B(K,K)=-W(I)/V(I)
      B(K,1)=1
      IF(I.EQ.1)GOTO 7
      DO 8 J=2,I
8     B(K,J)=B(I,J)+B(K,K)*B(I,K-J+1)
7     S(K)=0.
      DO 9 J=1,K
9     S(K)=S(K)+B(K,J)*R(J)
5     CONTINUE
      DO 10 I=1,Q
```

FORTRAN IV Data Analysis Routine (continued)

```
      Q2=Q-I+1
      CHI2=0.
      DO 11 K=1,6
11    CHI2=CHI2+Q2**(.5*FLOAT(3-K))*G(K)
      STAT=FLOAT(N)*(1.-S(Q1)/S(I))
      IF(STAT.LT.CHI2)GOTO 12
      IF(Q2.GT.1)GOTO 10
      WRITE(6,22)
22    FORMAT(30H NO AUTOREGRESSIVE ORDER FOUND )
      RETURN
10    CONTINUE
12    BSUM=0.
      DO 13 J=1,I
13    BSUM=BSUM+B(I,J)
      VXBAR=S(I)/(FLOAT(N)*BSUM*BSUM)
      LZ=0.
      DO 14 J=1,I
14    LZ=LZ+(I-1-2*(J-1))*B(I,J)
      DF=FLOAT(N)/(1.+2.*LZ/BSUM)-1.
      IF(DF.LT.1)DF=1
      QU=0.
      DO 15 J=1,4
15    QU=QU+H(J)/DF**J
      SVX=SQRT(VXBAR)
      QU=QU+Y
      CPL=XBAR-QU*SVX
      CPU=XBAR+QU*SVX
      WRITE(6,16)
16    FORMAT(1H1)
      I1 =I-1
      ALPHA=ALPHA/2
      WRITE(6,17)XBAR,ALPHA,CPL,ALPHA,CPU,R(1),VXBAR,N,DF,QU,I1,S(I)
17    FORMAT(75H STATISTICAL RESULTS OF SIMULATION EXPERIMENT USING AUTO
     *REGRESSIVE APPROACH//
     *36X,12HSAMPLE MEAN= F12.4//18X,F6.3,24H LOWER CONFIDENCE POINT=
     *F12.4//18X,F6.3,24H UPPER CONFIDENCE POINT= F12.4//
     *21X,27HSAMPLE POPULATION VARIANCE= F12.4//
     *17X,31HSAMPLE VARIANCE OF SAMPLE MEAN= F12.4//
     *30X,18HFINAL SAMPLE SIZE= I6//
     *18X,30HEQUIVALENT DEGREES OF FREEDOM= I6//
     *22X,26HCOMPUTED CRITICAL T VALUE= F6.2//
     *27X,21HAUTOREGRESSIVE ORDER= I6//
     *23X,25HSAMPLE RESIDUAL VARIANCE= F12.4////)
      IF(DF.EQ.1)WRITE(6,18)
18    FORMAT(44H COMPUTED EQUIVALENT DEGREES OF FREEDOM < 1 /
     *44H SET = 1 FOR COMPUTATION OF CONFIDENCE LEVEL ////)
      IF(I.LE.1)GOTO 20
      WRITE(6,19)
19    FORMAT(40X,34HSAMPLE AUTOREGRESSIVE COEFFICIENTS//)
      I1=I-1
      IZ=0
      WRITE(6,21)IZ,((B(I,J),J),J=1,I1),B(I,I)
21    FORMAT((6(2X,3H B(I2,2H)=E13.4))/)
      WRITE(6,27)STAT,Q2,CHI2
27    FORMAT(18H0TEST STATISTIC = F8.2,3X,7HD.F. = I4,3X,23HCRITICAL 0.0
     *25 VALUE = F6.2//////)
      WRITE(6,23)
23    FORMAT(44X,23HSAMPLE AUTOCORRELATIONS ///)
      DO 24 J=2,Q1
24    R(J)=R(J)/R(1)
      R(1)=1.
      WRITE(6,25)R
25    FORMAT(10E13.4/)
20    CONTINUE
      WRITE(6,26)
26    FORMAT(///////74H AUTOREGRESSIVE ORDER DETERMINED AS IN FISHMAN, G
     *.S., CONCEPTS AND METHODS/46H IN DISCRETE EVENT DIGITAL COMPUTER S
     *IMULATION ///)
      RETURN
      END
```

REFERENCES

Chapter 1

1. R. W. Conway, B. M. Johnson, and W. L. Maxwell, "Some Problems of Digital Systems Simulation," *Management Sci.*, Vol. 6, 1959, pp. 92–110.
2. P. J. Kiviat, *Digital Event Simulation: Modeling Concepts*, The Rand Corporation, RM-5378-PR, Santa Monica, Calif., 1967.
3. P. J. Kiviat, *Digital Computer Simulation: Computer Programming Languages*, The Rand Corporation, RM-5883-PR, Santa Monica, Calif., 1969.
4. M. R. Leavitt, "ALLY 1—Description of the Model," Project on Simulated International Processes, Northwestern University, Evanston, Ill., 1970.
5. Keith Tognetti, "A Discrete Bio Simulation—the Population and Regulation of Turtles," *Digest of the Second Conference on the Applications of Simulation*, New York, 1968, pp. 346–363.

Chapter 2

6. G. Carter and E. J. Ignall, *A Simulation Model of Fire Department Operations: Design and Preliminary Results*, The New York Rand Institute, R-632-NYC, New York, November 1970.
7. *Digest of the Second Conference on the Applications of Simulation*, New York, December 1968.
8. R. B. Fetter and J. D. Thompson, "A Decision Model for the Design and Operation of a Progressive Patient Care Hospital," *Medical Care*, Vol. 7, No. 6, November-December 1969, pp. 450–462.
9. M. A. Geisler and W. A. Steger, "The Use of Man-Machine Simulation in the Design of an Operational Control System," *Proceedings WJCC*, Vol. 19, 1961, pp. 51–62.
10. J. M. Hammersley and D. C. Handscomb, *Monte Carlo Methods*, Methuen, London, 1964.
11. H. H. Harman, *Simulation: A Survey*, Systems Development Corporation, SP-260, Santa Monica, Calif., 1961.
12. P. J. Kiviat, *Discrete Event Simulation: Modeling Concepts*, The Rand Corporation, RM-5378-PR, Santa Monica, Calif., 1967.

13. T. H. Naylor, "Bibliography 19: Simulation and Gaming," *Computer Revs.*, January 1969, pp. 61–69.
14. *Proceedings of the First Conference on the Applications of Simulation*, New York, December 1967.
15. *Proceedings of the Third Conference on the Applications of Simulation*, Los Angeles, December 1969.
16. *Proceedings of the Fourth Conference on the Applications of Simulation*, New York, December 1970.
17. *Proceedings of the Summer Simulation Conference*, Vols. I and II, Boston, 1971.
18. M. A. Shubik, *Preliminary Bibliography on Gaming*, College on Simulation and Gaming of the Institute of Management Science, March 1971.
19. M. A. Shubik, "On the Scope of Gaming," *Management Sci.*, Vol. 18, No. 5, Part II, January 1972, pp. 20–36.

Chapter 3

20. H. I. Ansoff and D. P. Slevin, "An Appreciation of Industrial Dynamics," *Management Sci.*, Vol. 14, No. 7, March 1968, pp. 383–396.
21. J. S. Hunter and T. H. Naylor, "Experimental Design for Computer Simulation Experiments," *Management Sci.*, Vol. 16, No. 7, March 1970, pp. 422–435.
22. E. J. Ignall, "On Experimental Designs for Computer Simulation Experiments," *Management Sci.*, Vol. 18, No. 7, March 1972, pp. 384–388.
23. P. J. Kiviat, *Digital Computer Simulation: Modeling Concepts*, The Rand Corporation, RM-5378-PR, Santa Monica, Calif., 1967.
24. J. R. Laski, "On Time Structure in (Monte Carlo) Simulations," *Operations Res. Quarterly*, Vol. 16, No. 3, 1965, pp. 329–339.
25. D. G. Malcolm, J. H. Rosenbloom, C. E. Clark, and W. Fazar, "Application of a Technique for Research and Development Program Evaluation," *Operations Res.*, Vol. 7, 1959, pp. 646–669.
26. C. Malkani and C. C. Pegels, *A Simulation Model for Evaluating Blood Utilization with Extended Life-Span*, State University of New York at Buffalo, January 1972.
27. G. H. Orcutt, M. Greenberger, J. Korbel, and A. M. Rivlin, *Microanalysis of Socioeconomic Systems: A Simulation Study*, Harper and Row, New York, 1961.
28. P-E Consulting Group Limited, *HOCUS Basic Manual*, Surrey, England.
29. A. R. Pai and K. L. McRoberts, "Simulation Research in Interchargeable Part Manufacturing," *Management Sci.*, Vol. 17, No. 12, August 1971, pp. B732–B743.
30. R. L. Petruschell, S. J. Benton, D. J. Dreyfuss, L. E. Knollmeyer, J. Y. Lu, and R. E. Stanton, *Reducing Costs of Stock Transactions: A Study of Alternative Trade Completion Systems*, Vol. 3: *The Trade Completion Simulation Model*, The Rand Corporation, R-552-ST, Santa Monica, Calif., 1970.
31. A. A. Pritsker, GERT: *Graphical Evaluation and Review Technique*, The Rand Corporation, RM-4973-NASA, Santa Monica, Calif., April 1966.
32. A. A. Pritsker and P. J. Kiviat, *Simulation with GASP II*, Prentice-Hall, Englewood Cliffs, N.J., 1969.
33. T. Sakai and N. Nagao, "Simulation of Traffic Flows in a Network," *Comm. ACM*, Vol. 12, No. 6, June 1969, pp. 311–318.

34. L. Takacs, *Introduction to the Theory of Queues*, Oxford University Press, New York, 1962.

Chapter 4

35. R. P. Bennett, et al., *SIMPAC User's Manual*, Systems Development Corporation, TM-602/000/00, Santa Monica, Calif., April 1962.
36. J. N. Buxton, "Writing Simulations in CSL," *Comp. J.*, Vol. 9, August 1966, pp. 137–143.
37. Consolidated Analysis Centers, Inc., *SIMSCRIPT II.5 Reference Handbook*, Santa Monica, Calif., 1971.
38. R. W. Conway, B. M. Johnson, and W. L. Maxwell, "Some Problems of Digital Systems Simulation," *Management Sci.*, Vol. 6, October 1959, pp. 92–110.
39. General Electric Company, *SIMCOM User's Guide*, Information Systems Operations, TR-65-2-149010, 1964.
40. M. Greenberger, et al., *On-Line Computation and Simulation: The OPS-3 System*, M.I.T. Press, Cambridge, Mass., 1965.
41. P. R. Hills, "SIMON—A Computer Simulation Language in ALGOL," in S. H. Hollingdall (ed.), *Digital Simulation in Operational Research*, American Elsevier, New York, 1967.
42. IBM, *Simulation Evaluation and Analysis Language (SEAL), System Reference Manual*, January 17, 1968.
43. IBM, *General Purpose Simulation System/360 OS and D05 Version 2 User's Manual*, SH20-0694-0, White Plains, N.Y., 1969.
44. IBM, United Kingdom Limited, *Control and Simulation Language Reference Manual*, London, reprinted 1966.
45. M. M. Jones and R. C. Thurber, *SIMPL Reference Manual*, M.I.T., Cambridge, Mass., 1971.
46. L. A. Kalinichenko, "SLANG—Computer Description and Simulation-Oriented Experimental Programming Language," in J. N. Buxton (ed.), *Proceedings of the IFIP Working Conference on Simulation Languages*, North-Holland, Amsterdam 1968.
47. H. W. Karr, H. Kleine, and H. M. Markowitz, *SIMSCRIPT 1.5*, Consolidated Analysis Centers, Inc., CACI 65-INT-1, Santa Monica, Calif., June 1965.
48. P. J. Kiviat, *Digital Computer Simulation: Computer Programming Languages*, The Rand Corporation, RM-5883-PR, Santa Monica, Calif., January 1969.
49. P. J. Kiviat, R. Villanueva, and H. Markowitz, *The SIMSCRIPT II Programming Language*, Prentice Hall, Englewood Cliffs, N.J., 1969.
50. D. Knuth, *The Art of Computer Programming*, Vol. 1: *Fundamental Algorithms*, Addison-Wesley, Reading, Mass., 1969.
51. D. C. Knuth and J. L. McNeley, "SOL—A Symbolic Language for General-Purpose System Simulation," *IEEE Trans. Electronic Computers*, August 1964.
52. H. S. Krasnow, "Dynamic Representation in Discrete Interaction Simulation Languages," in S. H. Hollingdale (ed.), *Digital Simulation in Operational Research*, American Elsevier, New York, 1967.

53. H. M. Markowitz, H. W. Karr, and B. Hausner, *SIMSCRIPT: A Simulation Programming Language*, Prentice-Hall, Englewood Cliffs, N.J., 1963.

54. H. Morgan and L. G. Siegel, "Synchronization Models for Simulation and List Processing," Technical Report 113, Department of Operations Research, College of Engineering, Cornell University, Ithaca, N.Y., June 1970.

55. R. E. Nance, "On Time Flow Mechanisms for Discrete System Simulation," *Management Sci.*, Vol. 18, No. 1, September 1971, pp. 59–73.

56. R. J. Parente, "A Language for Dynamic System Description," IBM Advanced System Development Division, Technical Report 17-180, 1966.

57. R. D. Parslow, "AS: An ALGOL Simulation Language," presented at the 1967 IFIP Working Conference on Simulation Languages, Oslo, Norway.

58. L. Petrone, "On a Simulation Language Completely Defined Onto the Programming Language PL/I," presented at the 1967 IFIP Working Conference on Simulation Languages, Oslo, Norway.

59. A. A. Pritsker and P. J. Kiviat, *Simulation with GASP II*, Prentice-Hall, Englewood Cliffs, N.J., 1969.

60. D. Teichroew and J. F. Lubin, "Computer Simulation: Discussion of Techniques and Comparison of Languages," *Comm. ACM*, Vol. 9, No. 4, October 1966.

61. D. Teichroew, J. F. Lubin, and T. D. Truitt, "Discussion of Computer Simulation Techniques and Comparison of Language," *Simulation*, Vol. 9, 1967, pp. 181–190.

62. K. D. Tocher and D. C. Owen, "The Automatic Programming of Simulations," *Proceedings of the Second International Conference on Operations Research*, 1969, pp. 50–68.

63. United States Steel Corporation, *Simulation Language and Library* (*SILLY*), Engineering and Scientific Computer Services, February 1968.

64. UNIVAC, 1106/1108, *SIMULA Programmer Reference*, UP-7556, 1971.

65. A. W. Wickham, "Time Flow Mechanisms for Discrete Event Simulation Models," unpublished Ph.D. thesis, Department of Computer Science, Southern Methodist University, Dallas, Tex., 1971.

66. J. W. J. Williams, "The Elliott Simulator Package (ESP)," *Comp. J.*, Vol. 6, No. 4, January 1964.

Chapter 5

67. Consolidated Analysis Centers, Inc., *SIMSCRIPT II.5 Reference Handbook*, Santa Monica, Calif., 1971.

68. O. Dahl and K. Nygaard, "SIMULA—an ALGOL-Based Simulation Language," *Comm. ACM*, Vol. 9, No. 9, September 1966, pp. 671–678.

69. O. Dahl, B. Myhrhang, and K. Nygaard, *Common Base Language*, Norwegian Computing Center, October 1970.

70. G. Gordon, *System Simulation*, Prentice-Hall, Englewood Cliffs, N.J., 1969.

71. IBM, *General Purpose Simulation System/360 User's Manual*, GH 20-0326, White Plains, N.Y., January 1970.

72. H. W. Karr, H. Kleine, and H. M. Markowitz, *SIMSCRIPT 1.5*, Consolidated Analysis Centers, Inc., CACI 65-INT-1, Santa Monica, Calif., June 1965.

73. P. J. Kiviat, R. Villaneuva, and H. Markowitz, *The SIMSCRIPT II Programming Language*, Prentice-Hall, Englewood Cliffs, N.J., 1969.

74. P. J. Kiviat and R. Villaneuva, *The SIMSCRIPT II Programming Language Reference Manual*, Prentice-Hall, Englewood Cliffs, N.J., 1969.

75. H. M. Markowitz, H. W. Karr, and B. Hausner, *SIMSCRIPT: A Simulation Programming Language*, Prentice-Hall, Englewood Cliffs, N.J., 1963.

76. J. McNeley, "Simulation Languages," *Simulation*, Vol. 9, No. 2, August 1967, pp. 95–98.

77. T. Schriber, *A GPSS Primer*, a preliminary printing, distributor: Ulrich's Books, Inc., 549 East University, Ann Arbor, Mich., 1972.

78. UNIVAC, 1106/1108, *SIMULA Programmer Reference*, UP-7556, 1971.

Chapter 6

79. T. W. Anderson, *The Statistical Analysis of Time Series*, Wiley, New York, 1970.

80. M. S. Bartlett, *An Introduction to Stochastic Processes*, Cambridge University Press, Cambridge, England, 1955.

81. D. R. Cox and H. D. Miller, *The Theory of Stochastic Processes*, Wiley, New York, 1965.

82. D. R. Cox and W. L. Smith, *Queues*, Methuen, London, 1963.

83. G. S. Fishman, *Spectral Methods in Econometrics*, Harvard University Press, Cambridge, Mass., 1969.

84. G. S. Fishman and P. J. Kiviat, "The Analysis of Simulation-Generated Time Series," *Management Sci.*, Vol. 13, No. 7, March 1967, pp. 525–557.

85. E. J. Hannan, *Multiple Time Series*, Wiley, New York, 1970.

86. P. Morse, "Stochastic Properties of Waiting Lines," *J. Operations Res. Soc. America* Vol. 3, No. 3, August 1955, pp. 255–261.

87. A. M. Yaglom, *An Introduction to the Theory of Stationary Random Functions*, translated by R. A. Silverman, Prentice-Hall, Englewood Cliffs, N.J., 1962.

Chapter 7

88. J. L. Allard, A. R. Dobell, and T. E. Hull, "Mixed Congruential Random Number Generators for Decimal Machines," *J. ACM*, Vol. 10, No. 2, 1966, pp. 131–141.

89. R. R. Coveyou, "Serial Correlation in the Generation of Pseudo-Random Numbers," *J. ACM*, Vol. 7, 1960, pp. 72–74.

90. R. R. Coveyou and R. D. MacPherson, "Fourier Analysis of Uniform Random Number Generators," *J. ACM*, Vol. 14, No. 1, January 1967, pp. 100–119.

91. D. Y. Downham and R. D. K. Roberts, "Multiplicative Congruential Pseudo-Random Number Generators," *Comp. J.*, Vol. 10, No. 1, 1967, pp. 74–77.

92. J. Durbin, "Tests of Serial Independence Based on the Cumulative Periodogram," 36th Session of the International Statistical Institute, 1967.

93. H. Felder, "The GPSS/360 Random Number Generator," *Digest of the Second Conference of Applications of Simulation*, New York, December 1968.

94. I. J. Good, "On the Serial Test for Random Sequences," *Ann. Math. Stat.*, Vol. 28, 1957, pp. 262–264.

95. S. Gorenstein, "Testing a Random Number Generator," *Comm. ACM*, Vol. 10, No. 2, February 1967, pp. 111–118.

96. M. Greenberger, "Notes on a New Pseudo-Random Number Generator," *J. ACM*, Vol. 8, 1961, pp. 163–167.

97. M. Greenberger, "An a Priori Determination of Serial Correlation in Computer Generated Random Numbers," *Math. Comp.*, Vol. 15, 1961, pp. 383–389; "Corrigenda," Vol. 16, 1962, p. 126.

98. M. Greenberger, "Methods of Randomness," *Comm. ACM*, Vol. 8, No. 3, March 1965, pp. 177–179.

99. T. E. Hull and A. R. Dobell, "Random Number Generators," *SIAM Rev.*, Vol. 4, No. 3, July 1962, pp. 230–254.

100. T. E. Hull and A. R. Dobell, "Mixed Congruential Random Number Generators for Binary Machines," *J. ACM*, Vol. 11, No. 1, January 1964, pp. 31–40.

101. D. W. Hutchinson, *A New Uniform Pseudo-Random Number Generator*, File 651, April 27, 1965, Department of Computer Sciences, University of Illinois, Urbana, Ill.

102. D. W. Hutchinson, "A New Uniform Pseudorandom Number Generator," *Comm. ACM*, Vol. 9, No. 6, June 1966, pp. 432–433.

103. IBM, *Random Number Generation and Testing*, Form C20-8011, 1959.

104. IBM, *General Purpose Simulation System/360 User's Manual*, GH 20-0326, White Plains, N.Y., January 1970.

105. B. Jansson, *Random Number Generators*, Almquist and Wiskell, Stockholm, 1966.

106. M. G. Kendall and A. Stuart, *The Advanced Theory of Statistics*, Vol. 3, Hafner, New York, 1966.

107. P. J. Kiviat, R. Villaneuva, and H. Markowitz, *The SIMSCRIPT II Programming Language*, Prentice-Hall, Englewood Cliffs, N.J., 1969.

108. D. E. Knuth, *The Art of Computer Programming: Seminumerical Algorithms*, Vol. 2, Addison-Wesley, Reading, Mass., 1969.

109. D. H., Lehmer, "Mathematical Methods in Large-Scale Computing Units," *Ann. Comp. Laboratory*, Harvard University, Vol. 26, 1951, pp. 141–146.

110. P. A. W. Lewis, A. S. Goodman, and J. M. Miller, "A Pseudo-Random Number Generator for the System/360," *IBM Systems J.*, Vol. 8, No. 2, 1969, pp. 136–145.

111. M. D. MacLaren and G. Marsaglia, "Uniform Random Number Generators," *J. ACM*, Vol. 12, No. 1, January 1965, pp. 83–89.

111a. G. Marsaglia, "Random Numbers Fall Mainly in the Planes," *Proc. Natl. Acad. Sci.*, Vol. 61, September 1968, pp. 25–28.

112. R. A. Olshen, "Asymptotic Properties of the Periodogram of a Discrete Stationary Process," *J. Appl. Prob.*, Vol. 14, No. 3, December 1967, pp. 508–528.

113. D. B. Owen, *Handbook of Statistical Tables*, Addison-Wesley, Reading, Mass., 1962.

114. W. H. Payne, J. R. Raburg, and T. P. Bogyo, "Coding the Lehmer Pseudorandom Number Generator," *Comm. ACM*, Vol. 12, No. 2, February 1969, pp. 85–86.

115. Rand Corporation, The, *A Million Random Digits with 1,000,000 Normal Deviates*, Free Press, New York, 1955.

116. A. Rotenberg, "A New Pseudo-Random Number Generator," *J. ACM*, Vol. 7, 1960.
117. *UNIVAC, 1106/1108, SIMULA Programmer Reference*, UP-7556, 1971.
118. A. Van Gelder, "Some New Results in Pseudo-Random Number Generation," *J. ACM*, Vol. 14, No. 4, October 1967, pp. 785–792.
119. J. Whittlesey, "A Comparison of the Correlational Behavior or Random Number Generators," *Comm. ACM*, Vol. 11, No. 9, September 1968, pp. 641–644.

Chapter 8

120. T. W. Anderson, Introduction to Multi-variate Statistical Analysis, Wiley, New York, 1958.
121. N. T. J. Bailey, *The Elements of Stochastic Processes with Applications to the Natural Sciences*, Wiley, New York, 1964.
122. M. B. Berman, *Generating Random Variates from Gamma Distributions with Non-Integer Shape Parameters*, The Rand Corporation, R-641-PR, Santa Monica, Calif., November 1970.
123. G. E. P. Box and M. E. Muller, "A Note on the Generation of Random Normal Deviates," *Ann. Math. Stat.*, Vol. 29, 1958, pp. 610–611.
124. B. Brown, *Characteristics of Demand for Aircraft Spare Parts*, The Rand Corporation, R-292 (AD 107426), Santa Monica, Calif., July 1956.
125. J. L. Doob, *Stochastic Processes*, Wiley, New York, 1953.
126. B. L. Fox, "Generation of Random Samples from the Beta and *F* Distributions," *Technometrics*, Vol. 5, 1963, pp. 269–270.
127. B. L. Fox, *Notes and Operations Research—*5, Operations Research Center, University of California, ORC 66-13, Berkeley, 1966.
128. D. P. Gaver, *Point Processes in Reliability*, U.S. Naval Postgraduate School, Monterey, Calif., December 1971.
129. P. G. Hoel, *Introduction to Mathematical Statistics*, 3rd Ed., Wiley, New York, 1962.
130. M. D. Jöhnk, "Erzeugung von Betaverteilten und Gammaverteilten Zufallszahlen," *Metrika*, Vol. 8, 1964, pp. 5–15.
130a. G. Marsaglia, "Generating Discrete Random Variables in a Computer," *Comm. ACM*, Vol. 6, January 1963, pp. 37–38.
131. A. W. Marshall and I. Olkin, "A Generalized Bivariate Exponential Distribution," *J. Appl. Prob.*, Vol. 4, 1967, pp. 291–302.
132. T. H. Naylor, J. L. Balintfy, D. S. Burdick, and K. Chu, *Computer Simulation Techniques*, Wiley, New York, 1965.
133. R. L. Plackett, *Principles of Regression Analysis*, Oxford University Press, Oxford, England, 1960.
134. E. Scheuer and D. S. Stoller, "On the Generation of Normal Random Vectors," *Technometrics*, Vol. 4, May 1962, pp. 278–281.

Chapter 9

135. K. O. Bowman and L. R. Shenton, *Asymptotic Covariances for the Maximum Likelihood Estimators of the Parameters of a Negative Binomial Distribution*, Union Carbide Corp., K-1643, Oak Ridge, Tenn., July 1, 1965.

136. K. O. Bowman and L. R. Shenton, *Biases and Covariances of Maximum Likelihood Estimators*, Union Carbide Corp., K-1633, Oak Ridge, Tenn. May 1965.

137. K. O. Bowman and L. R. Shenton, *Biases of Estimators for the Negative Binomial Distribution*, Union Carbide Corp., ORNL-4005, UC-32, Oak Ridge, Tenn., November 1966.

138. K. O. Bowman and L. R. Shenton, *Small Sample Properties of Estimators for the Gamma Distribution*, Union Carbide Corp., CTC-28, Oak Ridge, Tenn., July 1, 1970.

139. G. E. P. Box and G. M. Jenkins, *Time Series Analysis: Forecasting and Control*, Holden-Day, San Francisco, 1970.

139a. S. C. Choi and R. Wette, "Maximum Likelihood Estimation of the Parameters of the Gamma Distribution and Their Bias," *Technometrics*, Vol. 11, No. 4, November 1969, pp. 683–690.

140. J. Durbin, "The Fitting of Time Series Models," *Rev. Intern. Inst. Stat.*, Vol. 28, 1960, pp. 233–244.

141. D. J. Finney, "On the Distribution of a Variate Whose Logarithm Is Normally Distributed," *Suppl. J. Roy. Stat. Soc.*, Vol. 7, No. 2, 1941, pp. 155–161.

142. R. Gnanadesikan, *Tables of Maximum Likelihood Estimates of the Beta Distribution*, Bell Telephone Laboratories, Murray Hill, N.J.

143. E. Hannan, *Multiple Time Series*, Wiley, New York, 1970.

144. N. L. Johnson and S. Kotz, *Discrete Distributions*, Houghton Mifflin, Boston, 1969.

145. N. L. Johnson and S. Kotz, *Continuous Distributions*, Vols. I and II, Houghton Mifflin, Boston, 1970.

146. M. G. Kendall and A. Stuart, *The Advanced Theory of Statistics*, Vol. 2, Hafner, New York, 1963.

147. H. W. Lilliefors, "On the Kolmogorov-Smirnov Test for Normality with Mean and Variance Unknown," *JASA*, Vol. 62, 1967, pp. 399–402.

148. H. W. Lilliefors, "On the Kolmogorov-Smirnov Test for the Exponential Distribution with Mean Unknown," *JASA*, Vol. 64, 1969, pp. 387–389.

149. A. Madansky, "The Use of Method-of-Moments Estimates in $\chi 2$ Tests for Goodness of Fit," The Rand Corporation, RM-4524-PR, Santa Monica, Calif., January 1967.

150. H. B. Mann and A. Wald, "On the Statistical Treatment of Linear Stochastic Difference Equations," *Econometrica*, Vol. 11, July-October 1943, pp. 173–220.

151. D. B. Owen, *Handbook of Statisiical Tables*, Addison-Wesley, Reading, Mass., 1962.

152. E. Parzen, "An Approach to Time Series Analysis," *Ann. Math. Stat.*, Vol. 32, No. 4, December 1961, pp. 951–988.

153. K. Pearson (ed.), *Tables of the Incomplete Γ-Function*, Biometrika Office, University College, Cambridge, England, 1934.

153a. R. S. Pinkham, R. Gnanadesikan, and L. P. Hughes, "Maximum Likelihood Estimation of the Parameters of the Beta Distribution from Smallest Order Statistics," *Technometrics*, Vol. 9, No. 4, November 1967, pp. 607–620.

154. H. S. Sichel, "New Methods in the Statistical Evaluation of Mine Sampling Data," *Trans. Inst. Mining Metallurgy*, Vol. 61, 1951–1952, pp. 261–288.

155. M. M. Siddiqui, "On the Inversion of the Sample Covariance Matrix in a Stationary Autoregressive Process," *Ann. Math. Stat.*, Vol. 29, 1958, pp. 585–588.

156. D. R. Thoman, L. J. Bain, and C. E. Antle, "Inferences on the Parameters of the Weibull Distribution," *Technometrics*, Vol. 11, No. 3, August 1969, pp. 445–460.

157. P. Whittle, *Prediction and Regulation by Linear Least-Square Methods*, The English University Press, London, 1963.

Chapter 10

158. G. E. P. Box and G. M. Jenkins, *Time Series Analysis: Forecasting and Control*, Holden-Day, San Francisco, 1970.
159. R. W. Conway, "Some Tactical Problems in Digital Simulation," *Management Sci.*, Vol. 10, No. 1, October 1963, pp. 47–61.
160. G. S. Fishman, "Problems in the Statistical Analysis of Simulation Experiments: The Comparison of Means and the Length of Sample Records," *Comm. ACM*, Vol. 10, February 1967, pp. 94–99.
161. G. S. Fishman, *Spectral Methods in Econometrics*, Harvard University Press, Cambridge, Mass., 1969.
162. G. S. Fishman, "Estimating Sample Size in Computer Simulation Experiments," *Management Sci.*, Vol. 18, No. 1, September 1971, pp. 21–38.
163. G. S. Fishman, "A Study of Bias Considerations in Simulation Experiments," to be published in *Operations Research*.
164. G. S. Fishman and P. J. Kiviat, "The Analysis of Simulation Generated Time Series," *Management Sci.*, Vol. 13, No. 7, March 1967, pp. 525–557.
165. H. J. Godwin, *Inequalities on Distribution Functions*, Griffin, London, 1966.
166. H. Goldberg and H. Levine, "Approximate Formulas for the Percentage Points and Normalization of t and χ^2," *Ann. Math. Stat.*, Vol. 17, 1946, pp. 216–225.
167. E. J. Hannan, *Multiple Time Series*, Wiley, New York, 1970.
168. G. M. Jenkins and D. G. Watts, *Spectral Analysis and Its Applications*, Holden-Day, San Francisco, 1968.
169. E. Parzen, "An Approach to Time Series Analysis," *Ann. Math. Stat.*, Vol. 32, No. 4, December 1961, pp. 951–988.
170. E. Parzen, "Mathematical Considerations in the Estimation of Spectra," *Technometrics*, Vol. 3, No. 2, May 1961, pp. 167–190.
171. A. I. Sarhan and B. G. Greenberg, *Contributions to Order Statistics*, Wiley, New York, 1962.
172. H. Sheffé, *The Analysis of Variance*, Wiley, New York, 1959.
173. L. Takács, *Introduction to the Theory of Queues*, Oxford University Press, Oxford, England, 1962.
174. D. R. Thoman, L. J. Bain, and C. E. Antle, "Inferences on the Parameters of the Weibull Distribution," *Technometrics*, Vol. 11, No. 3, 1969, pp. 445–460.
175. M. Tin, "Comparison of Some Ratio Measures," *JASA*, Vol. 60, 1965, pp. 294–307.
176. P. Treuenfels, "Upper Bound for the Abscissa of the χ^2 Distribution," *SIAM Rev.*, Vol. 13, January 1971, pp. 136–137.

Chapter 11

177. G. E. P. Box and N. R. Draper, "A Basis for Selection of a Response Surface Design," *JASA*, Vol. 54, 1959, pp. 622–654.

178. G. E. P. Box, *Evaluationary Operation: A Statistical Method for Process Improvement*, Wiley, New York, 1969.

179. G. E. P. Box and J. S. Hunter, "Experimental Designs for Exploring Response Surfaces," in V. Chew (ed.), *Experimental Designs in Industry*, Wiley, New York, 1953, pp. 138–192.

180. G. E. P. Box and J. S. Hunter, "Multi-factor Experimental Designs for Exploring Response Surfaces," *Ann. Math. Stat.*, Vol. 28, 1957, pp. 195–241.

181. K. Brownlee, *Statistical Theory and Methodology in Science and Engineering*, 2nd Ed., Wiley, New York, 1965.

182. D. Burdick and T. Naylor, "Design of Computer Simulation Experiments for Industrial Systems," *Comm. ACM*, Vol. 9, No. 5, May 1966, pp. 329–338.

183. J. M. Burt, Jr., and M. Garman, "Monte Carlo Techniques for Stochastic Network Analysis," *Proceedings of the Fourth Conference on the Applications of Simulation*, December 9-11, 1970, pp. 146–153.

184. J. M. Burt, Jr., D. P. Gaver, and M. Perlas, "Simple Stochastic Networks: Some Problems and Procedures," *Naval Res. Logistics Quarterly*, Vol. 17, No. 4, December 1970, pp. 439–460.

185. G. Carter and E. Ignall, "A Simulation Model of Fire Department Operations," *IEEE-SSC*, Vol. 6, No. 4, October 1970, pp. 282–292.

186. G. Carter and E. Ignall, "Virtual Measures for Computer Simulation Experiments," The Rand Corporation, P-4817, Santa Monica, Calif., April 1972.

187. W. G. Cochran and G. M. Cox, *Experimental Designs*, 2nd Ed., Wiley, New York, 1957.

188. W. G. Cochran and G. M. Cox, *Experimental Designs*, 2nd Ed., Wiley, New York, 1957.

189. N. R. Draper and H. Smith, *Applied Regression Analysis*, Wiley, New York, 1960.

190. G. S. Fishman, "The Allocation of Computer Time in Company Simulation Experiments," *Operations Res.*, Vol. 16, 1968, pp. 280–295.

191. G. S. Fishman, "Variance Reduction in Simulation Studies," *J. Stat. Comp. Simulation*, Vol. 1, 1972, pp. 173–182.

192. G. S. Fishman, "Variance Reduction for Normal Variates in Monte Carlo Studies," Yale University, Department of Administrative Sciences, New Haven, 1972.

193. D. P. Gaver, Jr., "Statistical Methods for Improving Simulation Efficiency," *Proceedings of the Third Conference on the Applications of Simulation*, December 8-10, 1969, pp. 38–46.

194. J. M. Hammersley and D. C. Handscomb, *Monte Carlo Methods*, Methuen, London, 1964.

195. W. J. Hill and W. G. Hunter, "A Review of Response Surface Methodology: A Literature Survey," *Technometrics*, Vol. 8, November 1966, pp. 571–590.

196. A. E. Hoerl, Jr., "Fitting Curves to Data," in J. W. Perry (ed.), *Chemical Business Handbook*, McGraw-Hill, New York, 1954.

197. J. S. Hunter and T. H. Naylor, "Experimental Design for Computer Simulation Experiments," *Management Sci.*, Vol. 16, No. 7, March 1970, pp. 422–435.

198. E. J. Ignall, "On Experimental Designs for Computer Simulation Experiments," *Management Sci.*, Vol. 18, No. 7, March 1972, pp. 384–389.

199. N. L. Johnson and F. C. Leone, *Statistics and Experimental Design in Engineering and the Physical Sciences*, Vols. I and II, Wiley, New York, 1964.

200. H. Kahn and A. W. Marshall, "Methods of Reducing Sample Size in Monte Carlo Computations," *Operations Res.*, Vol. 1, No. 5, 1953, pp. 263–278.

201. M. G. Kendall and A. Stuart, *The Advanced Theory of Statistics*, Vol. 3, Hafner, New York, 1966.

202. J. P. C. Kleijnen, *Variance Reduction Techniques in Simulation*, doctoral disseration at Katholieke Hogeschool, Tilburg, The Netherlands, January 1971.

203. T. Naylor, *Computer Simulation Experiments with Models of Economic Systems*, Wiley, New York, 1971.

204. E. S. Page, "On Monte Carlo Methods in Congestion Problems, II: Simulation of Queueing Systems," *Operations Res.*, Vol. 13, No. 2, 1965, pp. 300–305.

205. A. K. Peng, *The Design and Analysis of Scientific Experiments*, Addison-Wesley, Reading, Mass., 1967.

206. H. Scheffé, *The Analysis of Variance*, Wiley, New York, 1959.

207. D. R. Thoman, L. J. Bain, and C. E. Antle, "Inferences on the Parameters of the Weibull Distribution," *Technometrics*, Vol. 11, No. 3, August 1969, pp. 445–460.

208. U.S. Department of Commerce, National Bureau of Standards, *Experimental Statistics*, Handbook 91, 1963.

INDEX

Activity, 24, 131
Activity scanning, 25, 38–40
Address, 7
Addressing, indirect, 104
Allard, J. L., 371
Alternatives, evaluation of, 6
  control, 7
  design, 7
ANALYS routine, 135, 364–366
Analysis of variance, 327, 330
ANALYSIS routine, 124, 127, 132, 133, 135, 162, 167, 362–364
Anderson, T. W., 371, 373
Ansoff, H. I., 368
Antithetic sampling, 319–322
Antle, C. E., 263, 342, 374, 375, 376
Assembler, 93
Association, linear, 142
Attributes, 4, 113
Autocorrelated sequences, 234–238
Autocorrelation function, 149–159
  sampling properties, 295–296
Autocovariance function, 149
Autoregressive representation, 128, 160–165
  autocorrelation function generation, 296
  estimation of sample mean variance, 286–288
  parameter estimation, 254–260
  variate generation from, 236–238

Bailey, N. T. J., 373
Bain, L. J., 263, 342, 374, 375, 376
Balintfy, J. L., 373
Bartlett, M. S., 371
Bennett, R. P., 369
Benton, S. J., 368
Berman, M. B., 373
Bernoulli trials method, 198
Beta distribution,201
  parameter estimation, 246, 359
  variate generation, 204–208, 209–211
Beta-binomial distribution, 218
  variate generation, 221–222
Bias, final conditions, 276–277
  in generation of autocorrelated sequence, 237
  initial conditions, 273
  in sample state time approach, 305
Binomial distribution, 218
  parameter estimation, 247
  variate generation, 220
Blood bank model, 64–66
Bogyo, T. P., 372
Bowman, K. O., 373, 374
Box, G. E. P., 373–374, 375, 376
Brown, B., 373
Brownlee, K., 376
Burdick, D. S., 373, 376
Burt, J. M., 322, 376
Buxton, J. N., 369

Carter, G., 314, 315, 367, 376
Cauchy distribution, variate generation, 240
CDC 6000 series computer, 129
Central limit theorem, 267, 290
Chain, current events, 105
  future events, 104
Chebyshev's inequality, 294
Chi-square distribution, variate generation, 294
Chi-square test, 252–253
Choi, S. C., 374
Chu, K., 373
Clark, C. E., 368
Classes, 72
Clock, 38
Cochran, W. G., 376
Coefficient of determination, 142
Comparison of experiments, 325–328
Compiler, 93
Computer, analog, 63
Conditions, initial, 90
Confidence intervals, 263, 288–293

nonparametric, 293–294
Confounding, 334
Congruence, 172
Congruential generator, linear, 172
mixed, 172–175
multiplication, 172, 175–180
Consolidated Analysis Centers, Inc., 113, 369, 370
Control variate, 319, 324–325
Conway, R. W., 1, 87, 367, 369, 375
Cornell Aeronautical Laboratory, 18
Correlation, 138–142
Correlogram, 156, 296
Covariance matrix, 142
Coveyou, R. R., 181, 371
Cox, D. R., 371, 376
Critical values, chi-square, 250, 299
Kolmogorov-Smirnov, 193, 360–361

Dahl, O., 129, 370
Data, analysis, 88–89
collection, 32–37, 88–89, 310
collection errors, 278
display, 88–89
structures, 71–82
crude chain, 76–77
ordered chain, 80–82
Debugging features, 92
Delay, conditional, 83
Delay statement, 40
Dependence, 138
Design of experiments, 310
Designs, fractional factorial, 334–335
Determination, coefficient of, 142
Deviate, normal, 212
uniform, 167
Distribution fitting, 249–254
Dobell, A. R., 371, 372
Documentation, 90–92
Doob, J. L., 373
Downham, D. Y., 371
Draper, N. R., 375, 376
Dreyfuss, D. J., 368
Durbin, J., 371, 374
DYNAMO, 63

Empirical distributions, 232–233
Entities, 4, 113
Epoch, 302
Ergodic, 153, 154
Ergodic properties, 148
Error messages, 90–92
Errors of measurement, 17
Estimator, consistent, 152
Event, 22, 24
conditional, 26, 42
next, 31
scheduled, 42
unconditional, 83
Event scheduling, 25, 26–37
Example, fire department, 5
maintenance shop, 5
outpatient, 5–6
theoretical queueing, 159
Exogenous economic processes, 235
Experiments, comparison of, 325–328
design of, 310
factorial, 331–334
full factorial, 311
Experimental, design, 136
layout, 310, 343
Exponential distribution, 201
bivariate generation, 215
chi-square test intervals, 252
modified Kolmogorov-Smirnov test, 361
parameter estimation, 246
variate generation, 168–203
Extrapolation, 336

F distribution, variate generation, 213
F test, 331
autoregressive order determination, 257
Facilities, 101
Facility, 105
Factor, 311, 325
Factorial experiments, 331–334
Factorial, fractional, 334
Faulkner, J., 131
Fazar, W., 368
Feedback, 7
Feld, H., 182, 371
Fetter, R. B., 367
Final conditions, 276–277, 310
Finney, D. J., 374
Fire department simulation, 314
Foresight, 341–344
Fishman, G. S., 371, 375, 376
Forrester, J., 63
FORTRAN, ANALYS routine, 364–366
Fourier transforms, 158

Fox, B. L., 206, 373
Fractional factorial designs, 334–335
Frequency domain, 155, 158

Gaming, operational simulation, 15
Gamma distribution, 201
  bivariate generation, 215
  parameter estimation, 246, 352–353
  variate generation, approximating method, 239
    integral shape parameter, 203–204
    nonintegral shape parameter, 208–209
GASP II, 71, 81
Garman, M., 322, 376
Gaver, D. P., 322, 323, 373, 376
Geisler, M. A., 367
General Electric Company, 369
Geometric distribution, 218
  parameter estimation, 248
  variate generation, 222–223
Gnanadesikan, R., 374
Godwin, H. J., 294, 375
Goldberg, H., 375
Good, I. J., 372
Goodman, A. S., 186, 372
Gordon, G., 98, 370
Gorenstein, S., 372
GPSS/360, 98–112
  extended computing capability, 108
  single server queueing problem, 108–112
Greenberg, B. G., 375
Greenberger, M., 64, 176, 368, 369, 372

Hammersley, J. M., 367, 376
Handscomb, D. C., 367, 376
Hannan, E. J., 371, 374, 375
Harman, H. H., 367
Hausner, B., 112, 370, 371
Header, 77
Hill, W. J., 376
Hills, P. R., 369
Histogram, 265
Hoel, P. G., 373
Hoerl, A. E., 376
Hughes, L. P., 374
Hull, T. E., 371, 372
Hunter, J. S., 344, 368, 376
Hunter, W. G., 376
Hutchinson, D. W., 372
Hypergeometric distribution, 218
  variate generation, 228

IBM, 369, 370, 372
Identification, 72
Ignall, E. J., 314, 315, 344, 367, 368, 376
Importance sampling, 317–318
Independence, 137–138, 169
Industrial Dynamics, 63
Initial conditions, 272–275, 310
Initialization, 89–90
Input parameter, determination, 136
  estimation, 242–260
Interaction point, 83
Interactions, 331
Interpolation, 336
Intervals, interevent, 23
Inventory problem, 49, 52, 56–58
Inverse transformation method, 168, 198
  discrete variate generation, 216

Jansson, B., 372
Jenkins, G. M., 374, 375
Job-shop simulation, 25
Johnk, M. D., 373
Johnson, B. M., 1, 87, 367, 369
Johnson, N. L., 374, 377
Jones, M. M., 369

Kahn, H., 377
Kalinichenko, L. A., 369
Karr, H. W., 112, 369, 370, 371
Kendall, M. G., 372, 374, 377
Kiviat, P. J., 1, 70, 112, 183, 367, 368, 369, 370, 372
Kleijnen, J. P. C., 377
Kleine, H., 369, 370
Knollmeyer, L. E., 368
k-tuples, 180–182
Knuth, D. E., 73, 184, 369, 372
Kolmogorov-Smirnov test, 254, 360–361
Korbel, J., 64, 368
Kotz, S., 374
Krasnow, H. S., 369
Kurtosis, excessive, 264

Language, assembly, 93
  basic machine, 93
  compiler, 93
  computer programming, 92
  interpretive, 94

list processing, 67
problem-oriented, 94
simulation programming, 70
object generation methods, 75
relationship concepts, 74
Laski, J. R., 368
Leavitt, M. R., 367
Lehmer, D. H., 173, 372
Leone, F. C., 377
Levine, H., 375
Lewis, P. A. W., 186, 372
Lilliefors, H. W., 254, 374
Linear association, 142
Logistics System Laboratory, 16
Logistics distribution, variate generation, 241
Lognormal distribution, 201
parameter estimation, 247
variate generation, 214
Lu, J. Y., 368
Lubin, J. F., 370

MacLaren, M. D., 170–171, 181, 372
McNeley, J. L., 131, 369, 371
MacPherson, R. D., 181, 371
McRoberts, K. L., 368
Madansky, A., 374
Main effects, 331
Malcolm, D. G., 368
Malkani, C., 64, 368
Mann, H. B., 374
Markowitz, H. M., 113, 369, 370, 371, 372
Marsaglia, G., 170–171, 181, 234, 372, 373
Marshall, A. W., 373, 377
Maxima generation, 39
Maximum likelihood estimation, 244–249
Maxwell, W. L., 1, 87, 367, 369
Mean, 264
conditional, 272
confidence interval, 267, 288–293
estimation example, 267–268
steady-state, 271
Mean-square error, 278
Mechanisms, relational, 72–73
Median, 264
estimation example, 270
Miller, H., 186, 371
Miller, J. M., 372
Minima generation, 239
Mnemonic, 93
Mode, 264
Models, 11–13
blood bank, 64–66
minimal, 18
queueing, 25–58
stock exchange, 66–67
Modulo notation, 172
Monte Carlo, 312, 321
Monte Carlo methods, 19–20
Morgan, H., 87, 370
Morse, P., 371
Moving average representation, 160
generation from, 235–236
Muller, M. E., 373
Multiple comparisons, 331
Miltitask-multiresource problem, 61–62
Multivariate normal distribution, 138
variate generation, 215–216
Myhrhang, B., 370

Nagao, N., 63, 368
Nance, R. E., 87, 370
Naval Electronic Warfare Simulator, 15
Naylor, T., 329, 344, 368, 373, 376
Negative binomial distribution, 218
parameter estimation, 248
variate generation, 226
Networks, PERT, 58–61, 263, 322
New York City Fire Department, 16
Normal distribution, 201, 291, 294, 304, 326
intervals for chi-square test, 252
modified Komogorov-Smirnov test of fit,360
multivariate generation, 215–216
parameter estimation, 247
variate approximation for binomial variate, 220
variate generation, 211–213
approximating method, 239
Normal process, 149
Norwegian Computing Center, 129
Nygaard, K., 129, 370

Object generation and manipulation, 73–74
Olkin, I., 373
Olshen, R. A., 372
Operational gaming, 15
Orcutt, G. H., 64, 368
Output analysis, 136
Owen, D. B., 370, 372, 374

Pai, A. R., 368

Page, E. S., 321, 376
Parente, R. J., 370
Parslow, R. D., 370
Parzen, E., 374, 375
Pascal distribution, 226
Payne, W. H., 372
P-E Consulting Group Limited, 368
Pearson, K., 374
Peng, A. K., 376
Pegels, C. C., 64, 368
Period, full, 173
Period modeling, 63–67
Periodogram, 192
Perlas, M., 322, 376
PERT networks, 58–61, 263, 322
Petrone, L., 370
Petruschell, R. L., 368
Pinkham, R. S., 374
Plackett, R. L., 373
Pointer, 77
Poisson distribution, 218
  parameter estimation, 247
  variate generation, 224–225
Preamble, 113
Predecessor, 81
Prior information, 312–318
Pritaker, A. A., 368, 370
Probability distribution function, cumulative, 265
Process, 24, 131
  interaction, 25, 40–44
  normal, 149
Program, object, 94
  source, 94
Pseudorandom number generation, 171–180
  GPSS/360, 182
  SIMSCRIPT II, 183
  SIMULA, 184
  testing, 184–193

Quantile, 265
Queue discipline, 31
Queueing example, theoretical, 159
Queueing models, one task-many servers, 44–46
  two tasks-many resources, 46–49
Queueing system descriptor, 34

Raburg, J. R., 372
Rand, 15, 16, 66, 112, 170, 372
  Systems Research Laboratory, 16
Random number generation, 88, 136, 167–196
Random variate generation, 136
Ratio estimate, variance of sample mean, 288
Reactivation point, 40, 83
Realization, 145
  finite, 147
Record, 28
Regeneration point, 302
Rejection method, 198, 230–231
Relationships, class, static, 5–6
  dynamic, 5–6
  logical, 23
  mathematical, 23
Replicate, 18
Replication, 147, 268, 310
Response surfaces, 335–341, 343–344
Rivlin, A. M., 64, 368
Roberts, R. D. K., 371
Rosenbloom, J. H., 368
Rotenberg, A., 373
Routine, timing, 182
Rule, job selection, 24
  logical operating, 24
Run length, 310
  specification, 34

Sakai, T., 63, 368
Sample size, determination, 297–299
Sampling, antithetic, 319–322
  importance, 317–318
  plan, 311
  stratified, 322–324
Sarhan, A. I., 375
Scheduling, 28
Scheuer, E., 373
Schriber, T., 98, 371
Seed, 172
SEED.V(1), 121
SEED.V(2), 121
Sets, 113
Sequencing set, 131
Sheffé, H., 375, 376
Shenton, L. R., 373, 374
Shubik, M. A., 368
Sichel, H. S., 374
Siddiqui, M. M., 374
Siegel, L. G., 87, 370
SIMSCRIPT, 112
SIMSCRIPT I.5, 112
SIMSCRIPT II, 112–129

ANALYSIS routine, 362–364
single server queueing problem, 121–128
SIMULA, 129
example, 131–133
SIMULA-67, 131
Simulation Associates, Inc., 113
Simulation, control program, 31, 82–85
discrete event digital, 22–69
gaming, 15
job-shop, 25
man-machine, 15
programming, 85, 92–96
restart, 306–307
Skewness, 264, 326
in sample state time approach, 305–306
Slevin, D. P., 368
Smith, H., 376
Smith, W. L., 371
Spectral density function, 149–159
Spectral distribution function, 192
Spectrum, 152, 154, 295
analysis, 163
estimation of sample mean variance, 283–286
sample, 259
Sperry Rand, 129
Standard numerical attributes, 101
State, 6
time, 301
Stationarity, 148–149
Stanton, R. E., 368
Statistical association, 137–143
Statistical reliability, 307
Steger, W. A., 367
Stock exchange model, 66–67
Stochastic sequences, 143–148
Stoller, D. S., 373
Storage, 105
units, 101
Stratified sampling, 322–324
Stuart, A., 372, 374, 377
Subsystems, 7
Successor, 77
System, 4
boundaries, 7
environment, 7
optimization, 11
performance, 10
response, 9
simulation, 4
computer, 16
identity, 14
laboratory, 15
quasi-identity, 14
state, 8–9

t distribution, 268, 293, 309, 332, 333
variate generation, 213–214
Tabled distribution, 233
Takacs, L., 369, 375
Teichroew, D., 370
Termination, 89–90
Test, chi-square, 182–187, 252–253
correlation, 191–193
F, autoregression order determination, 257
gap, 181, 190
independence and uniformity, 185–193
Kolmogorov-Smirnov, 187–188, 254, 360–361
poker, 181, 190
serial, 181, 188–190
Thoman, D. R., 263, 342, 374, 375, 376
Thompson, J. D., 367
Thurber, R. C., 369
Time, advance, fixed, 86
variable, 85
domain, 155, 158
flow, 85
series, 147
TIME.A, 117
TIME.V, 117
Timing routine, 31
Tin, M., 375
Tocher, K. D., 370
Tognetti, Keith, 367
Trailer, 77
Transactions, 101
creation, 103–104
Treatment, 311, 325
Treunenfels, P., 375
Triangular distribution, variate generation, 202
Truitt, T. D., 370
Truncated distributions, 228–232

Uniform distribution, 201
variate generation, continuous, 200–201
discrete, 219
Uniformity, 169
United States Steel Corp., 370
UNIVAC, 129, 131, 370, 371, 373
1107 and 1108 computers, 129
U.S. Department of Commerce, 376

Validation, 262, 328–330
Van Gelder, A., 373
Variance, 264
  analysis of, 327, 330
  estimation example, 268–269
  generalized, 339
  reduction, 310, 326
  sample mean, autoregressive estimate, 286–288
    estimation, 279–288
    ratio estimate, 288
    spectrum analysis estimate, 283–286
    subsamples estimate, 282–283
Villanueva, R., 11, 371, 372

Wait statement, 40
Wald, A., 374
Watts, D. G., 375
Weibull distribution, 201
  example, 342–343
  parameter estimation, 246, 354–358
  variate generation, 211
Wetté, R., 374
Whittlesey, J., 373
Whittle, P., 259, 375
Wickham, A. W., 87, 370
Wiener-Khintchine theorem, 155
Williams, J. W. J., 370

Yaglom, A. M., 371
Yule-Walker equations, 161
  estimates, 255